The Impact of Tourism on the Marine Environment

Part of the Responsible Tourism Series

Edited by Harold Goodwin, Director of Responsible Tourism, Institute of Place Management at Manchester Metropolitan University and John Swarbrooke, Associate Dean-International, University of Plymouth.

Sustainability is a necessity, climate change, biodiversity loss, the loss of cultural heritage and local economic development are challenges for the tourism sector. Too often sustainability is used as 'greenwashing'. Responsible Tourism requires transparency in reporting and respect for local people and their cultural and natural heritage. We need to leave more than footprints, to fund conservation and to compensate local communities for the opportunity cost of maintaining their heritage for our enjoyment. Too often tourism has just used destinations and this needs to be reversed. Responsible Tourism is about using tourism to make better places to live in and better places to visit, in that order.

The Impact of Tourism on the Marine Environment

John Swarbrooke

(G) Goodfellow Publishers Ltd

(G) Published by Goodfellow Publishers Limited,
26 Home Close, Wolvercote, Oxford OX2 8PS
http://www.goodfellowpublishers.com

British Library Cataloguing in Publication Data: a catalogue record for this title is available from the British Library.

Library of Congress Catalog Card Number: on file.

ISBN: 978-1-911635-57-4

Design and typesetting by P.K. McBride, www.macbride.org.uk

Cover design by Cylinder

Printed by Lightning Source

Contents

Dedication

This book is dedicated in the first instance to our son, John Michael Richard Swarbrooke, for it will be his and later generations which will pay the price if we do not tackle climate change and global warming.

It is also dedicated to my wife, Susan Horner, for her constant support and to my parents, John Wilfred Samuel Swarbrooke and Maureen June Bate, without whose sacrifices I would have never have been able to write books and teach students, which have been two of the greatest privileges and pleasures in my life.

Acknowledgements

I think that I first need to acknowledge that my decision to write this book was largely influenced by where I live. It is a small village in Cornwall, in the UK, on the wild Atlantic coast, by a small road which each year carries hundreds of thousands of tourists between various visitor attractions and beaches. From my window I see pleasure craft, fishing boats, and freighters going up and down the coast and the occasional cruise ship. And from the nearby cliff tops, at different times of the year, I can spot seals, dolphins, basking sharks and even whales, along with thousands of seabirds, both permanent residents and migrants. Nearby beaches are packed over the summer months with surfers, sunbathers, and sea anglers, too many of whom leave rubbish behind on the beach, and the receding tide sweeps this into the sea and thus into the marine ecosystem and the food chain. Six kilometres above my head are the contrails of jet airliners that are contributing to global warming, which in turn is affecting the water temperature of the ocean that I look at from my window. Given all these things, writing a book such as this one seemed an obvious thing to do!

A further influence on my decision has been my experience of working at the University of Plymouth, which is a centre of excellence, with a global reputation, in the fields of oceanography, marine science, environmental science and marine biology. Colleagues at the university have played a leading role in bringing the scandal of plastic pollution in the oceans to the attention of the world. My discussions with these colleagues helped stimulate me to write this book, as they acknowledged that they knew of no text that looked at the impact of tourism on the marine environment from the point of view of the oceans, rather than tourism!

I certainly owe a debt of gratitude to Professor Harold Goodwin and Professor Michael Hall, two giants in the world of tourism academia, for their contributions to the book. Although both are very busy, they readily and generously agreed to my request for short essays so that mine was not the only voice coming from the pages of the book. Both have written about subjects which are close to their hearts.

While I did make grateful use of academic journal papers and texts when writing this book, I would like to acknowledge the debt I owe to the many online sources I used to gather the latest information. These websites belonged to governments, the United Nations and the European Commission, university research centres, professional bodies, tourism organisations, individual researchers and non-governmental organisations and activist groups. In a subject area where things are happening every day and change is constant, these sources are vital resources because with books and refereed academic papers having such long gestation periods, they are overtaken by subsequent events. Indeed, I sadly have to acknowledge that in the few months between the completion of this manuscript and the publication of the book, some of the content will already seem out-of-date.

Finally, I would like to thank Tim Goodfellow and Sally North of Goodfellow Publishers, for being so supportive and encouraging, something that is rare in the academic publishing world today. The idea for this book was born during a conversation with Tim Goodfellow over lunch in Taunton and his enthusiasm for the subject spurred me on, for which I am very grateful.

Preface

I have written this book for the simple reason that I believe such a book needs to be written and that it needs to be written now.

This view has been strengthened during the period spent writing this book, a six-month period in which Greta Thunberg has been stirring consciousness around the world and the Intergovernmental Panel on Climate Change has issued its report on the impact of climate change on our oceans.

We have long neglected the well-being of our oceans, but have now become painfully aware of their fragility, not least due to the media with television documentaries such as the recent 'Blue Planet II' series, presented by Sir David Attenborough and seen by audiences in many countries.

As we, hopefully, begin to recognise the need for radical action to save the planet and safeguard the interests of those yet to be born, we are beginning to understand that the marine environment will play a major role in that struggle. If we can save our oceans, we may yet be able to save the planet as a whole.

Many of the challenges facing our marine environment have little or nothing directly to do with tourism, so why write a book specifically about the impact of tourism on our oceans?

The answer is a simple one in principle. Tourism adds another layer of challenges for our marine environment, and the rapid growth of marine tourism worldwide means that these challenges have grown dramatically in recent years. The pace of change has exceeded our ability to manage marine tourism.

Oceans are complex places in which to manage tourism. While no one owns the oceans themselves, the right to exploit them is a complex issue vested, usually, in national governments, and is often highly contested.

Unlike on land, there is no destination marketing organisation to promote tourism in the oceans but, nevertheless, they are at the heart of tourism and the tourist experience, whether it be a beach vacation or a cruise on the high seas.

With the exception of communities that live along their fringes, they have no resident communities to protect them or influence political decision-making. They are used in many different ways by humans, ways which often conflict with each other. Furthermore, we have little understanding yet of the carrying capacity of different marine environments in relation to tourism. Likewise, we do not really, as yet, have any idea of what the fashionable concept, sustainability, means in the context of the marine environment.

Of course, we also need to recognise that the only true residents of the oceans, the wildlife, have no vote, so just have to passively accept whatever tourism throws at them!

Tourism industry interest in the oceans has tended to be focused almost exclusively on the land that borders them, and really just the beaches at the interface between sea and land. The sea itself beyond the distance from shore the tourist wishes to swim has generally remained unknown to most tourists except to cruise passengers and ferry users and the odd sport angler, perhaps. Where it exists, the sea has been something to be exploited by tourism, a vital asset that has too often been taken for granted.

However, in the past 20 years we have seen a number of trends which have thrust oceans into the forefront of the tourism offer aimed at those who are, perhaps, looking for new experiences and are bored with conventional land-based attractions. Let us now look at just a few of these trends.

First, there has been a huge increase in cruising after decades in which it declined, and commentators predicted its imminent demise. Cruise ships are no longer a means of transport alone, they are now floating all-inclusive resorts, almost destinations in their own right. And cruise ships are getting bigger all the time, but their passengers want to be able to go anywhere, so we see bizarre pictures of huge ships doing ungainly three-point-turns in fjords, or monster vessels towering above the cityscape of Venice. And, while attention has focused on the carbon footprint of airliners in debates over climate change and the environment, it is only recently that attention has begun to be paid to the environmental impacts of cruise ships.

Second, we have seen an explosive growth in marine wildlife-watching fuelled by nature programmes on television. Whales and dolphins are being watched, or perhaps more accurately chased, in pretty well all the world's oceans. Tourists, no longer satisfied with seeing marine life, now want to interact with the creatures by swimming with them.

Third, more and more water-based activities are being developed or invented for the enjoyment of tourists, and some of these activities, such as sailing, ocean kayaking and wild swimming can take participants well beyond inshore waters. Those who follow such sports naturally wish to pursue their interest all over the world and in places as yet undiscovered by other tourists.

Fourth, seafood has become increasingly popular in recent years with tourists thanks to the culinary media and the guide-book writers, travel bloggers and the growing healthy eating media. This has put increased pressure on already threatened stocks.

These four points illustrate quite clearly that trends in the tourism market are becoming increasingly challenging for the well-being of our oceans.

It is worth making the point at this stage that the ocean is under threat from tourism particularly because tourists are mobile and not tied to any location. They can travel to pretty well any marine environment in the world if they have the money, and more and more do. There are therefore very few places of sanctuary, where the marine environment can exist without interference from tourists.

Yet at the same time the tourism industry has more reason than most industries to protect the oceans because tourists really only want to visit places where the sea is clean and there is a lot of wildlife to see. Unhealthy, unsustainable marine environments can easily destroy local tourism industries.

There are bigger issues to worry about in relation to the relationship between oceans and tourism. With global warming causing increased sea temperatures and rising sea levels we face the real prospect of some well-established coastal tourism destinations simply disappearing over time. Some tasteless and somewhat defeatist tourism promoters are already sometimes selling places like the Maldives with slogans such 'see it before it disappears!'.

And we cannot ignore the contribution which tourism itself makes to global warming across the planet, not least through the carbon emissions from air travel.

At the same time, we have many examples of where irresponsible marine tourism can destroy beautiful and irreplaceable places; the impact of diving on many coral reefs around the world demonstrates this very clearly.

This book will endeavour to take a holistic approach to tourism and marine environments exploring a wide range of issues from a range of perspectives.

While it may appear at times that the tone of the book is rather negative about the impacts of tourism on the oceans, I have tried hard to present a balanced overview of the key issues. However, this can be quite difficult. With tourism on land negative environmental impacts can often be set against economic benefits such as jobs or social benefits like services and facilities that exist due to tourism demand but can also be used by local residents. Frankly, it is hard to see how the oceans and their wildlife gain from tourism unless one believes that marine tourism provides a motivation and resources for marine conservation

I am grateful to Professor Michael Hall and Professor Harold Goodwin for sharing their thoughts on issues relating to the impact of tourism on the marine environment, which are of particular interest to them.

What we need is a truly inter-disciplinary approach to this subject but that is beyond my expertise or ability, notwithstanding the attempts which have been made to offer perspectives on the issues from scientists who are not tourism academics. What is actually needed is research by multi-disciplinary teams with tourism academics involved but not dominating.

Although I know relatively little about oceans from a scientific point of view, I do understand one thing and that is that the term 'marine environment' in the title of the book is not really accurate. There are many different and diverse marine environments and the impacts of tourism will be different for each of them. I hope that the reader will forgive the use of the singular version, which has been used purely for convenience. Of course, the same is true in relation to the use of the term 'impact' in the title when in fact there are a myriad of different impacts

However, we also need to place tourism in a wider context in relation to the impact of human beings on the marine environment. The issue of plastics in the ocean shows this clearly; most are not the result of tourism, but some certainly are and as a discretionary, non-essential activity that could be seen as a luxury, tourism should be able to solve this problem easier than some other economic sectors.

At the same time, we need to acknowledge that what happens in and above our oceans themselves has an impact that extends over the land masses of our planet as well. Most of our weather, wherever we live, is influenced by the oceans, and the majority of the extreme weather events on earth also have their origins in or above the seas of the world.

As with all my books, my main hope is that reading it will stimulate researchers, students and practitioners to want to find out more about the impact of tourism on our oceans and how tourism might contribute positively towards their conservation.

Certainly, I have learned a lot while researching this book and I can honestly say it has been a labour of love.

John Swarbrooke

Cornwall – overlooking the Atlantic Ocean!

January 2020

Postscript to the Preface

In the four months between writing the Preface in January and working on the final proofs of the book in May 2020 the world has been ravaged by the COVID-19 Pandemic. It is clear that this is having a devastating impact on the tourism industry and at the time of writing it is unclear how long the Pandemic will continue and over what period of time its effects will be felt within the tourism industry.

However, it is clearly a major issue facing tourism and the planet so, at the last possible moment, right at the end of the book production process, I have added a Postscript looking at what COVID-19 might mean for tourism and the marine environment.

John Swarbrooke

May 2020

1 Introduction – 'All At Sea!'

In this chapter, I will endeavour to outline the context for the rest of the book, setting the scene for what is to come.

The relationship between tourism and the sea

For centuries now, the sea has been at the very heart of tourism. For hundreds of millions of people worldwide, going on vacation still means going to the coast, particularly for their annual summer break. This magnetic pull of the sea motivates millions of people every week, from Stockholm to Sydney, New York to New Delhi, to head to their favourite beaches and seaside resorts. The lure of the sea makes them willing to put up with the hassles of modern airline travel and being stuck for hours in huge traffic jams so they can spend a few days each year by the ocean.

What they do when they arrive at the coast, however, varies dramatically from tourist to tourist. For some they are content just to drink in the views, take photos and sit in their car viewing the sea through their windscreens. For others it means lying on the beach soaking up the sun, people-watching or playing ball games. Some tourists come to enjoy the man-made attractions that develop wherever tourists make their annual seasonal migration to the ocean. This can mean everything from casinos to theme parks, gift shops to theatres. In these cases, the sea is simply a backdrop to the vacation, with little real interaction with it on the part of the visitor.

But, of course most visitors to the seaside will, at some point, at least dip their toe in the water and most will go further and swim in it. Depending on where they are, they will bask in the sensual pleasure of being in lovely warm clear water or they will grit their teeth as the cold water grips them like a vice and console themselves with the idea that swimming is good for them! Tourist engagement with the sea itself can range from a quick 'paddle' from the shore to a 'wild swim' for many kilometres beyond the sheltered inshore waters.

For an increasing number of tourists, though, the sea is a playground and being in it or on it may well be the main purpose of their vacation. We have seen, in recent years, large increases in the popularity of a range of marine-based activities, from surfing to paddle-boarding, windsurfing to jet-skiing, scuba diving to sea kayaking, sea angling to sailing.

In a world where an increasing number of tourists are urban dwellers who live away from the sea, there is another attraction of visiting the coast, which is the opportunity to enjoy fresh seafood. For many tourists being able to eat fish straight from the sea provides an element of authenticity for their trip

In recent years we have also seen a massive increase in marine wildlife tourism, mirroring what we saw earlier on land with the growth of safari tourism and wildlife experiences. For many coastal destinations, lists of their main selling points now usually include the opportunity to see whales, dolphins, stingrays or whatever happens to live in or migrate through, 'their' piece of ocean.

It is clear from this brief outline of how tourism relies on the sea as a resource, that it is the interests of both tourists and the tourism industry that the marine environment is in good health. This has been widely recognised by tourism interests which have supported initiatives such as the 'Blue Flag Beach' scheme within Europe which sets standards for the quality of bathing water. However, most industry lobbying and public policy in relation to tourism and marine environments has focused upon safeguarding the quality of beaches and inshore waters so they can be exploited by both the tourism industry and tourists.

In contrast there has been much less interest paid to the possible impacts of tourism on the well-being of oceans. It has always been assumed that there are no such impacts and that the oceans are resilient enough to deal with any impacts that may occur. However, in recent years, both of these assumptions have begun to be questioned and shown to be false.

As we will see in detail in a later chapter, the massive global growth of cruise tourism has been shown to have a negative environmental impact, which now can be seen across the planet. At the same time, the media has highlighted the damage which divers have done to some of the most iconic coral reefs in the world.

Furthermore, there has been growing public awareness around the issue of declining fish stocks, and the consumption of fish by tourists on vacation has served to exacerbate this problem. At the same time the desire of tourists for fish dishes when they vacation by the sea often leads to some destinations, where fish stocks are limited already, importing fish from considerable distances with the resulting carbon emissions.

However, concern over the state of our oceans has become much broader in focus than just tourism, not least due to television programmes such as the 'Blue Planet' series by David Attenborough that has been shown all over the world.

This has been set in the context of the broader issue of climate change and rising sea levels. Here the emphasis is on how human activity as a whole – including tourism but covering all aspects of our lives as consumers – is having a devastating impact on the oceans of the world.

This has been seen very clearly in relation to the issue of plastics in oceans which has become a high-profile issue in just a few years. Again, tourism is only part of the cause of the problem but given that tourism is not a necessity of life nor something available to everyone on the planet, it is understandable that there should be focus on how tourism is adding to Man's impact on the oceans.

Even in terms of enlightened self-interest rather than morality, there is good reason for the tourism industry and tourists to seek to act to reduce the impacts of tourism on the ocean, as well as getting involved in the broader issue of tackling global change. It is now well known that sea levels are continuing to rise and if that continues due to global warming, then popular destinations may simply disappear from the map. The case of the Maldives is well known in this respect, but by the middle of this century places such as the coasts of Florida and parts of the Mediterranean will also be under threat.

This idea that some places will disappear altogether is fuelling a depressing kind of 'see it while you can' type tourism. Farbotko explored this issue in relation to the Tuvalu Islands in 2010 (Farbotko, 2010) and it is regularly noted as a motivator for trips to Antarctica. In a few extreme cases this is leading to a frightening new way of promoting trips to a few destinations based on the idea of visit 'before it disappears'.

Less dramatically, the tourism industry continues to relentlessly market the ocean as an attraction and a commodity whether it be for cruises, water-sports activities or lazing on perfect beaches fringing a turquoise sea. Unlike tourism on land, tourist industry brochures seem to say very little about sustainable tourism or responsible tourism in relation to the oceans.

I reviewed six brochures from leading brands selling in the UK market between 2019 and 2021, three tour operators and three cruise companies.

The three tour operation brands were all owned by TUI, Europe's largest tour operator and a company with a strong track record in sustainable tourism. It was disappointing, therefore, to see that:

- The long-haul brand with a 156-page long brochure devoted just one paragraph to responsible tourism together with a small quarter page feature on a charity to which it made financial donations. The charity was concerned with providing clean water for local communities, clearly a very worthwhile cause. Some hotels were highlighted with a sustainability symbol although it was unclear what the symbol meant. Many of the resort descriptions focused on beaches, marine wildlife and water-sports but there was nothing said about the impact tourism might have on marine environments.

- The all-inclusive winter sun brand brochure of 112 pages had nothing specifically to say about sustainable tourism except a paragraph in the small print section saying the company was committed to helping make aviation more sustainable. Again, it talked quite a lot about water-based activities and marine wildlife-watching but said nothing about the need to protect the ocean environment. It specifically promoted whale-watching in Tenerife and diving in Malta but without any acknowledgement of the need to do both responsibly.
- A further long-haul brand had a paragraph about sustainable aviation at the back of a 220-page brochure. All of the products in this brochure were coastal resort-based vacations. Again, water-sports and beaches featured heavily but there was no information about responsible tourism and marine environments.

The suggestion appears to be either that tourism does not cause problems for marine environments or that the oceans can cope with any tourism impacts without any risk of long-term damage. As we will see throughout this book both ideas are false.

Perhaps even more worrying though were the brochures of three leading cruise industry brands. In the past few years there has been an increase in criticism, fair or otherwise, of the impact of cruise ships on marine environments, particularly in term of waste disposal, plastics, pollution. There have even been criminal cases against cruise companies relating to pollution issues, some of which have resulted in hefty fines. Questions have also been raised about the carbon footprint of cruise ships as the ships get bigger and bigger.

Although some companies are belatedly taking action on sustainability challenges, others seem to be in denial, not wanting to talk about it or even acknowledge that there are such challenges. This may be because the cruise sector, which has been very successful in recent years, sells a product based on conspicuous consumption, the concept of luxury and self-indulgence, and the idea that they can make all their customers dreams come true. In such a cocoon of escapist pleasure it probably seems unwise to prick the balloon by raising issues around the impact of these dream vacations on the marine environment.

Three leading cruise brand brochures were reviewed, and the following observations made:

- A 172-page brochure from a brand owned by Carnival Cruises had two paragraphs near the end about sustainability and the environmental management systems used on its ships, which offered some detail about what is being done to reduce negative environmental impacts. It also had a small feature about a new ship powered by liquefied natural gas which is designed as a step towards reducing carbon and nitrogen emissions and achieving zero sulphur emissions in the future. Disappointingly, though, issues such as wildlife impacts and issues such as plastic were not discussed.

- In the 148 page brochure produced by a leading US-based cruise brand, that is popular in the UK market, contained nothing about sustainable cruising except five lines of text about the introduction of solar power and new technology which allows the ship to remain stationary without the need for an anchor which might damage the seabed.
- Another US-based cruise brand, popular in the UK, offered a 114-page brochure which appeared to make no reference to sustainability or responsible cruising at all.

At a time when programmes such as David Attenborough's 'Blue Planet' series have highlighted the huge challenges facing the world's oceans, it is surprising that the cruise industry has not been more proactive. While tour operators still need to do far more, they, some time ago, publicly acknowledged their responsibility towards the environment and the host communities in the destinations to which they send their customers. Airlines, which can never be truly sustainable while they are reliant on current fuels, have instituted measures to ameliorate the impact of their activities. And over the past thirty years the hotel industry has worked holistically as an industry to share knowledge that helps them reduce their adverse impacts on the environment

It seems time that the cruise industry, although a successful modern industry, has been slow to respond to sustainability issues. Most attention to date has focused on aviation and tourist activities on land. Now, though, the oceans are beginning to take centre stage in the sustainability debate, in a time of rising sea levels, depletion of marine wildlife and plastic pollution in the oceans. The time has come, therefore, for the cruise industry to 'step up to the plate' as it begins to attract greater attention and increasing criticism for its impacts.

The model of the relationship between tourism and the sea

Let us now look at the broad relationship that exists between travel, tourism and the sea which I have endeavoured to portray in Figure 1.1. While not claiming to be comprehensive, it does illustrate a range of ways in which tourism and the sea are linked.

Of course, the relationship between 'tourism' and the sea is not new, it dates back many centuries. Since history began to be recorded, the sea has taken travellers on long journeys as traders, explorers, pilgrims, emigrants and conquerors. However, in terms of leisure and tourism the story is a shorter one, beginning with sea bathing and moving through the era of luxury liners to the modern era.

Over time the sea has changed from being largely a means of transport for travellers, a necessity, to representing a place of leisure and fun for tourists; in essence, that is as true of a seaside beach as it is of a cruise ship.

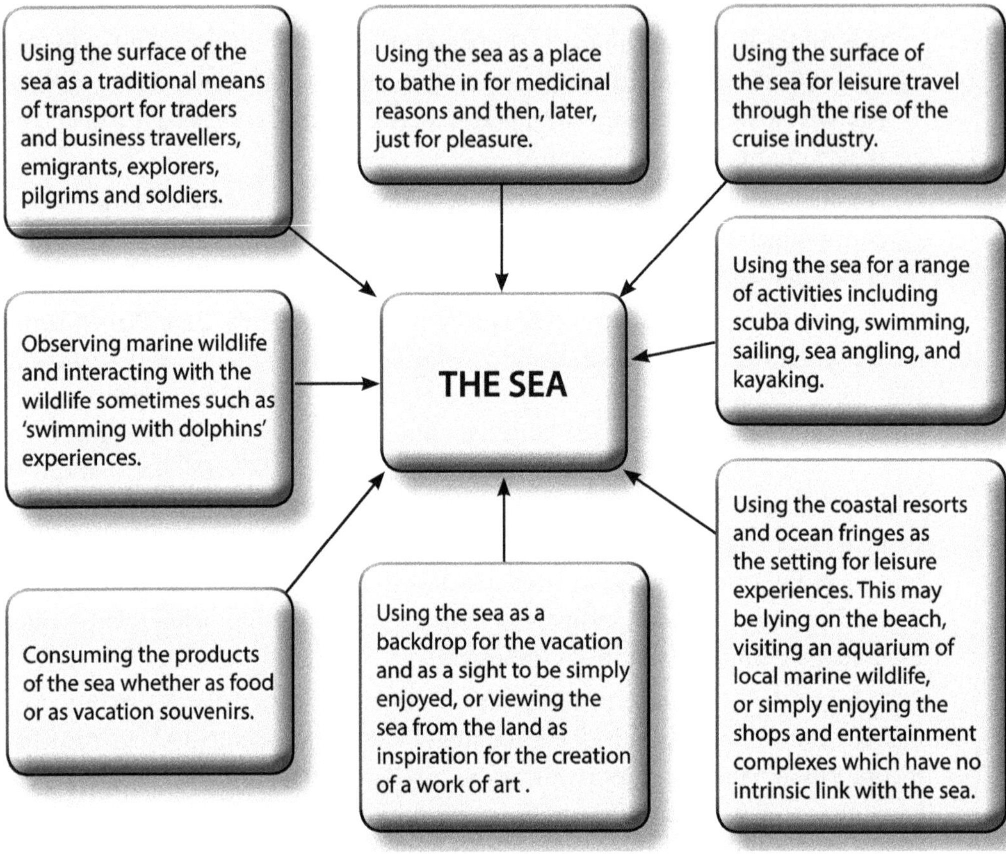

Figure 1.1: The relationship between tourism and the sea

Perhaps we can make a few more general points about the relationship between tourism and the sea.

- First, it is obviously a rather one-way relationship in which people gain a great deal from the sea, but the sea seems to gain little, if any, benefit in return. Indeed, as we will see the relationship is harmful to the sea, in a number of ways.
- Second, some tourism is based on being on the sea and separated from contact with it, not least for reasons of safety. Other tourists want to fully immerse themselves in the sea even if this is risky.
- Third, the sea is intrinsically an alien environment for human beings, so the relationship involves both risk and fear for the tourist. Every year many tourists die in the sea, a high price to pay for participating in a leisure activity.
- Fourth, most direct contact with the sea by tourists is very limited in both time and space so most tourists never experience anything more than a superficial interaction with a particular part of the vast marine environment. We may

swim for half an hour in a small area of sea off a certain beach or it may be a three-hour whale-watching cruise to a specific location known to be popular with migrating whales. Meanwhile, cruise passengers pass across an area of the ocean in a matter of minutes.

- Fifth, the tourism industry is constantly inventing new marine-based activities for those who love the sea but are always looking for activities they and their peers have not tried before. In recent years these have included ways of launching yourself into the air from the sea, coasteering, and taking submersible trips under the water. There is apparently a trend towards activity rather than just passive enjoyment, and towards interaction and co-creation in marine leisure. Some of this has become very controversial when it involves marine creatures as 'co-creators', as we will discuss in Chapter 4.
- Finally, as we will discuss in Chapter 2, there is a difference between 'the sea' and 'the marine environment'. The latter is an ecosystem, or rather a myriad of ecosystems, that also include the sea-bed geology and flora and fauna as well as the sea, which is the milieu in which the ecosystem exists and is itself also a component of the ecosystem. It is important to bear this distinction in mind when we use these different terms.

Tourism impacts in marine environments

We will now look at the types of impacts which tourism has, or can have, on marine environments. Some of these are presented in Figure 1.2. Again, the diagram does not aim to be wholly comprehensive but, rather, it is an attempt to illustrate the range of impacts.

There are several interesting observations to be made about the impact of tourism on marine environments as follows:

- Tourism impacts in coastal areas is not a new issue; nearly a quarter of a century ago Gormsen offered a general overview of the historical evolution of coastal tourism and its impacts. (Gormsen, 1997). However, in the early days of coastal tourism the focus was often more on the impacts of tourism on the coastal destinations than on the marine environment and the oceans.
- The nature and scale of the impacts is not the same in all marine environments, for these environments themselves are not homogeneous. Some are more resilient or fragile than others and the type of tourism and scale of the impacts also varies from one location to another.
- Some impacts are intrinsically more destructive of marine environments than others and some do irreversible damage while others do harm which can heal over time.

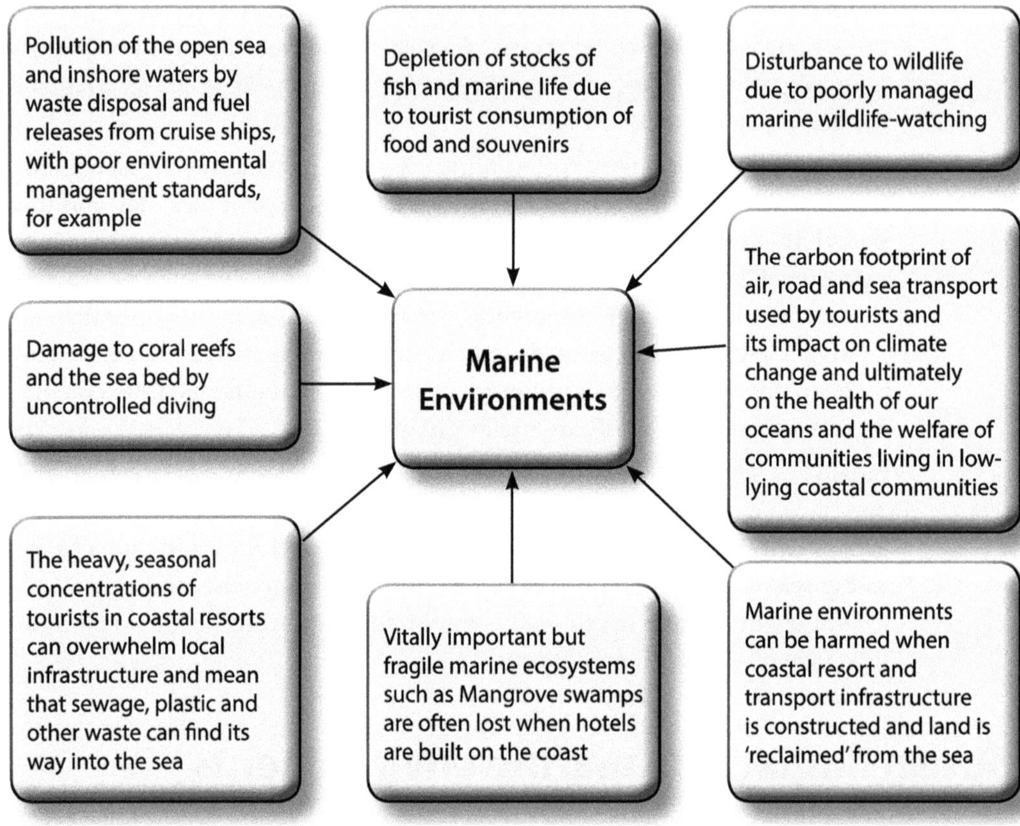

Figure 1.2: A typology of tourism impacts in marine environments

- Many of the impacts seen from tourism are not inevitable; they are often the result of poor environmental management practices within industry and/or a lack of governmental regulation.
- What is unprecedented about these impacts is that they are truly global for tourism now reaches every corner of the earth. Indeed, cruise companies seem intent on making every square kilometre of the seas accessible to tourists including those which surround both the Arctic lands and Antarctica. Indeed, one of the greatest areas for concern today is the impact of tourism in Antarctica. Amelung and Lamers noted in 2007 that '*the global impacts of Antarctica tourism have been largely ignored. Emissions are shown to be considerable, with aviation and cruises being the main sources*'. (Amelung and Lamers, 2007)
- While some impacts are specific to a particular location and immediate, others take effect more slowly or have impacts that are felt over a much larger geographical area. The ones which are ultimately the most damaging are probably those which occur over a long period with few visible clues until it is too late.
- While we tend to see, rightly, most tourism impacts on the oceans as negative it can be argued that there is one positive impact, potentially at least. It is

possible that if tourists can be made aware of the problems facing our oceans and income from marine tourism can be used for conservation purposes, then it is possible that tourism might also become a force for good in terms of our marine environments.

Finally, while our focus is upon the marine environment it is worth briefly noting that tourism also has an impact on those who work on the sea or live by the sea. It provides employment for millions of people, but can it can have downsides for them as well including:

- The alleged exploitation of the labour of those, often less well off, who work on cruise ships or in the 'informal' tourism economy, who may sometimes endure poor working conditions and low pay, denial of their human rights, and possibly danger when trying to earn a living.
- The development of coastal resorts in some countries, which may lead to local people losing their traditional rights to certain beaches which are fenced and available only to the tourists.
- Coastal zone dwellers suffering flooding because of global warming and sea level rise, partly caused by the tourism industry, particularly air travel and the use of private cars worldwide by those on vacation.

'All at sea!' – the problems of managing and planning sustainable tourism in marine environments

Over the years a copious literature has grown up around tourism planning and management, but this has all been concerned with tourism that is land-based. There is no such literature to guide the planning and management of tourism in the oceans. Furthermore, marine tourism poses challenges for such activities.

First, on land, there are various tiers of government that are responsible for virtually every hectare of land. These governmental bodies have a duty to manage their areas in the public interest and are responsible for planning the futures of their areas as well as managing them on a day-to-day basis. They may not own all the land, but they have influence over it through established systems of regulation and legislation. By contrast most of the world's oceans are outside the remit of any governmental organisations, and governments usually only have jurisdiction over the sea around their coasts for a number of kilometres. Thus, large areas of open ocean are owned by no governmental body so there is no one with the right or duty to plan or manage them.

This lack of governmental oversight of marine environments means that there are no destination management organisations in place that could coordinate the planning and management of tourism in marine areas.

However, increasingly perhaps, we are seeing some governments seeking to claim ownership or control over areas of the ocean, particularly when the ocean bed is thought to contain oil, natural gas or minerals. These claims can often lead to conflict, which then makes sensible management of the oceans more difficult or even impossible.

Second, oceans have no residents – at least not human residents – and that means they also have no voters! Whereas most areas on land have people who live on them and most of these have a vote and therefore can influence public policy, the oceans have no such resident population. As a result, the oceans have no political power of their own and can be ignored by politicians except where they provide a livelihood for a large number of land-based voters.

Third, in most areas of open sea international law says that all vessels should have free and unhindered passage. This would make it very difficult to restrict access to certain particularly fragile or endangered marine environments. This is in contrast to the land where it is possible to legislate to control access to protect particular areas, even if this power is used rarely in reality.

Fourth, there are powerful vested interests involved in the exploitation of ocean resources which mean that tourism impacts on the oceans and the need to plan marine tourism are a low priority for most governments and supra-governmental organisations. These interests include the fishing industries of many European, African and Asian nations, which employ large numbers of people, and the oil, natural gas and mineral corporations, which generate significant export earnings for countries and tax revenue for governments.

Fifth, we have a general lack of knowledge about the impacts of tourism in marine environments and how tourism interacts with different economic users of the oceans. While I am sceptical about the value of the idea of 'carrying capacity', in practice it remains a valuable concept in tourism planning. However, whereas researchers have been seeking to find ways of applying this on land in different geographical milieu, there is very little experience of trying to do the same thing in respect of marine environments. There is a clear need for urgent research in this area which will be discussed further in the final chapter when I offer a research agenda for the future.

Sixth, marine environments are not really destinations in the conventional sense. They are rarely a place where tourists come to spend their vacation. They may visit them for a short time to undertake a leisure activity or pass through a number of them while in transit on a ferry or cruise ship, or spend time on their fringes in land-based destinations. In most cases the marine environment itself has no physical infrastructure for tourism apart from the odd accommodation units built over water or the occasional underwater restaurant. Infrastructure, such as airports, can be built on land reclaimed from the sea, but once built they are no longer marine environments, obviously. Thus, marine environments cannot be conceptualised in the same way as normal destinations, which adds a further layer of complexity to their planning and management.

Finally, in tourism planning and management on land we are used to balancing the various impacts of tourism. The generally negative impacts on the destination environment are usually offset against the normally positive economic impacts for the host community, with the social impacts being a mixture of positive and negative. These impacts are generally felt in the same geographical area. This is not the case for tourism in marine environments. There, the economic benefits are largely enjoyed by coastal communities on the fringes of the oceans, whereas the environmental impacts are largely felt in the open sea some distance from the coastal communities. The idea of some kind of balancing of the two types of impacts within a single destination entity, which underpins most land-based tourism planning and management, is simply not applicable to marine environments.

These points and others will be explored further in Chapter 10.

'The perfect storm!' – the growing pressures on marine environments

As we have noted, the relationship between tourism, the sea and marine environments is not a new one. Nor is the impact of mankind's activities on the oceans a new story. However, what is new is that the oceans are facing unprecedented pressures due to the effects of climate change and global warming, as well as the growth of tourism and increasing exploitation of the ocean's resources, whether that be fish or oil or minerals. The idea of a 'perfect storm' has been used because:

- We have no previous experience of some of the changes, notably climate change and global warming.
- The pace at which the changes are taking place seems to be getting faster.
- The impacts of the factors are being felt on a truly global scale and are affecting every kind of marine environment.
- The growing pressures on the marine environment are occurring at a time when there appears to be a lack of political leadership around sustainability and a rise of populism which seems to accord a low priority to environmental issues in general and the plight of the oceans in particular.
- They are also taking place at a time when governments keen to deliver growth and prosperity for their populations are eyeing the resources of the oceans as their best chance of achieving this, and this is leading to competition and conflict between nations.

In this situation it seems fair to say that the threats and pressures facing our oceans have never been greater. While tourism is not the main cause of these pressures it is as a significant contributor and as something that is not essential to human survival it could, perhaps, be argued it should be controlled more to help us tackle the threats facing the planet and humanity.

■ The relationship between marine environments and other industries and human activities

While in no way denying the adverse impacts which tourism have on marine environments, it is important to place these in context to see them in proportion. Although comprehensive empirical data is not available, it seems clear that tourism is, by no means, the largest single cause of negative man-made impacts on our oceans. Agriculture, perhaps, has the largest impact through the runoff of pesticides from this massive industry, which seems addicted to the use of chemicals in almost every country. Most of these chemicals eventually reach the sea at some point in the hydrological cycle. There are other industries which can, and often do, have a devastating impact on the well-being of our seas, such as shipping and the extractive industries, especially oil and mining. At the same time, across the world, we are seeing fish stocks being reduced by unsustainable fishing practices together with the decimation of certain species to meet the demand for 'delicacies' such as shark's fin soup. And when we try to solve the depletion of wild fish stocks through fish-farming we cause further problems for our seas through pollution.

As well as industries, our modern lifestyles also pose a threat to marine environments. Perhaps the most significant of these is the recently recognised scourge of plastic waste and micro-plastics, which is increasingly affecting even the most remote marine environments on the planet. While tourism is responsible for creating some of this pollution, far, far more is the result of the daily lives of most people, even those who live hundreds of kilometres from the sea, due to its transportation by rivers and the wind.

However, while recognising that tourism is not the major cause of the problems facing our marine environments it is important to add two caveats. First, unlike food and shelter, tourism is not a necessity of life, so its impacts are the result of our pursuit of pleasure in our leisure time rather than something we have to do to survive. We could, therefore, reduce some of man's physical impacts on the oceans, by reducing the exploitation of the oceans by tourism. However, that would have an adverse economic impact as millions of jobs worldwide depend on cruise tourism or tourism in coastal resorts.

Second, we need to recognise that the tourism industry is also playing a major, if often indirect, part in the biggest challenge facing our oceans, namely climate change and specifically global warming. While estimates vary, everyone is agreed that air travel is a significant factor in climate change and all air travel, by definition, relates to tourism, whether it be leisure travel or business tourism.

Some of these issues will be discussed further in Chapter 11.

The structure of this book

The purpose of this chapter is to set the scene for the rest of the book and the chapters that will follow. This first chapter focuses on the relationship of tourism and marine environments from the point of view of tourism.

In Chapter 2 we will look at different types of marine environment and explore the various types of challenges which our oceans are facing today from a range of sources. It is followed by an 'opinion piece' by Professor C. Michael Hall of the University of Canterbury, who examines the issue of marine litter and waste and tourism.

Chapter 3 looks at the cruise industry, whose rapid growth and global reach means that its effects are being felt increasingly in virtually every marine environment on the planet. At the end of the chapter you will find an 'opinion piece' by Professor Harold Goodwin from the Responsible Tourism Partnership, on the subject of responsible tourism and the cruise industry.

In Chapter 4 we will explore the phenomenon of marine wildlife-watching and try to evaluate its impacts, both positive and negative.

The focus of Chapter 5 is on the uses which tourists make of the sea for leisure activities, and how that impacts different types of marine environments.

The emphasis in Chapter 6 is on the 'coastal strip', the interface between the sea and the land, looking specifically at the impacts of man-made resorts and tourism infrastructure as well as how tourists use the beaches and foreshore.

Chapter 7 looks at the consumption of marine resources by tourists while recognising that such consumption is not just by tourists. Here we will look at issues such as over-fishing and souvenirs sourced from the marine environment.

In Chapter 8, we look at the two-way links that exist between marine environments, natural disasters and tourism.

Chapter 9 focuses on the tourism industry and its vested interest in ensuring that our oceans are healthy and conserved. We will look at the concrete actions industry could and should take to play its part in protecting our marine environments.

Chapter 10 examines the public policy dimension of the relationship between tourism and the sea. It explores the real challenges that exist in the planning and management of marine environments.

In the final chapter I endeavour to bring everything together, by offering some conclusions and making some predictions about the future. I also offer suggestions for a future research agenda. At the same time, the final chapter seeks to set tourism in the context of broader issues, such as climate change and global warming as well as the impact of other economic sectors and consumer lifestyles on marine environments.

Having focused on tourism it is now time to turn our attention to the oceans by looking at marine environments, their characteristics and their diversity, together with the challenges they face from climate change and global warming.

References

Amelung, B and Lamers, M (2007) Estimating the greenhouse emissions from Antarctic tourism. *Tourism in Marine Environments*. **4**(2-3) 121-133.

Farbotko, C (2010) 'The global warming clock is ticking so see these places while you can': voyeuristic tourism and model environmental citizens on Tuvalu's disappearing islands. *Singapore Journal of Tropical Geography* **31**(2) 224-238.

Gormsen, E (1997) The impact of tourism on coastal areas. *Geojournal* **42**, 39-54.

Questions and exercises

1. Select a coastal tourist destination with which you are familiar. Look at its origins as a destination, its development over time and its situation today. For each stage in its development, seek to identify two things, namely:

   - ☐ What role did the sea play in its tourism industry?
   - ☐ What impact did tourism in the destination have on the marine environment?

2. Collect several holiday or vacation brochures featuring a range of destinations, preferably on more than one continent. Look at the words used to describe the ocean in each destination. To what extent are the same words or phrases used for the different destinations? What conclusions can you draw from your findings?
3. Look at the websites and other marketing materials produced by DMOs (Destination Management Organisations). Record what each one says about the impacts of tourism on marine environments. Do they offer advice or guidance on what tourists can do to help minimise the negative impacts of tourism on the sea around the destination?
4. In a group, set up a group discussion where each group member represents a particular economic sector such as tourism, agriculture, transport, oil and natural gas, construction and retailing. Each member should research their sector and produce a five-minute presentation in which they seek to persuade the rest of the group that their sector is less harmful to the marine environment than the others
5. Write an essay answering the question, 'Why is so much less written about sustainable tourism in the marine environment than there is about sustainable tourism on land?'

2 The Marine Environment

Introduction

The fact that open ocean covers two-thirds of the surface of our planet dramatically illustrates the importance of the marine environment to life on Earth. But the importance of the oceans goes far beyond their sheer size for it is the oceans that largely determine our climate for the weather around the world is heavily influenced by what happens in our seas.

'Weather patterns are primarily controlled by ocean currents which are influenced by surface winds, temperature, salinity, the Earth's rotation and ocean tides....Ocean currents bring warm water and rain from the equator to the poles and cold water from the poles towards the equator' (www.greentumble.com, 2016).

Every schoolchild knows that the sun evaporates water from the sea which then become clouds that then produces almost all of the rain and snow which falls on every land mass in the world. The oceans also absorb heat from the sun and from human activities; this heat is then carried to the land in those places where the prevailing winds blow from the sea to the land. At the same time, the oceans play a vital role in the carbon cycle by absorbing carbon dioxide that is in the air.

Our oceans, of course, are also a major source of food for people on every inhabited continent with fish being an important source of protein. It is also seen as a very healthy food, being low in fat and an excellent source of Omega 3. This, of course, has led to the problem of over-fishing, while dishes such as shark's fin soup have contributed to the depletion of shark populations in some parts of the world.

Increasingly our oceans are also being exploited because of the resources found in the seabed, resources that are currently essential pre-requisites for modern economies including oil, natural gas and minerals.

Of course, the seas of the world continue to be the main conduit for the transportation of the goods which consumers demand, whether that be consumers in the USA who want to buy German cars or the European consumer who buys 'summer fruits' from New Zealand or South Africa when it is winter in Europe. The oceans also play a role in the transportation of people and are particularly important in the life of island communities.

And lastly, there is the role of our seas in the modern industry of tourism. It is hard to imagine what tourism would be like without the sea. The sea means beaches and sunbathing, a place to go swimming or scuba-diving and wildlife-watching. In recent years it has also become highly fashionable to take your whole vacation on the ocean, on a cruise ship, being pampered.

So, it is clear that marine environments play a crucial role in the life of virtually everyone on Planet Earth. But what are they and how much do we know about them? This chapter will try to answer these two questions, at least in part. We will also look at the relationship between marine environments and tourism but this time from the point of view of the oceans.

Key definitions

As you would expect there are a myriad of definitions of the term 'marine environment' so let's begin with a simple one which seems to capture its essence.

The **marine environment** consists of '*the oceans, seas, bays, and estuaries, and other major water bodies, including their surface interface and interaction, with the atmosphere and with the land seaward of the high water mark*' (www.definitions.net, 2019)

The UK government Joint Nature Conservation Committee, quoting Connor et al. in 2002, states that '*the marine environment can be described or characterised at a number of different scales, ranging from ocean-level processes through to those that occur at species and genetic level*'. They go on to describe these three elements in a little more detail as follows:

- *Species provide the globally accepted classification of biological diversity, with well established rules of taxonomy, to distinguish between different types. Their classification is arranged in a hierarchy of genera, families, orders, classes and phyla.*
- *Habitats comprise suites of species communities or assemblages that consistently occur together, but which are derived from different parts of the taxonomic hierarchy (e.g. kelps, molluscs and fish in a kelp forest habitat). Their classification can also be structured in a hierarchy (biotopes, biotope complexes, broad habitat) reflecting degrees of similarity.*
- *Marine landscapes comprise suites of habitats that consistently occur together, but which are often derived from different parts of the habitat classification hierarchy (e.g. saltmarsh, intertidal mudflats, rocky shores and subtidal mussel beds in an estuary).*

The term 'marine habitat' is clearly an important one in this context and the same source goes on to define this word as a term which encompasses *'the sub-stratum rock (rock, sediment or biogenic reefs such as mussels), its topography, and the particular conditions of wave exposure, salinity, tidal currents and other water quality characteristics(e.g. turbidity and oxygenation) which contribute to the overall nature of a place on the shore or seabed.'* (www.jncc.defra.gov.uk, 2007)

There are other key concepts that we need to discuss before we go any further, starting with **marine ecosystems**. This term covers the interaction of plants, animals and the marine environment, in other words in bodies of water where the water is saline. For the planet these ecosystems are vitally important given that seas cover some 70% of the surface of the Earth.

In 2017 the online Biological Dictionary made several points about marine ecosystems that the reader may find of interest in the context of this book, including:

- *Marine ecosystems are easily the most diverse of all the ecosystems on the planet. Coral reefs alone are home to over 25% of all marine life, despite occupying less than 1% of the ocean floor*
- *Like all ecosystems, marine ecosystems are finely balanced and highly complex. There are many different parts that make up an ecosystem, and each part plays a role in maintaining balance within the system. Organisms depend on, and are highly influenced by, the physiochemical environmental conditions in their ecosystem*
- *The marine environment ... can be divided into smaller, distinct ecosystems upon closer inspection.* (www.biologydictionary.net, 2017)

This final point is important as it reinforces the idea that we must not think about the marine environment as a single homogenous entity but rather recognise that there are many different and diverse marine environments around the world.

Typologies of marine environments

This point about the heterogeneity of marine environments will now be explored further as we look at typologies of marine environments, under three main headings namely, scientific, geographical and tourism usage.

Scientists have developed numerous typologies and classifications of marine environments over a period of many decades, including important contributions such as the work of Joel Hedgbeth in the 1950s. The most basic typology divides ocean environments into two types, *Benthic* and *Pelagic*. *Benthic* includes everything on the ocean floor while *Pelagic* includes all of the mass of water.

Scientists also sub-divide marine environments based upon the depth of water as this obviously has a major impact on marine ecosystems due to variations in temperature, pressure and the amount of light which permeates the environment.

This sub-division includes the Inner Shelf, the Outer Shelf, the Bathyl Zone, the Abyssal Zone and the Hadal Zone, the latter with water depth of up to 11,000 metres. Unsurprisingly, scientific classifications of marine environments tend to be strongly related to typologies of marine ecosystems. For example, as far back as 1984, Hayden et al. had divided marine environments up in several ways in relation to their ecosystems. They identified seven types of *ocean realms* and thirteen types of *coastal margin realms.* They identified '*the defining characteristics for these realms* [to be] *the seasonal variations in ocean surface currents and the companion seasonal variations in the main wind-currents of the atmosphere.*' They went on to identify and map forty *faunal provinces.* (Haydn et al., 1984).

Modern developments such as GIS have also been used to classify marine environments, including a major project undertaken by NIWA with the Ministry for the Environment in New Zealand.

The importance of these scientific classifications is that they illustrate that there are many types of marine environment which means that the impacts of tourism are likely to be different in different places and that the systems required to manage these environments will vary between locations.

This brings us to the question of geography and the geographical classification of marine environments. This can be based upon a number of criteria including, for example:

- Their size and shape and whether their boundaries are clear.
- Their geographical location and whether they are in the southern or northern hemisphere or both.
- Their relationship with land masses, such as they are mostly 'open sea' like the Pacific Ocean or relatively narrow strips of water such as the English Channel.
- Those areas of sea where marine resources are exploited, whether that be fish or minerals or oil and gas.
- Their relationship with the human settlement on their fringes.
- The prevailing winds and currents and the directions in which they flow and how this change with the seasons.
- The climatic zones or zones in which the area of sea is located.
- The extent to which the area of ocean represents a transportation route.

This consideration of geography links very nicely to the tourism usage of oceans and how this can be classified. An attempt is made to produce a basic typology of tourism usage of oceans in Figure 2.1.

This simple outline of these three approaches to classifying marine environments clearly shows the diversity and complexity of marine environments. This needs to be borne in mind when reading the chapters which follow.

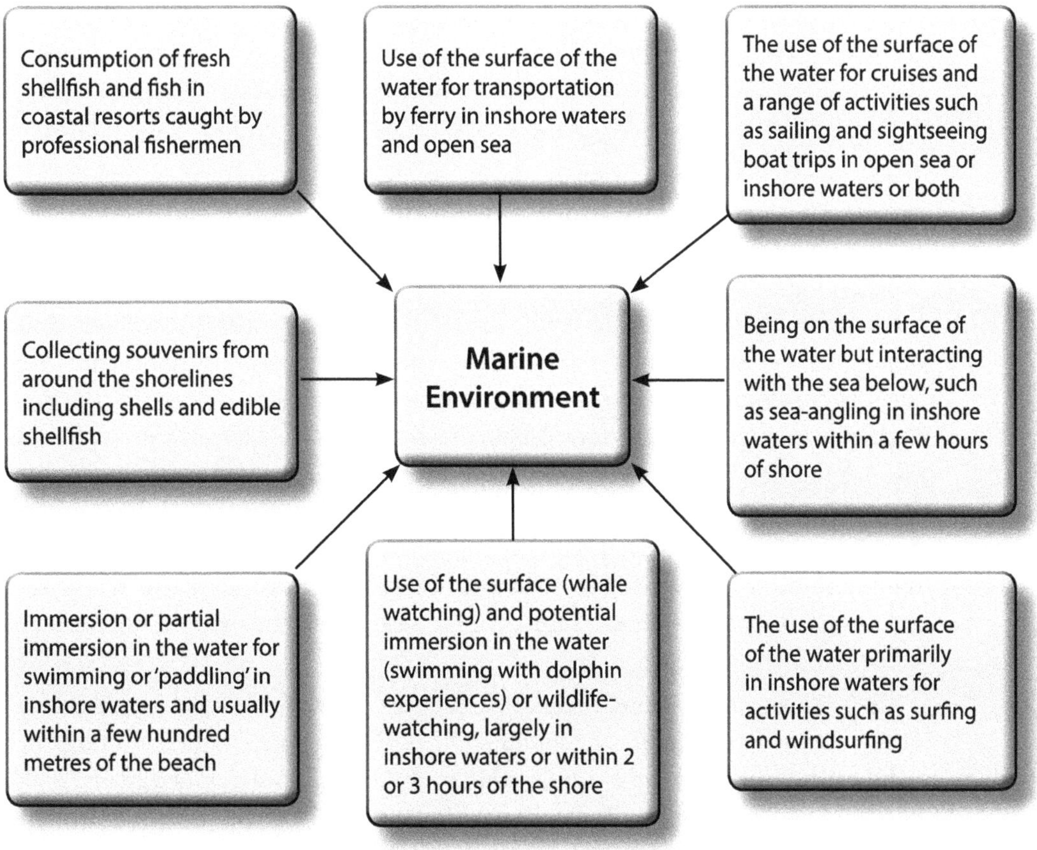

Figure 2.1: A basic typology of tourism usage of marine environments

It is clear from this simplistic diagram that tourism makes use of the marine environments in many ways. This, together with the fact that, as we have seen, there are many different types of marine environment, this give us clear warning that any attempt to manage the impacts of tourism on the oceans is going to be a complex matter.

Let us now move on to look at an interface which is crucial to any study of the impacts of tourism on marine environments, namely the relationship between the sea and the land.

The relationship between the marine environment and the land

There is obviously a two-way relationship between the ocean and the land, and more particularly the settlements on land where people live and work. As Sir David Attenborough noted at the launch of the Blue Planet 2 series, '*we may think that we live a long way from the oceans, but we don't. What actually we do here'* (wherever we live on land) *'has a direct effect on the oceans and what the oceans do then reflects back on us'*. (David Attenborough, reported by the *Independent*, 15 October 2017).

We recognised earlier that most of the weather experienced on land, even places many kilometres from the coast, largely originates from, and above, the oceans. This is especially true of extreme weather phenomena such as hurricanes or typhoons. They are created when warm and moist air from the surface of the sea rises rapidly and comes into contact with cooler air forming rain drops and large storm clouds. These extreme storms tend to have most impacts on coastal areas, but their effects are also felt further inland, usually most noticeable in the form of heavy rain. And while they tend to occur in tropical or sub-tropical areas their effects spread much wider, albeit on a smaller scale. I live in Cornwall in the UK, on the Atlantic Coast, and after a Caribbean hurricane the remnants of the storm usually bring unusually strong winds and heavy rain to the village where I live. It appears that hurricanes and typhoons are becoming a more regular occurrence and there are increasing instances of them occurring in places where they are not traditionally seen or at times of the year when they would not normally occur.

In 2004 we also saw the tragic consequences for those who live on the fringes of the oceans when there is a tsunami. On 26 December in that year a huge tsunami caused by an underwater volcano in the Indian Ocean killed over 220,000 people in eleven countries. Many coastal settlements have yet to recover from the devastation, which was made worse by the fact that many poorer people were living in areas at risk from flooding and hotels had been built on beaches at sea level.

Of course, the oceans influence the land and the population on land in less dramatic ways too, not least through providing resources to sustain human life such as fish and minerals, natural gas and oil to literally power our economies. And, given the subject of this book, they also provide the resources on which coastal tourism and seaside resorts depend.

So, what does the land, and primarily the human population that lives on the land, give the ocean in return?

The first answer to this question is water! The rivers that flow into the sea are a vital part of the hydrological cycle on which life on earth depends. Unfortunately, this increasingly means water that is polluted by agriculture and other industries and water that carries plastics and microplastics into the oceans.

The second main answer is not something given to the oceans but rather something done to the oceans namely, development. In places such as the United Arab Emirates and Hong Kong, for example, the sea has been 'reclaimed' and built upon for infrastructure such as airports or homes. The term 'reclaimed' here is odd because the areas of sea where it has happened have never been land! This construction has a substantial impact on marine environments.

Human coastal building projects have damaged some types of marine habitat particularly severely, perhaps most notably mangrove swamps. This is bad news as mangroves play a vital role in preventing coastal erosion and thus reducing the impact of storms. They also filter out pollutants originating from the land and serve as nurseries for fish species.

On a more positive note there has been some attempts to conserve marine environments and some protected marine areas have been established such as the Marine Protected Areas (MPAs) in the USA, National Marine Conservation Areas (NMCAs) in Canada, and a number of marine national parks in countries such as India, Kenya, Sri Lanka and Australia. In the UK, in summer 2019, a major new initiative was announced, the Plymouth Sound National Marine Park. However, most of these parks, while a step in the right direction, tend to be limited to inshore waters rather than open sea, have limited funding and few legally enforceable regulatory powers.

However, as we will now see, the health of the oceans on which so much human existence and economic activity depends is at greater risk than ever before.

The crisis in our oceans

In recent years, and even months at the time of writing, public awareness of the crisis facing our oceans has risen dramatically, worldwide, largely as a result of the media, in terms of news media and television programmes such as the 'Blue Planet 2' series fronted by Sir David Attenborough. While made in the UK for the BBC, this series has been shown throughout the world, although not everyone will be aware that it is the second such series; the original 'Blue Planet' series was produced in 2001. Much of the second series was concerned with how the plight of the oceans had deteriorated dramatically since the first series was made.

With apologies for the use of the title of a well-known movie, and for a pun in a book about the sea, it is fair to say that our marine environments now face a set of challenges that might fairly be termed, a 'perfect storm'. Global warming is affecting sea water temperatures as well as land temperatures, leading to a melting of the ice caps and a significant rise in sea levels globally. This in turn is affecting every organism that lives in the sea.

At the same time, a new and frightening form of pollution in our oceans has been recognised, namely plastics, and more especially microplastics that are found in many everyday household items. Except now, they are not just found in suburban homes in Melbourne or inner-city apartments in New York; they are turning up in the oceans, including the Arctic, thousands of kilometres away from any town or city. And they are getting into the food chain which means, ironically, that human beings who threw away items that contain microplastics are likely to get them back when they digest their fish-and-chip suppers!

The impact of plastics on the oceans was nicely summarised by Attenborough during the launch of the Blue Planet 2 series in 2017 when he said that during the filming '*the Albatross parent has been away for three weeks gathering stuff for her young and what comes out? You think it is going to be squid but it's plastic. The chick is going to starve and die*'. (Sir David Attenborough, reported by the *Independent*, 15 October 2017). Of course, plastic pollution is not just an issue for our oceans, but the scale of marine pollution has come as a surprise to most people including politicians.

Man is also contributing to the crisis in the marine world through over-fishing and the depletion of fish stocks and continued destruction of marine wildlife habitats, whether that be through major construction projects or the extraction of oil, natural gas and mineral resources.

We will now look at each of these points in a little more detail.

■ The impact of climate change and global warming

It would be virtually impossible for anyone on the planet to still be unaware of the concepts of climate change and global warming. While a few lone voices continue to deny both are reality, the vast majority of scientists now are just debating the scale of the change and how quickly it will cause massive changes to life on earth. Most governments, except those desperate to protect polluting industries and ensure re-election through increases in economic growth, seem to have accepted, at least in principle, the need to act on climate change and global warming.

Public pressure also appears to be growing, and at the time of writing we are seeing disruptive demonstrations around the world by loosely organised campaigns such as Extinction Rebellion. At the same time a Swedish teenager, Greta Thunberg, is energising people around the world to get involved in action to tackle climate change and global warming.

This is not the place to go into details about the causes of climate change, and most people will already be aware of the main causes, given how widely the subject has been covered by the mainstream media in most countries. Our focus must be on what this climate change will mean for the world's oceans and this is worrying enough!

The first key point is that global warming is leading to rising sea levels. According to the US National Oceanic and Atmospheric Administration, '*Sea level has been rising over the past century and the rate has increased in recent decades. In 2018 global mean sea level was 81 millimetres* above the 1993 average. *It was the …23rd out of the last 25 years in which global mean sea level had increased relative to the previous year…The rising water level is mostly due to a combination of meltwater from glaciers and ice sheets and thermal expansion of seawater as it warms….As global temperatures continue to warm, sea levels will continue to rise'*. (www.climate.gov, 2019)

This article goes on to explain that rising sea level is a threat to coastal communities in the USA and points out that eight of the ten cities in the world are on or near a coast. The article went on to report that their research suggested that they were confident that global mean sea level would rise by between 0.2 metres and 6.6 metres by 2100!

Other recent research seems to indicate that the Greenland and Antarctica ice sheets are melting faster than forecast, which will potentially speed up the rise in sea levels. At the time of writing the Inter-Governmental Panel on Climate Change published a report which also forecast significant changes in sea level due to various factors including the melting of the ice sheets.

The rises in sea level are likely to have two main impacts, namely:

- Problems for coastal communities in terms of everyday life with more frequent flooding that may render some coastal settlements uninhabitable. These coastal communities will also face difficulties with infrastructure, such as transport and safe waste disposal due to flooding. It will also significantly increase coastal erosion.
- Higher storm surges causing greater damage over a wider area than previously.

Furthermore, the warming of the oceans is likely to also lead to more frequent and more destructive weather events, particularly typhoons and hurricanes. It will also lead to increases in rainfall in those numerous land masses whose climate is heavily affected by the oceans, even considerable distances inland.

Readers will be very familiar with the highly publicised actions of politicians in the Maldives, who held a cabinet meeting underwater to highlight the fact that a modest increase in sea level would mean that the Maldives could disappear altogether. The coast of Florida is also at risk, Venice could become a 'no-go' area for tourists, and it is predicted that the Pacific islands of Tuvalu could disappear by 2050! Already, we are seeing 'last chance to see' type tourism in Antarctica (Eijgelaar et al., 2010)

Scott et al. noted in 2012 that *'an estimated 29% of resort properties* (in the Caribbean region) *would be partially or fully inundated by a one metre sea level rise while between 49% and 60% of resort properties would be at risk of beach erosion damage …with the same sea level rise* (Scott et al., 2012)

And global warming is also having an adverse effect on marine ecosystems as the water becomes too warm for some species or for the organisms on which they feed, while melting ice takes away breeding sites for some marine animals and birds. Given the inter-dependence of all elements in the marine ecosystem, a small change in water temperature or level can have a significant knock-on effect on every element of the ecosystem.

Pollution

The first thing to note is that marine pollution is not new but the scale of it and the diversity of the sources of marine pollution are new. The seas have always been polluted directly from vessels traversing the oceans through the disposal of waste and later through pollution from the ship's engines and catastrophic oil spillages following oil tankers and other vessels sinking or running aground. Disasters such as the wrecks of the Torrey Canyon in the UK in the 1960s and the Exxon Valdez in Alaska in 1989 have shown us what can happen. And we have also seen the risk posed to the marine environment by deep sea oil exploration, with incidents such as the BP Deepwater Horizon spillage in the Gulf of Mexico in 2010.

However, most marine pollution originates from the land. As the Worldwide Fund for Nature (WWF) has noted: '*Over 80% of marine pollution comes from land based activities. From plastic bags to pesticides - most of the waste we produce on land eventually reaches the oceans, either through deliberate dumping or from run-off through drains and rivers.*' (www.wwf.panda.org, 2019).

In 2019, in a piece in *National Geographic,* Jenny Howard discussed several aspects of marine pollution and stated that:

Before 1972, humans around the world spewed trash, sewage sludge and chemical, industrial and radioactive wastes into the oceans with impunity... The London Convention ... in 1975 ...was the first international agreement to spell out better protection for the marine environment. An updated agreement, the London Protocol, went into effect in 2006... Many...pollutants sink to the ocean's depths or float far distances from their original source, where they are consumed by small marine organisms and introduced into the global food chain. (www.nationalgeographic.com, 2019).

So, while we have been aware of the pollution of seas from the land, the situation is just getting worse and worse each year. Perhaps the biggest worry today is the growth of agricultural pollution that is getting into the oceans through run-off. This comes from the use of fertilisers to bring unproductive land into agricultural use and increase yields, and in the run-off from intensive livestock production units. Agricultural pollution is also being generated within the oceans themselves through the boom we have seen in recent years in 'fish-farming' in many parts of the world.

The *National Geographic* article mentioned above also drew attention to two other important forms of marine pollution namely light pollution and noise pollution.

Since the invention of the lightbulb, light has spread across the globe ...Light pollution penetrates under the water, creating a vastly different world for fish living in shallow reefs near urban environments. Light disrupts the normal cues associated with Circadian rhythms to which species have evolved timing of migration, reproducing and feeding. Artificial light at night can make it easier for predators to find smaller fish prey and can affect breeding in reef fish.

The increased presence of loud or persistent sounds from ships, sonar devices and oil rigs disrupt natural noises in the marine environment. Such unnatural noises interrupt communications disrupting migration... hunting and reproduction patterns for many marine mammals. (www.nationalgeographic.com, 2019)

While all marine pollution is to be regretted the issue has been dominated over the past two or three years by the issue of plastic pollution in the oceans.

Plastics

Within the past few years, scientists and the media have drawn attention to plastic pollution on land. For the media it is a highly visual and easy story to cover as it just needs a few shots of mountains of plastic waste or fences encrusted in migrating plastic carrier bags to vividly illustrate the problem. As a result, public pressure and a desire by governments and companies to show they are in tune with public opinion has led to significant initiatives, such as those targeted at reducing the use of plastic shopping bags by retailers.

However, it was left to a few academics, committed broadcasters such as Sir David Attenborough, and concerned citizens, to draw attention to the pollution of the oceans by plastic. I am proud to say that no small part is this awareness-raising activity was played by my colleagues at the University of Plymouth in the UK.

We should not be surprised by the problem of plastic pollution in the oceans as we know that run-off from the land deposits all kinds of undesirable materials into the oceans on the world. But it is almost certainly fair to say that the scale of the problem has shocked most people.

According to the International Union for Conservation and Nature or IUCN website in 2019 '*at least 8 million tons of plastic end up in our oceans every year and make up 80% of all marine debris. ...Marine species ingest or are entangled by plastic debris, which causes severe injuries and deaths ... Plastic pollution threatens food safety and quality, human health, coastal tourism and contributes to climate change.*' (www.iucn.org, 2019).

A UK-based pressure group, Plastic Oceans UK, has highlighted the impact of plastic pollution of the seas on marine wildlife today as well as making predictions about the future. Their website stated in 2019 that:

- *100% of the seven species of sea turtle are known to ingest plastic and all seven species are known to become entangled in plastic*
- *Ingesting just a single item of plastic has a 22% chance of killing a turtle*
- *99% of seabird species could ingest plastic by 2050*
- *100% of seabird chicks in some regions can have plastic in their stomachs*
- *450 pieces of plastic waste were found in a single Laysan Albatross chick*
- *20.4% is the lifetime chance of mortality (for a seabird) from ingesting just a single piece of plastic*
- *47 species of whales are known to ingest plastic*
- *57.9% of North Atlantic Right Whale deaths in 2003-2018 were caused by entanglement in plastic fishing gear*
- *Microplastics have been found in the intestines of a Humpback Whale*

(www.plasticoceans.uk, 2019)

The final line of the quote above talks about the latest crisis to be identified in relation to the pollution of our oceans and the marine food chain, namely microplastics.

According to the internationally renowned Woods Hole Oceanographic Institution in the USA:

Marine microplastics are small fragments of plastic debris that are less than five millimetres long. Some …are micro by design… microbeads are tiny plastic spheres manufacturers add to body washes, toothpastes and other products to give them extra scrubbing power. Secondary microplastics are fragmented pieces of plastic resulting from the 'weathering' of larger pieces of plastic debris.

Ocean currents and circulation patterns move microplastics around like confetti. Scientists have only recently begun looking for microplastic 'hot spots' in the ocean.

Marine organisms at the base of the food chain including plankton and fish larvae are known to consume microplastics, and filter-feeding animals which humans eat, such as oysters and scallops, ingest the particles as they filter seawater.

Plastics could represent a risk to both marine animals and humans since they may contain toxic chemicals like phthalates and bisphenol A and others used in the manufacturing process…..Plastics can also become toxic by soaking up harmful chemicals from the environment such as polychlorinated biphenyls, exposure to which has been directly linked to cancer.

(www.whoi.edu, 2019).

In the long term these micro-plastics may turn out to be a major threat to both the health of human beings and the well-being of marine wildlife, and it is to the topic of the welfare of marine wildlife that we will now turn.

At the end of this chapter readers will find one of two 'opinion pieces' generously contributed to the book by two world renowned academics, an excellent essay on marine litter and waste and tourism by Professor Michael Hall. It provides another perspective on the issue of waste pollution of the oceans and the relationship between ocean waste and tourism

Loss of marine wildlife habitats

According to the Ocean Health Index, '*Habitat destruction is one of five global ecological pressures affecting the ocean, along with fishing pressure, climate change, water pollution and the introduction of alien species. Damage or destruction of habitats kills the plants and animals responsible for the habitat's ecological functions, and in some cases, its survival and regeneration.*' (www.oceanhealthindex.org, 2019).

In addition to the causes of marine habitat destruction we have already looked at, or will look at shortly, in this chapter, the same source, the Ocean Health Index, identified other causes of marine habitat destruction, including:

- Development where habitat is destroyed to facilitate building on the foreshore, roads and bridges, marinas, reclamation for agricultural production and aquaculture.
- Logging and vegetation removal which damage habitats and/or introduce sediments which can damage habitats.
- Dredging and extractive operations which destroy habitats and create sediment.
- Dunes and foreshores can be eroded and damaged by activities such as the driving of off-road vehicles and the wash from boats.
- Tidal barrages can change the salinity of a particular habitat and can also make migration more difficult for marine species.

We have already talked about the issue of agricultural run-off polluting the oceans, but it also causes *eutrophication*. The introduction of fertilisers into our seas increases the quantity of chemicals like phosphate and nitrates in the marine environment. These stimulate the growth of algae, which reduces the amount of sunlight which reaches marine flora, and many plants die as a result. This, in turn, reduces oxygen levels in the water, which can leave areas of the sea devoid of life.

In 2016, Greentumble, a pressure group, published an article on the causes of marine habitat destruction. Quoting a number of studies, they claimed that:

- *30-35 % of critical marine habitats such as seagrasses, mangroves and coral reefs are estimated to have been destroyed.*
- *There are close to 500 'dead zones' covering more than 245,000 square kilometres globally, equivalent to the UK's surface*

(www.greentumble.com, 2016).

The same source looked at the impact of shipping on the destruction of marine habitats and stated that *'With around 90% of world trade being conducted via shipping, it is reasonable to suspect that it is also a major cause of marine habitat destruction.... shipping impacts our aquatic environment in several ways and has a disproportionate impact on parts of our seas which are busy shipping lanes. Beyond air emissions, ships can pollute the marine environment and cause disturbances of those ecosystems by releasing into the water biocides, used in anti-fouling paints applied to the exterior of ships and through the discharge of garbage and sewage into the sea. Further damage is caused through the dropping of anchors, noise and wave disturbances, and striking of whales and other marine mammals.'*

Of course, while most of the damage done to marine habitats by thousands of ships is the responsibility of ships carrying cargoes, the cruise industry also contributes to these impacts to some degree, as we will see in a later chapter. Likewise, tourism development on the coast and leisure activities undertaken on the foreshore also contribute to the destruction of marine habitats.

Depletion of fish stocks

The United Nations Food and Agriculture Organization (FAO), in its latest report on global fish stocks paints a frightening picture of the impacts of over-fishing on global fish stocks as can be seen from the data which follows:

- *Of the 600 marine fish stocks monitored by FAO:*
  - *3% are underexploited*
  - *20% are moderately exploited*
  - *52% are fully exploited*
  - *17% are over exploited*
  - *7% are depleted*
  - *1% are recovering from depletion*

  (www.fao.org, 2019).

In other words, the FAO considers that three-quarters of fish species are currently fully exploited, over-exploited or depleted.

Their report goes on to note that *'amongst the fish stocks identified by FAO as falling into its worst category of depleted'* are species in virtually every ocean of the world including atlantic cod, salmon and haddock in the Atlantic, various species of tuna in the Pacific and Indian Ocean, hake in the Pacific Ocean and sardines, whiting and bonito in the Mediterranean and the Black Sea.

Most countries which have a coastline are involved in fishing, although there are great differences in the degree to which the industry is regulated and sustainable fishing is practiced, between these countries. Fishing rights are also the subject of international agreements in some places, and are one of the most contentious areas of policy in the European Union. Fishing is a big business and is crucial for the diet of many nations and so is a high priority for many governments.

Much of the over-fishing we see is thought to be the results of the subsidies paid by some governments, which tend to favour large fleets which catch a disproportionately high percentage of the fish caught globally. Indeed, UN Sustainable Development Goal 14, which is about the ocean *'calls on the World Trade Organisation members to prohibit certain forms of fisheries subsidies which contribute to over-capacity and overfishing, eliminate subsidies that contribute to illegal, unreported and unregulated fishing and refrain from introducing new such subsidies by 2020'* (www.sustainabledevelopment.un.org,2019). Sadly, there seems no prospect of this happening at the time of writing!

At the same time, it is the type of fishing practiced which is at the heart of debates over over-fishing, most typically trawling which can damage the ocean bed and scoop up marine creatures which will never appear on any dinner plate.

Interestingly, the reduction in fish stocks hits developing countries most heavily, as this is where there are a high number of small-scale fishermen who are wholly dependent on fishing for their living.

There are two other controversial issues relating to fishing and the health of our oceans, namely:

- The damage caused to the marine environment by the deliberate or accidental dumping of fishing gear in the sea
- The targeting of species such as shark in connection with gastronomic delicacies such as shark's fin soup, although similar issues exist with tuna and abalone. Even where bans on consumption exist, they are often subverted by illegal fishing in some parts of the world.

Tourism plays a small but significant role in the depletion of fish stocks through the demand of tourists for fresh fish in coastal resorts but, in general, over-fishing is a result of the demand for fish from consumers in their everyday lives.

Conclusions

It would be very easy to become rather depressed after reading this chapter, but it is important that we draw out the key points before we go on to look at the various ways in which tourism interacts with, and has impacts upon, the marine environment.

First, and important, we have seen that there is no such thing as the 'marine environment' but rather there are a wide variety of different 'marine environments'. When we begin to look later at ways of managing and planning tourism and its impacts on the oceans, we will need to bear this fact in mind. It shows that 'one size fits all' solutions will not work, and we will need to develop different approaches in different locations.

Second, as we noted in Figure 2.1, tourism interacts with marine environments in many ways and often different tourism uses are taking place in the same geographical area of sea.

Third, we have identified the range of ways in which the oceans interact with land masses and have recognised that much of what happens on the land, particularly in relation to weather, has its origins in the oceans.

Finally, and most important, we have seen that marine environments across the world are in crisis due a 'perfect storm' of factors that all appear to be gaining strength and whose effects are changing oceans at an ever-faster pace. These factors include climate change and global warming, pollution including the scourge of plastics, destruction of marine wildlife habitats and depletion of fish stocks.

The effect of each of these factors is magnified by the fact that they are often inter-related and reinforce each other.

While this chapter has not been about tourism but rather about the oceans it is impossible to escape from two crucial truths, namely that:

- Tourism is, in various ways, contributing to each of the factors that are causing a crisis in our marine environments.
- With the dependence on oceans which many tourist destinations and tourist businesses have, the current crisis of the oceans poses a direct threat to their future well-being or perhaps even their survival.

With both points in mind it is now time to turn our attention to the cruise industry, a sector which depends on the oceans for its success but which has a considerable impact on the oceans and on the land masses that fringe them and the people who live in the places they visit.

References

www.biologydictionary.net (2017) www.biologydictionary.net>Botany [20 Sept 2019]

www.climate.gov, (2019) www.climate.gov?news-features/understanding-climate/climate-change-global-sea-level [2 Feb 2020]

www.definitions.net (2019) www.definitions.net>definition>marine+environment [14 Sept 2019]

Eijgelaar, E, Thaper, C and Peeters, P (2010) Antarctic cruise tourism; the paradoxes of ambassadorship, 'last chance tourism' and greenhouse gas emissions. *Journal of Sustainable Tourism* **18**(3) 337-354.

www.fao.org (2019) www.fao.org/newsroom/commonecg/1000505/en/stocks.pdf [21 Nov 2019]

www.greentumble.com (2016) www.greentumble.com>how-does-the-ocean-affect-climate [14 Sept 2019]

Hayden, BP, Carleton Ray, G and Dolan, R (1984) Classification of Coastal and Marine Environments. *Environmental Conservation* **11**(3) 199-207.

www.iucn.org (2019) www.iucn.org/news/protected-areas/201912/iucn-green-list-gets-down-earth-and-sea-Malaysia. [5 Jan 2020]

www.jncc.defra.gov.uk (2007) www.jncc.defra.gov.uk>...>MarineHabitat Classification>Shallow classification [18 Sept 2019]

www.nationalgeographic.com (2019) www.nationalgeographic.com/environment/oceans/critical-issues-marine-pollution.html [29 Oct 2019]

www.oceanhealthindex.org (2019) www.oceanhealthindex.org 2019 [6 Dec 2019]

www.plasticoceans.uk (2019) www.plasticoceans.uk/the-facts-plastic-pollution [22 Dec 2019]

Scott, D, Simpson, MC and Sim, R (2012) The vulnerability of Caribbean coastal tourism to scenarios of climate change related sea level rise. *Journal of Sustainable Tourism* **20**(6) 883-898.

www.sustainabledevelopment.un.org (2019) www.sustainabledevelopment.un.org/sdg14 [21 Dec 2019]

www.whoi.edu (2019) www.whoi.edu/know-your-ocean/ocean-topics/pollution/marine-plastics [3 Jan 2020]

www.wwf.panda.org (2019) www.wwf.panda.org/our_work/oceans/problems/tourism/tourism_pressure [28 Nov 2019]

Questions and exercises

1. Produce a short presentation to tourism industry leaders to persuade them that it is their interests to contribute to improving the state of our oceans. Prepare a single power point slide containing the main points you would like to make.
2. Using a number of examples, identify at least ten ways in which human activities on land affect the well-being of the oceans.
3. Select a country or an area of sea, then look at Figure 2.1. Try to develop your own diagram, based on Figure 2.1, showing the ways in which tourism takes place in your chosen country or area of sea. You might also want to do the same for a different country or area of sea and see how they compare to each other.
4. Discuss the relationship between the phenomenon of global warming, marine environments and weather patterns across the world. Try to present this relationship in diagrammatic form.
5. In groups, design a questionnaire and use it to survey a small number of people to find out their views on the problems facing oceans that were outlined in the chapter. Your aim is to see how well-informed people are, whether they feel positive or negative about the state of our oceans and whether their views vary based on age or gender.
6. Choose **one** of the following statements which you believe is the most accurate and then justify your selection, using examples and published data and academic references to support your view:

   - ☐ 'The current crisis in the oceans has little or nothing to do with tourism specifically'
   - ☐ 'The current crisis in the oceans is partly caused by tourism and could harm the tourism sector'
   - ☐ 'The so-called crisis of the oceans is not as bad as we have been led to believe'

Opinion Piece:
Marine Litter and Waste and Tourism

C. Michael Hall

Introduction

Marine litter and waste have been found at all ocean depths (Pham et al. 2014) and in the most remote islands (Lavers & Bond, 2017) and form one of the most important contemporary issues facing the marine environment. Growing awareness of the impacts of marine plastics (Bergmann et al., 2015), together with concerns over the environmental effects of cruise ships, especially in relation to sewage discharge (Hall et al., 2017), has meant that marine waste has gained a high media profile and generated substantial public awareness and concern. Three-quarters of all marine litter is composed of plastic and this has become a focus for government and consumer action (UNEP, 2019). However, it is important to note that although there is growing awareness of plastics in the marine environment, and particularly their impact on wildlife and beach composition, the reality is that the predominance of plastics in marine litter is not necessarily the result of relatively more plastics compared to other waste in the oceans, but its durability (Andrady, 2015).

Marine litter or debris is any anthropogenic item deliberately discarded or unintentionally lost in the sea and along the coastline, or transported into the marine environment by rivers, drainage, sewage systems, coastal processes or winds. Marine litter, is defined by the United Nations Environment Programme (UNEP) (2005: 3) as 'any persistent, manufactured or processed solid material discarded, disposed of or abandoned in the marine and coastal environment' and includes a range of materials including plastics, metals, sanitary waste, paper, cloth, wood, glass, rubber and pottery (Galgani et al., 2013). Other marine waste includes oils, sewage and chemicals. Marine litter is usually categorized according to its source. Land-based litter enters the marine environment through rivers, wastewater and sewage inputs, as well as from wave action on the coast. Sea-based sources include fisheries, recreational boating, commercial shipping and ferries, energy production, aquaculture, and dumping activities, some of which are illegal (UNEP, 2019). However, a complicating matter in differentiating between sea and land-based waste in tourism is that many cruise ships have to deposit waste in ports, which may then go to land-fill sites.

Tourism both contributes to waste in the marine environment and is also affected by it. Examples of contributions include waste from cruise ships and recreational fishers, as well coastal tourism activities. Examples of the impacts of waste on tourism include lower levels of beach and water quality in coastal areas, decline in quality of marine ecosystems used for nature-based activities, the fouling of recreational boats by plastic and effects on destination or attraction image.

Waste in the marine environment

There is an enormous amount of waste entering the marine environment. At the global scale, land-based sources contribute approximately 80% of maritime litter, with the remaining 20% from sea-based sources (Strand et al., 2015). Sources of land-based marine litter can generally be correlated with the efficiency of solid waste management and wastewater treatment (Schmidt et al., 2017). Eighty per cent of all wastewater produced globally is discharged into the environment without any treatment, with the result that pharmaceuticals enter the oceans affecting marine ecosystems and human faecal matter contributes to coastal eutrophication and algal blooms (UNEP, 2019).

Jambeck et al. (2015) estimate that 275 million tons of plastic waste were generated in coastal countries in 2010, of which 4.8-12.7 million tons (Mt) may have ended up into the ocean. The middle range of their estimates would suggest that eight Mt of plastics alone enters the ocean each year from land-based sources with the UNEP (2019) suggesting that without substantial interventions, the quantity of plastic in the ocean is expected to increase to 100-250 Mt by 2025. This is a dramatic increase from the approximately 6.4 Mt of litter, plastic and other waste, that the UNEP suggested were being deposited into the oceans in 2005 (UNEP, 2005). The amount of waste in the oceans is therefore enormous and growing. Eriksen et al. (2014) estimated that there was a minimum of 5.25 trillion plastic particles weighing 268,940 tons afloat in the sea, but this figure does not include debris on beaches or on the sea floor. Galgani et al. (2015, p.29) suggest that plastics 'typically constitute the most important part of marine litter sometimes accounting for up to 100% of floating litter'.

Impacts on tourism

Recreational boating is threatened by the hazard that marine litter, such as fishing nets, poses to navigation and the damage it can cause. For example, approximately 3.5% of all call outs by the Royal National Lifeboat Institution (RNLI) in the UK are for fouled propellers/impellers (RNLI, 2017). Costs are also incurred by harbours and marinas in clearing marine litter, although no systematic work is available Mouat et al. (2010) reported one UK marina had an annual bill of €39,000 for litter removal.

Tourism is impacted by the aesthetic damage that marine waste can cause in coastal areas (Polasek et al., 2017; Rangel-Buitrago et al., 2017) which can affect destination image and branding, as well as tourism resources such as beaches, coral reefs and marine wildlife, and charismatic fauna – such as whales, dolphins, seals and marine turtles - in particular (Eagle et al., 2016). The cost of removing beach litter may be substantial, in the UK the cost to coastal municipalities was estimated to be in the region of €18–19 million per annum and €10.4 million per year in Belgium and The Netherlands (Mouat et al., 2010).

However, although the overall economic impact on tourism is one of the potential effects of marine littering, it can be difficult to ascertain. For the 21 economies of the Asia-Pacific Rim, McIlgorm et al. (2011) estimated that marine debris-related damage to marine industries, including tourism, cost US$1.26bn per annum in 2008 terms, equivalent to 0.3% of the gross domestic product for the marine sector of the region. Some forms of marine based tourism, and especially cruise ships, are also potentially impacted by the reputational costs of marine waste, if they are also perceived as being a potential contributor to the waste problem (Hall et al., 2017).

Impact of tourism

One of the most difficult issues in managing the environmental impacts of tourism-related, e.g. cruise ships and ferries, ship-based waste is being able to determine just how much waste shipping is responsible for. Waste production and disposal statistics are usually not provided in the annual reports of cruise lines. International Maritime Organization (IMO) statistics do not distinguish the specific waste contribution of cruise operations. Waste generation statistics are also not very accurate in some jurisdictions because of poor record keeping, and illegal dumping and incineration may mean that some waste contributions are not accounted for (Seas At Risk, 2011)

In one of the few assessments, TRT Trasporti e Territorio Srl (2007) in a report for the European Parliament estimated that at a global level, only 27% of ship waste production appeared to have been given to reception facilities, to which cruise and passenger ships are a major contributor, with the majority being discharged or incinerated. However, this figure changes substantially if ships were operating in special regulated areas (e.g. Mediterranean, North Sea, Baltic), and particularly the EU fleet of ferries and passenger ships, because of different waste management requirements, leading to 51% of solid waste in the EU fleet going to reception facilities instead of being incinerated or discharged at sea and therefore the overall contribution of cruise and passenger ships falls.
In terms of smaller craft, in the early 1990s recreational leisure and fishing craft were identified as being responsible for approximately 52% of all rubbish dumped in US waters (Derraik, 2002). Recreational fishing is also a source of marine litter. García-Rivera et al. (2017) reported that fishing activity was the source of 29.16% of macro-marine litter, almost 68.1% of the plastics, and 25.1% of the metal, but they noted that it can be hard to differentiate the relative contributions from merchant ships and recreational and fishing vessels in coastal waters.

Under international maritime law, sewage (black water) is covered under Annex IV of the *International Convention for the Prevention of Pollution from Ships*, 1973, as modified by the

Table 1: Solid waste produced on board cruise and passenger ships by category and disposal (ships > 100 GT) All figure in cubic metres per year.

Impact Type	Total Ship generated solid waste (before disposal)	Cat. 1 Plastics	Cat. 2 Floating packaging covering materials	Cat. 3 Paper, rags, glass, metals, bottles, & other similar residues	Cat. 4 Triturated paper, rags, glass, metals, bottles, etc.	Cat. 5 Food waste	Cat. 6 Other waste mixed with dangerous substances	Total -Solid waste discharged overboard or incinerated	Total solid waste to reception facilities
World fleet									
Cruise & passenger ships[1]	1,576,481	230,545	22,157	108,376	686,380	487,691	41,333	632,191	944,291
Total	10,490,239	1,448,371	343,766	680,858	4,312,102	3,063,862	641,280	7,609,832	2,880,408
Cruise & passenger ships contribution (%)	15.0%	15.9%	6.4%	15.9%	15.9%	15.9%	6.4%	8.3%	32.8%
EU fleet									
Cruise & passenger ships[1]	360,335	52,223	6,147	24,549	155,478	110,471	11,467	144,691	215,644
Total	3,142,268	450,606	64,592	211,823	1,341,548	953,205	120,493	1,539,652	1,602,616
Cruise & passenger ships contribution (%)	11.5%	11.6%	9.5%	11.6%	11.6%	11.6%	9.5%	9.4%	13.5%

1. excludes cargo/passenger ships and roll on – roll off passenger ships

Source: TRT Trasporti e Territorio Srl (2007)

Protocol of 1978 (MARPOL). Because of the large numbers of people they carry, sewage management is a major problem for the cruise ship sector that has grown so rapidly in the recent past (Hall et al., 2017). Butt (2007) estimated that approximately 50 tonnes of sewage per day are produced by an average cruise ship, equating to between 20 and 40 litres per person per day. In addition, cruise ships also discharge considerable quantities of grey water, which is other waste water such as from sources such as kitchens, laundries, and showers (Gössling, Hall & Scott, 2015). It has been estimated that a cruise ship of 3,000 passengers can generate 340–960 m^3 of grey water daily (Guilbaud et al., 2012). Like sewage, grey water also contains organic matter which, because of the often coastal nature of much cruise traffic, can have considerable impacts on algal growth and eutrophication, as well as being a potential vector for invasive species (Hall et al., 2010).

Environmental impacts of waste

Marine ecotourism depends on the environmental wellbeing of species and ecosystems (Gallagher & Hammerschlag, 2011; Silva, 2015). However, while ecotourism can help justify marine conservation and ecological restoration programmes, marine waste and litter pose substantial threats to marine environmental health (Hall, 2019). Perhaps the most high profile effect of plastic waste on marine organisms is the entanglement of wildlife in marine litter, often plastic bags or discarded or lost fishing nets and ropes (what is referred to as 'ghost fishing') which can kill organisms for years after entering the ocean (Kühn et al., 2015). Entanglement can affect the mobility of animals and their capacity to eat and breathe, while ingestion of plastic debris by animals affects reproduction and survival, even if they do not directly cause mortality. There are no accurate estimates available for the actual number of animals that become entangled in waste and litter. However, the number of species known to have been affected by either entanglement or ingestion of plastic debris has doubled between assessments conducted by Laist (1997) and Kühn et al. (2015), from 267 to 557 species among all groups of marine wildlife (Table 2). Accurate assessments of entanglement for fish, invertebrates and small species of reptiles, are not available as they are not recorded for the many thousands of species that exist. However, the number of marine bird, turtle and mammal species with known entanglement almost doubled, from 89 to 161 species, between 1997 and 2015 (Kühn et al., 2015).

Table 2: Number of species known to have been affected by either entanglement or ingestion of plastic debris

Marine Wildlife	Number of species	% affected species reported affected 1997	% affected species reported affected 2015
Marine turtles	7	86%	100%
Marine mammals	123	43%	66%
Seabirds	406	44%	50%

Source: Kühn et al. 2015

Marine litter, and especially plastics, can also transport hitch-hiking 'alien' species to ecosystems across the surface of the ocean (Kiessling et al., 2015) or vertically from the sea surface to the seafloor (Kuhn et al., 2015), and this is a significant factor in biological invasion and marine ecosystem change (Molnar et al., 2008; Hall, 2015). For example, Kiessling et al. (2015) report that 387 taxa, including microorganisms, seaweeds and invertebrates, have been found rafting on floating litter in all major oceanic regions with 335 taxa associated with plastic substrata, such as domestic waste, plastic fragments or plastic buoys.

Responses to ship waste, litter and sewage

There is a range of regulatory and behavioural interventions available to supranational, national and regional actors to enable reductions in ship waste. These are closely interconnected to changes in technology to reduce the amount of ship waste. Figure 1 outlines some of the regulatory and behavioural responses to waste reduction. Educational programmes, while significant, may only have a limited impact on behaviours as while seafarers and shipping companies and owners may be aware of waste management issues, knowledge alone may not overcome other incentives to retain existing behaviours, such as increased financial returns by dumping waste at sea. Therefore, there is increased interest in the use of social marketing approaches to changed embedded social and economic practices (Chen, 2015; Newman et al., 2015). Such measures are significant as the development of education and promotional campaigns to change waste management behaviours may act to limit regulatory interventions as well as provide a basis for collaborative action between different stakeholders (Hall, 2014). For example, international regulations under MARPOL, the *International Convention for the Prevention of Pollution from Ships*, govern the what and where of waste that can be discharged overboard. MARPOL prohibits the disposal of plastics and most other garbage at sea, and requires signatories to ensure that adequate reception facilities for ship-generated waste are available (Newman et al., 2015). MARPOL reflects the main focus of international, governmental, and non-governmental organisation actors in ship waste management policy in the creation of incentives to encourage ships to use port waste facilities, rather than dump waste at sea and is the approach favoured by the International Maritime Organization and the European Union (Pérez et al., 2017). Nevertheless, while legislative and regulatory actions are significant what is hugely important but often lacking, because of the inherent difficulties of policing the marine environment, is the capacity of agencies and governments to implement interventions.

The future

Writing over 40 years ago Hees (1977) observed that ship sewage discharges in the open ocean are predicted to be completely harmless. Such predictions reflect a common attitude with respect to the oceans as a dumping ground for anthropogenic litter and waste. Unfortunately, they are also profoundly mistaken and wrong. What was once 'out of sight, out of mind', is clearly no longer.

Figure 1: A continuum of behavioural interventions in altering ship-based marine littering behaviour with examples

Behavioural Intervention	Education	Social Marketing	Legislation & Regulation
Behavioural state of target audience	Unaware of impact of littering / Considering behavioural change / Maintaining change behaviour	Aware of need to change / Not considering change	Entrenched littering and waste behaviours / No desire to change
Rights	Rights of companies and individuals stronger		State / societal rights stronger
Locus of power	Resides in individuals and ship owners and managers	Relatively balanced between change agency, individuals and shipping companies	Resides in change agency (although overall extent of power depends on implementation capacity)
International interventions	2006 IMO Marine Environment Protection Committee Action Plan on tackling the inadequacy of Port Reception Facilities; UNEP Regional Sea Programme & Global Programme of Action; UNEP Guidelines on the Use of Market-Based and Economic Instruments; UNEP/IOC Guidelines on Surveying and Monitoring of Marine Litter; UNEP Global Partnership of Marine Litter, Clean Seas campaign	Seas At Risk campaigns; Friends of the Earth Cruise Report Campaign	UN Convention on the Law of the Sea (UNCLOS); Annex IV (with respect to sewage) and Annex V (with respect to garbage) of MARPOL 73/78; London Protocol
Supranational interventions	Commission for the Conservation of Antarctic Marine Resources (CCAMLR) Marine Debris Program; KIMO (Kommunenes Internasjonale Miljøorganisasjon) educational initiatives	Fishing for Litter (FFL) projects; KIMO (Kommunenes Internasjonale Miljøorganisasjon) campaigns	1992 Helsinki Convention & HELCOM (Helsinki Commission) strategy for Port Reception Facilities (Baltic Strategy); EU Port Reception Facility Directive; EU Marine Strategy Framework Directive
National / Regional interventions	Virginia Department of Environmental Quality's Virginia Eastern Shorekeeper program; GhostNets Australia	Fishing for Litter Scotland; Gestes Propres ('Clean Habits') 'I Sail, I Sort' campaign	US Marine Debris Research, Prevention, and Reduction Act; UK Merchant Shipping (Prevention of Pollution by Sewage and Garbage from Ships) Regulation 2008

Source: After Hall, 2014

Land and sea-based waste and sewage technologies are improving, although substantial challenges in nitrogen and micro-plastic removal in particular remain (Köster et al., 2016). New technologies will play a part in reducing waste, but they take time to have an effect, especially while older ships, for example, remain in use. Prior to the COVID-19 pandemic in 2020 the cruise ship sector, which is a major producer of sewage and food waste because of the number of passengers carried, was one of the fastest growing tourism sectors, with increases in the amount of per passenger waste being outstripped by the absolute growth in the sector (Hall et al., 2017). The sheer size of many of the cruise vessels at 3,000-5,000 passengers and crew also means that they produce as much waste and sewage as that of a small town. The increasing size of cruise ships increase pressures for improved waste management and pollution control. For example, Royal Caribbean's *Harmony of the Seas* carries 6,780 passengers and a crew of 2,100 (Vidal, 2016). However, the growing opposition to cruise ships in some destinations prior to COVID-19 (Giuffrida, 2017), may end up being tempered by economic desperation post COVID-19. Nevertheless, the cruise sector will continue to face long-term challenges to brand integrity given the amount of pollution that is produced and the association of cruising with disease outbreaks.

Regulatory and economic measures to reduce the extent of ship-based marine littering have grown, but their application, along with research on ship waste and sewage is primarily focussed on the fleets and regions of the developed world. Many areas, particularly in developing countries, lack analysis and the provision of adequate port waste disposal facilities and the incentives to use them and reduce dumping at sea. The marine waste situation therefore reflects a classic trans-boundary pollution problem given that the impacts of waste at sea can spread beyond where it is deposited.

Adoption rates and the size of the global marine waste problem means that the greatest possibility for rapid waste reduction lies in behavioural change, whether as a result of regulatory developments and/or financial incentives. However, such initiatives require action and investment from governments and port authorities in the provision of adequate waste management facilities and enforcement. Campaigns against single-use plastic bags do not solve the problem. Indeed, successive neoliberal market-oriented policies have meant that often much of the cost of behavioural interventions are borne by the public rather than by the polluter, although the shipping industry, and the fisheries sector in particular, are increasingly aware of the economic impacts of marine litter on their own industries. Nevertheless, a much wider global public backlash to the level of marine pollution is needed in order to force governments to reduce both land and ship-based contributions. Effective policies and accompanying reductions will therefore only occur when global monitoring systems are established together with world-wide available of waste disposal facilities. Without effective public pressures the likelihood of this occurring continues to remain low.

References

Andrady, A.L. (2015). Persistence of plastic litter in the oceans. In M. Bergmann, L. Gutow & M. Klages (Eds.), *Marine Anthropogenic Litter* (pp. 57-74). Springer.

Bergmann, M., Gutow, L., & Klages, M. (Eds.) (2015). *Marine Anthropogenic Litter*. Springer.

Butt, N. (2007). The impact of cruise ship generated waste on home ports and ports of call: A study of Southampton. *Marine Policy*, **31**, 591-598.

Chen, C.-L. (2015). Regulation and management of marine litter. In M. Bergmann, L. Gutow & M. Klages (Eds.), *Marine Anthropogenic Litter* (pp. 395-428). Springer.

Derraik, J.G.B. (2002). The pollution of the marine environment by plastic debris: a review. *Marine Pollution Bulletin*, **44**, 842-852.

Eagle, L., Hamann, M., & Low, D.R. (2016). The role of social marketing, marine turtles and sustainable tourism in reducing plastic pollution. *Marine Pollution Bulletin*, **107**, 324–332.

Eriksen, M., Lebreton, L.C., Carson, H.S., Thiel, M., Moore, C.J., Borerro, J.C., ... & Reisser, J. (2014). Plastic pollution in the world's oceans: more than 5 trillion plastic pieces weighing over 250,000 tons afloat at sea. *PloS one*, **9**(12), e111913.

Galgani, F., Hanke, G., & Maes, T. (2015). Global distribution, composition and abundance of marine litter. In M. Bergmann, L. Gutow & M. Klages (Eds.), *Marine Anthropogenic Litter* (pp. 29-56). Springer.

Gallagher, A.J., & Hammerschlag, N. (2011). Global shark currency: the distribution, frequency, and economic value of shark ecotourism. *Current Issues in Tourism*, **14**(8), 797-812.

García-Rivera, S., Lizaso, J.L.S., & Millán, J.M.B. (2017). Composition, spatial distribution and sources of macro-marine litter on the Gulf of Alicante seafloor (Spanish Mediterranean). *Marine Pollution Bulletin*, **121**, 249-259.

Giuffrida, A. (2017). 'Imagine living with this crap': tempers in Venice boil over in tourist high season. *The Guardian*, 23 July. https://www.theguardian.com/world/2017/jul/23/venice-tempers-boil-over-tourist-high-season

Gössling, S., Hall, C.M., & Scott, D. (2015). *Tourism and Water*. Channelview.

Guilbaud, J., Massé, A., Andrès, Y., Combe, F., & Jaouen, P. (2012) Influence of operating conditions on direct nanofiltration of greywaters: Application to laundry water recycling aboard ships. *Resources, Conservation and Recycling*, **62**, 64-70.

Hall, C.M. (2014). *Tourism and Social Marketing*. Routledge.

Hall, C.M. (2015). Tourism and biological exchange and invasions: a missing dimension in sustainable tourism?. *Tourism Recreation Research*, **40**(1), 81-94.

Hall, C.M. (2019). Tourism and rewilding: an introduction–definition, issues and review. *Journal of Ecotourism*, **18**(4), 297-308.

Hall, C.M., James, M., & Wilson, S. (2010). Biodiversity, biosecurity, and cruising in the Arctic and sub-Arctic. *Journal of Heritage Tourism*, **5**, 351-364.

Hall, C.M., Wood, H., & Wilson, S. (2017). Environmental reporting in the cruise industry. In R. Dowling & C. Weeden (Eds.), *Cruise Ship Tourism* (2nd ed.) (pp. 441-464). CABI.

Hees, W. (1977). Sewage discharges from ships transiting coastal salt waters. *Journal of the American Water Resources Association*, **13**, 215-230.

Jambeck, J.R., Geyer, R., Wilcox, C., Siegler, T.R., Perryman, M., Andrady, A. et al. (2015). Plastic waste inputs from land into the ocean. *Science*, **347**(6223), 768-771.

Kiessling, T., Gutow, L., & Thiel, M. (2015). Marine litter as a habitat and dispersal vector. In M. Bergmann, L. Gutow & M. Klages (Eds.), *Marine Anthropogenic Litter* (pp. 141–181). Springer.

Köster, S., Westhof, L., & Keller, L. (2016). Stand der Technik der Abwasserreinigung an Bord von Kreuzfahrtschiffen [The state of the art of wastewater treatment on cruise ships]. *GWF, Wasser - Abwasser*, **157**, 528-537.

Kühn, S., Bravo Rebolledo, E.L., & van Franeker, J.A. (2015). Deleterious effects of litter on marine life. In M. Bergmann, L. Gutow & M. Klages (Eds.), *Marine Anthropogenic Litter* (pp. 75-116). Springer.

Laist, D.W. (1997). *Impacts of Marine Debris: Entanglement of marine life in marine debris including a comprehensive list of species with entanglement and ingestion records.* Springer.

Lavers, J. & Bond, A. (2017). Exceptional and rapid accumulation of anthropogenic debris on one of the world's most remote and pristine islands. *Proceedings of the National Academy of Sciences*. **114**(23): 201619818.

McIlgorm, A., Campbell, H.F., & Rule, M.J. (2011). The economic cost and control of marine debris damage in the Asia-Pacific region. *Ocean and Coastal Management*, **54**, 643–651.

Molnar, J.L., Gamboa, R.L., Revenga, C., & Spalding, M.D. (2008). Assessing the global threat of invasive species to marine biodiversity. *Frontiers in Ecology and the Environment*, **6**, 485–492.

Mouat, J., Lopez Lozano, R. & Bateson, H. (2010). *Economic Impacts of Marine Litter*. Kommunenes Internasjonale Miljøorganisasjon (KIMO) International Secretariat.

Newman, S., Watkins, E., Farmer, A., ten Brink P., & Schweitzer, J.-P. (2015). The economics of marine litter. In M. Bergmann, L. Gutow & M. Klages (Eds.), *Marine Anthropogenic Litter* (pp. 367-394). Springer.

Pérez, I., González, M.M., & Jiménez, J.L. (2017). Size matters? Evaluating the drivers of waste from ships at ports in Europe. *Transportation Research Part D*, **57**, 403–412.

Pham, C.K., Ramirez-Llodra, E., Alt, C.H., Amaro, T., Bergmann, M., Canals, M., ... & Huvenne, V.A. (2014). Marine litter distribution and density in European seas, from the shelves to deep basins. *PLoS One*, **9**(4), e95839.

Polasek, L., Bering, J., Kim, H., Neitlich, P., Pister, B., Terwilliger, M., Nicolato, K., Turner, C., & Jones, T. (2017). Marine debris in five national parks in Alaska. *Marine Pollution Bulletin*, **117**, 371–379.

Rangel-Buitrago, N., Williams, A., & Anfuso, G. (2017). Killing the goose with the golden eggs: Litter effects on scenic quality of the Caribbean coast of Colombia. *Marine Pollution Bulletin*, **125**, 22-38.

Royal National Lifeboat Institution (RNLI) (2017). *RNLI 2016 operational statistics*. RNLI.

Schmidt, C., Krauth, T., & Wagner, S. (2017). Export of plastic debris by rivers into the sea. Environmental Science & Technology, **51**(21), 12246-12253.

Seas at Risk (2011). *Ship waste dumping and the clean ship concept. How an improved EU PRF Directive can play a key role in cleaning up the seas.* http://www.seas-at-risk.org/images/pdf/Seas_At_Risk_Position_Paper160911.pdf

Silva, L. (2015). How ecotourism works at the community-level: the case of whale-watching in the Azores. *Current Issues in Tourism*, **18**(3), 196-211.

Strand, J., Tairova, Z., Danielsen, J., Hansen, J.W., Magnusson, K., Naustvoll, L-J., & Sørensen, T. K. (2015). *Marine Litter in Nordic Waters*. Nordic Council of Ministers.

TRT Trasporti e Territorio Srl (2007). *External Costs of Maritime Transport*. European Parliament.

United Nations Environment Programme (UNEP) (2005). *Marine Litter, an analytical overview*. UNEP.

United Nations Environment Programme (UNEP) (2019). *Global environment outlook – GEO-6*. UNEP.

Vidal, J. (2016). The world's largest cruise ship and its supersized pollution problem. *The Guardian*, 21 May. Available at https://www.theguardian.com/environment/2016/may/21/the-worlds-largest-cruise-ship-and-its-supersized-pollution-problem

3 The Cruise Industry

In many ways the cruise industry is the success story of modern tourism. Less than 30 years ago commentators were predicting its demise after years of slow and steady decline. It had become an object of ridicule with the oft repeated joke that its customers consisted just of the 'newly-wed, overfed, and the nearly dead'. The hit US TV show, 'Love Boat', set on a cruise ship, and broadcast from 1977 to 1986 seemed to mirror the decline of traditional cruising; highly popular with audiences in its early days its ratings fell and it was axed after nine years in 1986.

And then, in the early 1990s, cruising went through a period of change that can only be described as revolutionary. American brands led the way in over-hauling the whole cruise experience in line with how consumer society was changing, making cruising both affordable and attractive to a wider range of demographic groups. It was no longer just for the rich and the elderly; the industry reached out to people of all ages and means. The market responded enthusiastically, and the cruise market has been growing steadily ever since. Today, cruise operators invest billions in new ships and developing new onboard experiences that will put them ahead of their competitors.

Most coastal tourist destinations with any kind of port or harbour facility seek to attract as many cruise ships as possible, fully aware of the spending power of cruise passengers when they come ashore.

For their part consumers are very loyal to the cruise experiences with an increasing number of people enjoying several cruises every year. There are now cruise channels on satellite and cable television and specialist cruise magazines that focus solely on cruise products and destinations.

However, in the past few years, some commentators have begun to view the cruise sector through more critical eyes. Its environmental impact has not come under scrutiny until recently, as most attention has focused on the more obvious target, aviation. However, several high-profile news stories have begun to shine a spotlight on the environmental impact of cruise ships. This scrutiny is likely to intensify as more and more public and media attention is focused on the plight of our oceans and the impacts on them of global warming.

Yet, today, the cruise industry seems to be much less engaged with sustainability concerns than other sectors of tourism such as the tour operation, hotels or the airline industry, for example. Sometimes it can seem as if both the industry and its customers are taking a view that if they do not talk about them these sustainability concerns might just go way. However, I believe that the industry is going to have to engage more, as an industry as well as individual brands, with sustainability issues in future. It seems certain that the cruise industry will attract more critical attention as cruise ships get bigger and bigger and as cruise destinations grow to include pretty well every area of sea on the planet.

The current debate over the hot topic of 'over-tourism' is bound to involve cruising which brings thousands of people to already popular coastal destinations for just a few hours before they are gone.

The pressure will be on for this dynamic and innovative industry to develop a mature and responsible approach to the broader issue of sustainable tourism and its part in it. It needs to develop models of good practice and work cooperatively as a sector to make cruising as a whole more sustainable. If not, it may face tougher regulation and more controls over its activities. After all, no industry has a greater vested interest in the health of our oceans than the cruise industry.

Historical perspective

Ships have been used for centuries to transport people across the seas but cruising itself is a relatively modern invention. Until the last hundred years or so very few people travelled by sea for leisure; they travelled by sea because they had somewhere, they needed to be, and there was no alternative. The passengers might have been businesspeople making a trip for commercial purposes, emigrants seeking a new life, pilgrims in search of spiritual fulfilment or colonial administrators taking up a new post.

In the days of sailing ships, journeys were long and dependent on the winds and there were few comforts on board. The risk of being shipwrecked was very real and it is not surprising that few people looked forward to a sea journey, and only a few adventurous travellers chose to take an ocean voyage when they did not have to. With the invention of steam things began to improve and journey times were reduced. Steam power meant vessels could become larger and more stable and over time amenities for passengers improved.

It is interesting to note that brands which are now known for luxury cruising trace their origins back a long way and started life as transporters of mails and a few passengers, such as Cunard, and P&O, which both came into existence in the 1830s. Purpose-built ocean liners started to be built at the beginning of the twentieth century. By common consent the first true ocean liner was the 'Prinzessin Victoria Luise' which was launched in 1900 by the Hamburg-America line.

However, even these ocean liners were a means of transport rather than cruise ships. They still largely carried people who needed to travel to a destination, rather than those who wanted to take a vacation on the sea. However, for those who could afford First Class tickets there is no doubt that travel on these liners was pleasurable with good food, comfortable cabins and entertainment. But for the much greater numbers travelling in Second or Third Class the liners were still, fundamentally, a means of transport to be endured rather than enjoyed.

The cruise business as such really began in the 1920s and 1930s with people starting to take sea voyages for pleasure and to discover new places and then return home with their photos and souvenirs. This was a golden era for luxury cruises because this type of vacation was really only available to the most affluent people – those with both money and time. Cruises were lengthy vacations at a time when few workers received paid holidays. Glamour and exotic adventures were what cruise lines sold, as can be seen from the beautiful advertising posters from that time.

It was during this era that the trans-Atlantic journey from European ports to New York became a byword for glamorous lifestyles, celebrity and ostentatious luxury. The companies competed to see whose ships could make the quickest passage across the Atlantic Ocean. For some people today there is still something special about this route.

At this time, of course, air travel was also growing with many of the world's most famous airlines being founded during this era. Now those who needed to travel had a choice and those travelling for business could get to their destination much more quickly by air, although not without a significant risk of accidents.

The Second World War put a temporary end to the rise of cruising, but the world was a different place after the war and the future of cruising began to look uncertain. Europe in particular was living through post-war austerity. Air travel was becoming the preferred means of travel for vacations particularly after the introduction of jet airliners in the latter half of the 1950s. The cruise ships there were ageing, and shipping companies were reluctant to invest in new vessels when the future of market did not seem to be particularly bright. The launch of the new super-liner, the SS France, in 1962 was the exception which proved the rule. She always relied on government subsidies, and while a beautiful ship, she was a commercial failure.

By the 1960s and 1970s the cruise industry was widely predicted to be in terminal decline. Its market was made up almost exclusively of old people with few younger people showing any real inclination towards cruising. Ships offered an old-fashioned vision of luxury and faded glamour. The atmosphere on board was often staid and their very formal style alienated most potential consumers in an age where informality was increasingly the order of the day.

Just when it looked as if the days of cruising were numbered, everything changed! New, largely US, players entered the market with new ideas. As we will see a new generation of cruise operators totally re-imagined and re-modelled the cruise experience in a deliberate attempt to grow the market and broaden the appeal of cruising to attract new segments. They did this by looking at how society was changing and anticipating how these changes might create new types of cruise consumer. Furthermore, unlike their predecessors, the cruise companies have continued to innovate and invest heavily, which largely explains the growth in cruising that we have seen, globally, over the past two decades or so.

The nature of modern cruising

Modern cruising was probably conceived in the 1980s and implemented from the 1990s. As well as investing in new and ever larger ships as we shall see later, the industry introduced a number of cruise concepts which perfectly reflected the evolution of consumer demand. These are illustrated in Figure 3.1

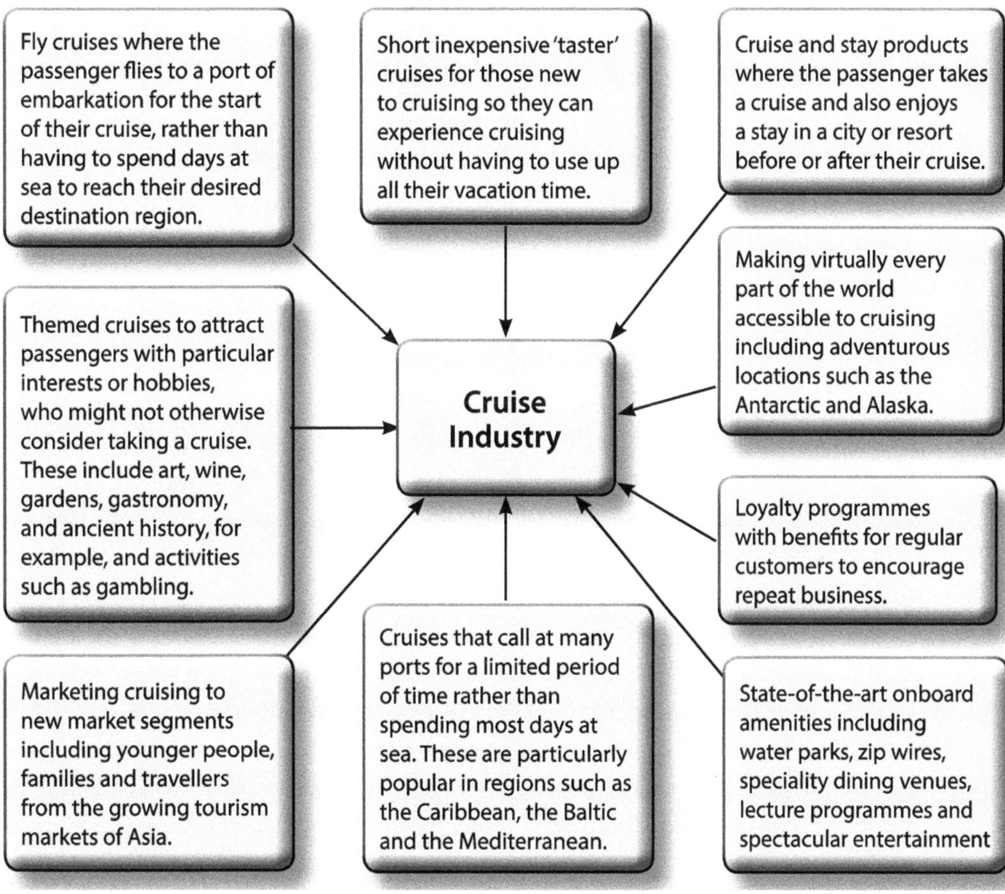

Figure 3.1: Key concepts in modern cruising

The developments shown in Figure 3.1 are just some of those which have been used by cruise companies to grow their markets. They have also addressed issues such as the traditional formality of cruising, so most cruise ships are now relaxed, although there are still formal nights on many ships. Companies have also introduced more flexibility into dining arrangements, and some have tackled that age-old source of embarrassment and irritation for some nationalities, namely tipping.

The industry has also recognised that different people prefer different types of vessels so there is now a wide choice available to cruisers including:

- Mega ships with thousands of passengers that are large enough to offer a bewildering range of leisure facilities, dining options, retail outlets and entertainment venues.
- Small ships with a few hundred passengers where the atmosphere are cosy and the service personalised.
- Large modern ships powered by engines, but which have huge sails that give a romantic appearance and makes them look like a historic vessel.
- Exploration ships, small sturdy vessels, that can take travellers into remote locations where the larger ships are unable to go.
- Yachts and small cruisers that accommodate no more than twenty or thirty cruisers and are like a traditional house part or chalet party in ski destinations.
- Working vessels which carry local people going about their business plus cargo, but which also carry passengers, such as the world famous Hurtigruten in Norway.
- River cruisers which explore waterways such as the Rhine, Nile, Danube, Amazon, Mekong, Yangtze or Mississippi but never venture into the open sea.

The cruise ships are like floating all-inclusive resorts, and many cruisers prefer 'sea days' when they are on the open sea, to the days when they are in port, so they fully take advantage of the facilities offered by 'the resort'. As with the land-based all-inclusive resorts, there is also the idea that as you have already paid for everything you need to consume as much as possible to ensure that you are getting full value-for-money.

With the larger ships, one could even argue that they function as floating destinations, offering a similar range of restaurants, bars, entertainment venues and activities as one might find in a small seaside tourist destination on land. Not surprising, therefore, that some passengers choose to stay on board rather than disembarking at the ports of call to look around. Unlike some ports of call, at least, the ship is clean, air conditioned and safe, with no unusual smells or pushy traders. For such consumers, the destinations the ships stop at are merely a backdrop to their photographs of them enjoying life on board

However, other people see the ship as a means to an end, with the end being the opportunity to visit a number of places without needing to pack and unpack. These people choose a cruise based primarily on the itinerary and the ports of call, and their preference is likely to be for cruises with the most ports of call or those which visit the most unusual or exotic ports of call.

A major feature in the modern cruising experience is ship size with a growth in both massive mega ships but also small ship cruising. However, it is the mega ships that have attracted the most media attention.

At the time of writing, in 2019, the largest cruise ship in the world is the 'Symphony of the Seas', part of the Royal Caribbean cruise line fleet. Launched in 2008, it carries 6,680 passengers and 2,200 crew and has a gross weight of 228,000 tons. It is around the length of three soccer pitches and is the height of a building of more than twenty stories. This single ship cost some $1.35 billion to build. However, I predict that by the time this book is published new even larger ships will have been built and launched.

This growth in the size of ships is causing problems though because passengers still expect these massive ships to be able to take them up the fjords of Norway, the waterways of Alaska and even into the heart of fragile places such as Venice. In this there is a clear tension between market demand and the desire of consumers, and the conservation of some very vulnerable landscapes and townscapes.

As we are about to see, another challenge arising from cruising is the sheer pace of growth in the cruise market globally, which shows no sign of slowing; and the wish of cruise lines and their passengers that nowhere on earth should be beyond the reach of cruise ships. Let us now look at this in a little more detail.

The global cruise market

The size of the global cruise market is impressive at an estimated 30 million in 2019, according to the Cruise Lines International Association. However, in terms of international tourism flows the cruise market only represents around 2% of annual international trips.

However, the growth rate of the cruise market has been spectacular over recent decades as can be seen from Table 3.1. Growth rates in the market in recent decades have often been above 10% p.a. and although growth has slowed a little in the last few years' they are still usually above 5% p.a. The market has grown particularly strongly in Europe where the market grew by nearly 350% between 2000 and 2014 according to data from CLIA, FCCA and CMW.

Table 3.1: Number of global cruise passengers

Year	Number of passengers
1970	500,000
1980	1,400,000
1990	3,800,000
2000	7,200,000
2010	18,400,000
2019 - estimate	30,000,000

Source: Baker, 2016, and data from Cruise Lines International Association (CLIA), Florida Caribbean Cruise Association (FCCA), and Cruise Market Watch (CMW)

The geographical origin of passengers is interesting and can be seen in Table 3.2.

Table 3.2: Geographical origin of cruise passengers

Country	Number of cruise passengers (millions)
USA	11.90
China	2.40
Germany	2.19
United Kingdom	1.93
Australia	1.34
Canada	0.92
Italy	0.77
Spain	0.51
France	0.50
Brazil	0.45

Source: CLIA, 2019

This data illustrates that the USA totally dominates the market, but also shows the growing interest in cruising in the Chinese market, as well as the appeal of cruising in several European national tourism markets.

In terms of trends in the market CLIA identified a number in their 2019 report on market trends including:

- *Achievement over experiences, as experiential travel has evolved into achievement travel as vacationers are looking for experiences beyond sightseeing. 'Bucket lists' have become goal-oriented.*
- *Instagram photos are driving interest in travel around the world. With onboard connectivity, cruise passengers are filing Instagram feeds with diverse travel experiences both on board and on land from several cruise destinations.*
- *Travelers use tech in daily lives and are expecting smart tech when vacationing as well. Cruise lines have adopted technology for cruise travellers in order to provide a highly personalized travel experience while on and off the ship.*

- *Conscious travel. Travelers want to see the world in a conscious, mindful way. The cruise industry is more conscientious than ever, working to .... minimise environmental footprints.*
- *Access is the new luxury. Travelers are setting sights on destinations that were previously out of reach.*
- *Total restoration. Stressed out from fast-paced lives, travellers are seeking ways to check out from daily responsibilities and rejuvenate more than ever before.*
- *Gen Z at sea. Generation Z is set to become the largest consumer generation by the year 2020.....This generation ...prefers experiences over material items and is seeking out travel. The appeal of multiple destinations and unique experiences, such as music festivals at sea, is attracting this new category of cruisers.*
- *Working nomads. Combining work with leisure time is on the rise. Straying far from the notion of device-free travel, many modern travellers ...are opting for trips where they can work remotely which cuts down on time off and lost wages.*

(CLIA, 2019)

3

While no empirical evidence is offered in the report to substantiate these ideas, they do seem to reflect the findings of researchers looking at the tourism market as a whole and seem convincing. However, within them are the seeds of some of the challenges that the industry is posing for marine environments, most notably the desire of cruise passengers to visit destinations that would not previously have been accessible to cruisers. Interestingly, CLIA term this phenomenon as '*access is the new luxury*' (CLIA, 2019), which, perhaps, illustrates the point we will look at later that cruising is marketed with a message of 'affordable luxury'. This messaging seems to resonate strongly with the cruise market.

The global cruise industry

Having looked at the market let us now turn our attention to the supply side, the cruise line industry. Several observations can be made about the industry as follows:

- There has been a proliferation of brands, but ownership has increasingly been concentrated within a small number of corporations. According to a report by Lock, based on data from Statista for 2018, two companies, the Carnival Corporation and Royal Caribbean controlled just over two-thirds of the global cruise market between them. The clear market leader is Carnival which has over 40% of the market through its brands which include: Carnival, P&O, Cunard, Princess; Holland America, Costa, and Seabourne.
- According to the CLIA report on the industry published in 2019, referred to earlier in the chapter, the economic impact of the sector in 2017 included the

employment of 1,108,676 full-time equivalent staff worldwide and total output amounting to 134 billion dollars.

- According to the same source, there were 272 cruise ships in operation with CLIA member companies in 2019 with a further 18 due to come into service during the year of 2019.
- CLIA also noted that a third of all cruise ships were deployed in the Caribbean with just over a quarter deployed in Europe. However, they also noted that cruise ships were deployed literally all over the world, based on the season, with many moving between continents depending on the time of year.

The Caribbean is an interesting case, where the tourism industry is heavily based on the cruise industry as the CLIA data above illustrates. In 2004, Lester and Weeden noted that the task of managing cruise tourism in the region was complicated by three factors, namely:

- The economic dependency of the Caribbean on tourism
- The inequality of power relations between the various stakeholder groups
- The lack of proven collaboration within this fragmented region of culturally diverse islands.

(Lester and Weeden, 2004)

Four years earlier, Wood suggested that the Caribbean was experiencing *manifestations of globalisation* through cruise tourism, including industry restructuring, changes in the product, changes to *deterritorialization, cultural theming and simulation.* (Wood, 2000). The role of cruising in the Caribbean is probably due to its location, in the 'back yard' of Florida, the current world cruising capital.

However, as CLIA recognised through its data, cruising is a global phenomenon so that as early as 2006 Stewart and Draper were able to report that *cruise tourism in Arctic Canada is increasing, albeit haphazardly.* (Stewart and Draper, 2006)

The marketing of modern cruising

The first thing to say on this subject is that the cruise lines are extremely professional in their marketing activities, which is one of the reasons why the cruise market continues to grow, worldwide.

They recognise that the market responds best to a message that emphasises luxury, but at an affordable price. They also realise that for many people cruising remains an aspirational leisure activity. This is clear from the strap lines used on the glossy brochures which include; 'This is the life!', 'This is how to holiday!', 'Welcome to extraordinary!', and; 'Spread your wings and depart from the expected!'.

At the same time, they are fully aware that cruising still carries status even if the golden age of luxury cruising is long gone. Furthermore, they realise that in addition to regular cruisers there is a huge market of potential cruisers who need to be encouraged to, if the reader will pardon the pun, 'dip their toe in the water' for the first time.

The brands are aimed at different market segments, but their sales messages tend to emphasise certain points regardless of the target market including:

- The all-inclusive nature of the cruise experience and the value-for-money it represents, particularly given the chance to eat as much as you want, when you want.
- The wide variety of cabins or staterooms available, from the inside cabins for the budget conscious traveller to huge luxurious suites at the top end.
- The opportunities to pay a little extra for an enhanced experience, whether that be through a 'drinks package' or modest supplements to dine in what are termed 'speciality dining' venues or treatments in the onboard spas.
- The quality and variety of onboard facilities especially the entertainment and leisure facilities.
- The level of service provided, with regular use of terms such as 'pampering' and 'luxury'.
- The onboard restaurants which carry the brand of well-known celebrity chefs such as Jamie Oliver.
- The range of ports of call included in each itinerary and the shore excursions available from each port.

The promotion and distribution of the cruise product involves several channels which generally operate in parallel to the mainstream tourism marketing system. In the UK, for example, these channels include:

- At least two satellite or cable television channels and further shows on other channels that focus just on promoting cruising and selling cruises, usually at what appear to be discounted prices.
- Promotional features about cruise brands, usually in the form of inserts, in newspapers and magazines.
- At least one glossy monthly magazine that is just about cruising and the latest products on sale from the industry.
- Specialist high street travel agencies whose business is based upon cruising.
- The use by the cruise brands of every form of social media to make potential customers aware of the joys of cruising and the latest offers.
- Online travel agent sites which specialise in cruises as well as more generic online travel agents which offer cruises alongside other vacation opportunities.

- The call centres and websites of the cruise brands themselves.
- Brand loyalty programmes that offer incentives to attract repeat customers.

The cruise sector also has its own influential customer review site called Cruise Critic, which has been part of TripAdvisor since 2007.

However, while recognising the effectiveness of cruise industry marketing we need to make some observations in the context of the subject of this book. These follow on from the mini survey of the content of cruise brand brochures that was reported on in Chapter 1. In most brochures there is little or no mention of the impacts of cruising on the oceans nor of the environmental challenges facing our marine environments. This is disappointing given that tour operators and airlines have started, albeit on a limited scale, to acknowledge and talk about impacts and sustainability in their brochures and on their websites.

That is not to say that the whole sector is doing nothing about sustainability, as there are some initiatives underway, but such initiatives are either not common or at least not talked about by the companies. By not talking about them they are failing to make their customers aware of the issues, at least. It is as if they either do not want these realities to burst the bubble of escapist enjoyment that they seek to create on their ships. Or, alternatively, the cruise companies are in denial about the negative impacts cruising can have but the growing evidence means that this stance will be hard to sustain going forward.

There is a real opportunity for cruise companies to play a positive role by developing more responsible cruising and encouraging their customers to think about the issues facing the oceans and behave responsibly while on vacation.

It is now time for us to move on to consider the impacts of cruising on the marine environment in some detail.

The environmental impact of cruising on marine environments

Traditionally the main scientific and media attention on the impacts of tourism on the environment has focused on the airline industry, the use of private cars for leisure travel, and the design and operation of hotels. However, in recent years the cruise industry has begun to come under increasing scrutiny on several fronts. At first, the focus was on the carbon footprint of cruising, but now the debate has widened to include fuel pollution, waste disposal at sea and even air pollution in the ports of call. Let us now look at these issues.

Carbon footprint

The first point to make is that the carbon footprint of cruise ships is not a new issue. More than ten years ago the media was drawing attention to the issue as can be seen from the following two excerpts from newspaper articles.

According to the *Daily Telegraph*, '*cruise ships emit three times more carbon emissions than aircraft, new research has revealed. Carnival, which comprises 11 cruise lines, said in its annual environmental report that its ships, on average, release 712 kilograms of CO_2 per kilometre. Carnival's ships carry, on average, a maximum of 1,776 passengers. This means that 401 grams of CO_2 is emitted per passenger per kilometre, even when the boat is entirely full. This is 36 times greater than the carbon footprint of a Eurostar passenger and more than three times that of someone travelling on a standard Boeing 747 or a passenger ferry.* (Daily Telegraph, 19 January 2008).

In fairness, the carbon emissions may well have been reduced since then through the introduction of new technologies.

Even earlier, in 2006, another UK newspaper was reporting on the issue has part of a report entitled *'Is cruising greener than flying?'*, this time presenting the views of travel industry experts. George Monbiot was reported as saying that researchers looking at the Cunard ship, Queen Elizabeth II believed that *'a tonne of shipping fuel contains 0.85 tonnes of carbon, which produces 3.1 tonnes of carbon dioxide when it is burnt. Every passenger is responsible for 9.1 tonnes of emissions. Travelling to New York and back on the QE II, in other words, uses almost 7.6 times as much carbon as making the same journey by plane'*. (The Guardian, 20 December 2006).

This is based on old data and new more fuel-efficient ships have been introduced since 2006, but the point of including these quotes is to show that concern over the carbon emissions of cruise ships is not something that has just arisen.

In the intervening years, the industry has taken measures to reduce the carbon footprint of its vessels, P&O claimed to have reduced its CO_2 emissions by 20% in 2014 alone (www.euractiv.com, 2018). Most cruise lines do seem to have made efforts to reduce their emissions but the challenge for the industry in this respect has been two-fold, namely:

- Ships are now much larger which makes to more difficult to reduce overall emissions although it will mean that emissions per passenger should fall substantially, of course
- Due to the growth of the cruise market more and more ships are being launched

Despite all the progress made, though, Euractiv was still reporting in 2018 that, '*daily emissions of cruise ships are the same as one million cars*'. (www.euractiv.com 2018). Thus, reducing the overall total emissions for any cruise company remains a major challenge. At the same time, a number of the other impacts of cruise ships have attracted attention, not least due to several high-profile legal cases.

Fuel pollution

A report on Forbes.com by James Ellsmoor in June 2019 included the following quote:

'Last year the German watchdog, Nabu, surveyed 77 cruise ships and found that all but one used toxic heavy fuel oil...The biggest issues with cruise emissions are the levels of nitrogen oxide which has ben linked to acid rain, higher rates of cancer and other forms of respiratory diseases. As such, cruise operators have been urged to switch to cleaner fuel alternatives with a lower sulfur content by 2020, but few have heeded these calls. Safer fuel, such as liquefied natural gas, is more expensive and operators have favoured using scrubbers, which have been called 'emission cheat' systems. These scrubbers wash cheap fuel in order to meet environmental standards, but then discharge the pollutants collected directly into the ocean.' (www.forbes.com, 2019).

Welcome developments have occurred with some new ships being introduced which use cleaner fuels, and P&O has announced that it has reduced fuel consumption 28% since 2005 (www.euractiv.com, 2018). Nevertheless, at the time of writing, it appears that there is still a long way to go before we can be satisfied that the fuels used by cruise ships no longer pose a threat to the marine environment.

At the same time, fuel spills from cruise ships continue to cause marine pollution, most recently in the case of the ship, 'Marella Dream', at Palma de Mallorca in Spain. We have also seen in spillages in fragile marine environments such as Antarctica when the MS Explorer sank in 2007, and the oil spill incident with the MS Nordkapp in the same year.

Water pollution

There are two further ways in which cruise ships can pollute the oceans as follows:

- 'Grey water' pollution that comes from showers, galleys and laundries onboard which may contain harmful chemicals.
- Chemical pollution from the 'black water' channels on board, which can come from things like cleaning products and batteries.

Waste disposal

Cruise ships, due to their size and purpose, inevitably generate a great deal of waste. For the purposes of this book our interest is in the disposal of this waste. In general, it is fair to say that cruise ships have efficient systems for recycling and handling the huge amounts of waste produced on board. However, as we will see shortly there have been some examples of waste being disposed of at sea which in some cases have led to legal action being taken against cruise companies.

Sewage disposal

Thousands of passengers and crew inevitably produces a large volume of sewage. Cruise ships have well established systems for managing this in ways which do not cause marine pollution. However, there have been cases where things have gone wrong, which have also led to legal action being taken against cruise companies. Claims have been made from time to time that some ships still have antiquated sewage treatment facilities that mean that some pollution still occurs after the sewage has been treated. This was the subject of a report by Friends of the Earth in 2014 although they acknowledged that the situation was improving.

Air pollution at sea

There is also growing concern about the air pollution generated by cruise ships when they are at sea. As Faig Abbasov, of Transport and Environment, said in June 2019, *'Luxury cruise ships are floating cities powered by some of the dirtiest fuel possible. Cities are rightly banning dirty diesel cars but they're giving a free pass to cruise companies that spew out toxic fumes that do immeasurable harm both to those on board and on nearby shores. This is unacceptable'*. (www.transportenvironment.org, 2019).

In 2011, Poplawski et al. published an interesting study looking at fine particulate matter (PM 2.5), nitrogen dioxide and sulphur dioxide associated with cruise ship operations in James Bay, Victoria, Canada (Poplawski et al., 2011)

Noise pollution

Although research evidence is patchy, it seems highly likely that the noise pollution caused by cruise ships has a negative impact on marine wildlife, particularly for those creatures which use sonar to help them navigate. In addition of the engine noise there is the noise pollution that comes from the onboard entertainment and leisure activities.

Wildlife habitat damage and injury to wildlife

Cruise ships can damage vulnerable marine habitats and to wildlife in several ways including:

- Ballast water pollution, where water collected at one point and carried onboard as ballast, is discharged somewhere else which may be hundreds or even thousands of kilometres from its point of origin. This water will contain microbes and organisms that, in this way, get introduced into ecosystems, sometimes causing major ecological damage
- Accidental damage to coral reefs and areas of the seabed which results from anchoring or from accidental groundings. For example, in 2017, the cruise ship

MS Caledonian crashed into coral reefs in Indonesia and destroyed nearly 2,000 square metres of coral reef. (www.marineinsight.com, 2019).

- Injuries to creatures such as dolphins and whales which are accidentally struck by cruise ships.

Over time, research will hopefully show us what impacts, if any, cruise ships have on marine life, including feeding, migrating and breeding, from any disturbance their presence causes in marine environments, especially in fragile environments in the Arctic and Antarctic regions.

The law, marine pollution and cruise ships

We have noted throughout this section that things are improving slowly as the industry takes steps to reduce its negative environmental impacts. However, there continue to be serious breaches of sound practice which have damaging impacts on the marine environment. Some of these have resulted in high profile legal cases against leading cruise companies.

In 2016, Princess Cruises was fined $40 million after pleading guilty to a few offences relating to the discharge of oily waste into the sea off the UK coast. The court also required the parent company, Carnival, to subject its 78-vessel fleet to a five-year environmental compliance programme. It appeared that the ship had been discharging contaminated water like this since 2005. Princess blamed employees for the incident. Then, in 2019 it was reported that the Carnival Corporation had agreed to pay a $20 million penalty because its ships were still polluting the oceans despite the outcome of the previous case. As part of this court case the company admitted that its ships had still been dumping 'grey water' in places where this is prohibited, including the Glacier Bay National Park in Alaska.

Even as I am writing this in October 2019 news is coming through that Carnival Corporation had been called back into court in Miami to explain what it was doing to reduce pollution of marine environments. These cases have shown two things namely that:

- Despite the introduction of better environmental protection systems on cruise ships pollution incidents continue to occur.
- There is sometimes an apparent lack of transparency from cruise companies about ocean pollution incidents.

The environmental impact of cruising on the ports of call

Cruise ships are also causing air pollution in the ports of the world according to many scientists. A report published in 2019 by Transport and Environment claimed that '*NOX (nitrogen oxide) emissions from cruise ships in Europe also heavily impact some cities. In Marseille, for example, 57 cruise ships emitted in 2017 almost as much NOX as one quarter of the city's 340,000 passenger cars. Along the coasts of Norway, Denmark, Greece, Croatia and Malta a handful of cruise ships are also responsible for more NOX than most of their domestic car fleet.* (www.transportenvironment.org, 2019). The same report claimed that 105 cruise ships created 560,000 kilograms of sulphur oxide in Barcelona, 245,000 kilograms in Palma de Mallorca and 110,000 kilograms in Venice.

Several studies have looked at the environmental impact of cruising on ports of call. In 2018, Dragović et al. published a study of the impact of cruise ships on air quality in Dubrovnik in Croatia and Kotor in Montenegro. They concluded that ‹*cruise ship traffic produces continuously increasing air pollution in both ports'* (Dragović et al., 2018). This followed on from a paper on the environmental impacts of cruising in the Adriatic Sea, in which both ports are located, which was published in 2014 (Carić and Mackelworth, 2014).

Some researchers have even attempted to put a number on the number of deaths in particular coastal cities which they claim have been caused by shipping pollution including cruise ships!

The negative impacts – literally – of cruise ships in their ports of call was vividly illustrated by a frightening event in Venice in 2019. In June of that year the MSC Opera crashed into a wharf and a tourist boat injuring five people. This incident re-ignited the debate about whether huge cruise ships should be able to enter the city of Venice where it is alleged that their wash damages the foundations of the buildings in this unique destination.

At the same time, considerable concern has been expressed in recent years about the environmental impacts of cruising in Antarctica as a 'port of call' albeit one without a host population. Amelung and Lamers stated in 2007 that not only did the impacts of cruising in Antarctica have a global effect, rather than just affecting the ecosystem there, but that it is hard to improve the environmental performance of Antarctic tourism because it inevitably relies on long-haul trips. (Amelung and Lamers, 2007)

Summary

Let us now try to bring together a few key points from this section on the environmental impacts of cruise ships. It is clear that improvements are being made to reduce the negative impacts of cruise ships, but it is equally clear that there continue to be major problems, not least the substantial carbon footprint of the cruise industry.

There seems little doubt that the marine environment suffers damage from cruising, also, in terms of fuel and other forms of water pollution as well as in terms of damage to wildlife habitats and harm to marine wildlife.

On land it can be argued that tourism generates revenue that can be used to fund conservation or projects to support local communities. There are no such benefits from cruise tourism for the oceans currently. The balance sheet from, therefore, is virtually totally one that tells a story of loss from the point of view of the marine environment. And, perhaps most worryingly, these negative impacts are being seen in some of our most fragile marine environments including the Arctic, the Antarctic and Alaska.

The reality is that things improving but very slowly and not nearly fast enough given the global growth in both the number of cruise ships and their size. And finally, we have to recognise that the impacts of cruise ships discussed above are true for all global shipping. However, in fairness it is important to note that freight shipping is largely essential to modern consumerism that helps maintain living standards across the world through trade. On the other hand, cruising is a leisure activity that is not essential to life; there are alternative ways of vacationing. That is why it is very important that the cruise industry dramatically improves its environmental performance before people start to suggest it should be more heavily regulated or even reduced.

While this book is focusing upon physical damage to the marine environment it is important for us to look at some other types of impacts when we are thinking about cruising. This is important because, as we all know, sustainability is about the balance between the environment, the economy and society. We will, therefore, now take a brief look at a few of these other types of impacts and some of the ethical challenges facing the modern cruise industry.

The economic impact of cruising

There is no doubt that cruising generates enormous revenues for the ports of call through both passenger expenditure and direct and indirect taxation, but a number of factors mean that this revenue does not come without 'strings' or disadvantages. These factors are as follows:

- Cruise ships require increasingly expensive infrastructure in terms of quays and berths as the size of ships increases. The cost of such infrastructure usually falls on local or central government which involves an element of opportunity cost. Every dollar spent on port facilities for cruise ships by the public sectors is a dollar that cannot be spent on health or education or other services that might bring more benefits for local people.
- Cruise ships tend to stay in port for just a few hours so that businesses have to make all their income in a very short time, which can encourage profiteering or corner cutting by businesses because the window of opportunity is so small.

- As cruise ships are all-inclusive, passengers are unlikely to spend much money ashore with restaurants and nothing with hotels. Their spending is likely to focus, therefore, on modest amounts spent on souvenirs and soft drinks.
- Shore excursions are likely to be organised through the ship so that local suppliers will only receive a proportion of the excursion price paid by the passenger as the cruise company will deduct its own costs and profit margin first.
- By necessity cruise ships need to carry most of their supplies with them which means they will buy relatively little from suppliers in their ports of call during their voyage.
- Many passengers do not actually disembark in a particular port of call, preferring instead to stay onboard enjoying the peace and quiet once the other passengers have gone ashore.

The social impact of cruising

The social impact of cruising has three the dimensions: namely impacts on the passengers, the crew and the ports of call. For the passengers the impact is certainly a positive one; cruise passengers generally are highly satisfied with the experience and return home refreshed and enriched. However, the picture for the crew is different even though the cruise industry provides them with a livelihood. The lifestyle is hard, and the crew are separated from their families. For many crew, promotion opportunities are very limited and certain nationalities appear to face discrimination in terms of such opportunities. As we will see in the next section, employment on cruise ships involves some significant ethical issues.

In terms of the ports of call the main one relates, perhaps, to the 'hot' topic of 'over-tourism', the phenomenon where communities in some tourist destinations are resisting tourism and suggesting that it has reached a tipping point in their community, where it adversely affecting the spirit of place, the sense of community and the well-being and quality of life of residents, despite the revenue it brings to the destination. While 'over-tourism' encompasses all of tourism activity, it appears to be particularly relevant to cruising because cruise ships disgorge huge numbers of passengers who visit the destinations for just a few hours. This can mean local people feeling swamped by the visitors and having their daily routine disturbed by the sheer volume of visitors. In 2011, Klein published an interesting paper on the way in which host communities in port cities viewed the impacts of cruising. (Klein, 2011)

It is now time to address several issues relating to cruising that might best be described in broad terms as ethical challenges.

The ethical challenges of cruising

Of course, all the issues above are ethical challenges too. These are less obvious ones, perhaps, in a book about the impacts of tourism on the marine environment, but they are significant enough to be worth discussing, in the context of cruising. Like much tourism there is the contrast between the conspicuous and often ostentatious consumption taking place on the ships and the poverty of many of the lower paid crew members and the deprived communities that live in some of the cruise ship ports of call. Cruise companies appear to engage much less in projects to support poorer communities within the destinations than tour operators do, for example.

Perhaps the greatest criticism directed against the ethics of the cruise industry is that focused upon the treatment of the labour. There is no real concern over the situation of officers or the higher paid skilled staff. Instead, the debates that rage are concerned with the lower paid staff, usually from poor countries, who work in the hospitality services onboard whether that be cooking, waiting on table or cleaning. This issue has been discussed for some time. In 2002 War on Want published a report entitled '*Sweatships – what its really like to work on board cruise ships.* Its highly emotive headline for the online version was entitled, '*A living hell below decks'*.

Since then several journalists have infiltrated cruise ships by getting jobs on board and then reporting back on their experiences. Few of these have painted a positive picture of working onboard cruise ships. The journalist Leo Hickman also produced a well-researched critique of employment in the cruise sector in a Florida based chapter, entitled '*All at sea'* in his 2007 book, *'The Final Call: investigating who really pays for our holidays'.*

The main issues raised by critics are as follows:

- The low salaries paid, particularly to the hospitality providers on board although the companies point to the fact that staff also receive food and accommodation.
- The long hours, with working hours regularly going up to 60 per week and bar staff, for example, often working shifts of 12 to 16 hours.
- The lack of days off when the ship is at sea.

Much cruise ship recruitment for lower paid positions is in the hands of agents. Some of the less scrupulous ones in developing countries can exploit the workers by lending them money which places them in debt and makes it difficult for them to leave their jobs.

Apologists for the cruise industry console themselves by saying they must be good jobs because lots of people apply for them. While sounding logical this rather naively ignores the fact that people who are desperate apply for any job, particularly when a recruitment agent paints a rosy picture of the job that may not always be realistic.

It is also hard to escape the idea that subtle and discrete forms of discrimination are at work on cruise ships, where particular trades seem to be allocated to people from particular cultures and countries or even ethnic backgrounds. Most of the highest paid jobs seem to be held by white Europeans or North Americans, while most of the people working long hours for low salaries in the galleys or cleaning cabins tend to come from very poor countries, particularly in Asia.

Of course, as always in the hospitality sector, low salaries are excused with the idea that salaries are enhanced by gratuities and tipping, although it is commonly known that these do not always reach the staff who performed the service, on land let alone at sea. It does raise the question of why guests should have to pay again, in terms of a gratuity, for a service they have already paid for, particularly as the gratuities are often added to the bill automatically, making a lie of the fact they are designed to reward exceptional service. There is also the fact that having to constantly seek gratuities to ensure a decent salary is quite demeaning for people who work hard and deserve their dignity.

Moving on to the subject of food waste, the industry is making big efforts to reduce food waste, perhaps more due to cost control issues rather than ethics. However, it remains an issue, particularly given the 'all you can eat' buffets' and the fact you can eat as often as you like on cruise ships. One cruise line which looked at this issue in 2017 estimated that with simple actions it could reduce wastage per day per passenger from over 200 grams to 98 grams. However, even this lower figure would give a total food waste of 7,500 kilograms, or seven and a half UK tons, on a 15-day cruise on a ship with 5,000 passengers, a not inconsiderable amount. Interestingly, a few cruise lines are now, where possible, making unused food available to charities in their ports of call.

The final two points in this section relate to destinations that cruise ships visit.

First, there is the trend for cruise lines to buy islands to use them exclusively as playgrounds for their customers. This reduces the economic benefits for normal destinations on those days when ships are visiting private islands. And it raises moral questions about whether cruise companies are the right people to own and manage parts of the marine environment. This phenomenon seems to be most common in the Caribbean region where there are at least seven such islands owned by companies including: Disney Cruise Line, Holland America Line, Royal Caribbean International, Norwegian Cruise Line and Princess Cruises.

Second, there is the question as to whether large cruise ships, or indeed any cruise ship at all, should visit some of the world's remotest and most fragile environments such as Antarctica. Their visits there, as we saw in 2007, can lead to accidents and damage to this unique marine environment. There are no real settlements there that need the economic benefits of cruising to flourish. In such an environment, no matter how responsibly they try to behave, cruise ships and cruisers can only cause harm. This raises the question of should we want or need to visit everywhere by cruise ship just because we can? We will take up this point towards the end of this chapter.

Accidents, sinkings and terrorism

It is a fact of life that even the best run cruise ships can have accidents or even sink or be the victim of terrorism. The last ten to fifteen years has seen a number of high-profile incidents involving cruise ships. While cruise ship sinkings, such as the MV Explorer in 2007 and the Costa Concordia in 2012, are mercifully rare, accidents and incidents involving cruise ships are more common than you might expect. A quick look through selected media over just three months, chosen at random, in 2019 revealed the following headlines:

- *'Two Holland America ships collide'* (in Canada)
- *'Cruise ship comes to help Oasis of the Seas'* (after a crane crashed on the cruise ship during dry dock in Freeport, Bahamas. Eight people were injured)
- *'Viking Sky engine failure caused by low oil level'* (The vessel was stranded over the weekend with 1,400 passengers on board off the coast of Norway)
- 'Terrifying *night on Norwegian Escape – crew and passengers injured'* (The ship was hit by a 100-knot wind which caused the ship to tilt heavily to one side)
- *'MSC cruise ships collide in Buenos Aires'*

The point is that any sinking or accident has the potential to result in damage to the marine environment, as well as injury to crew and passengers.

Terrorism is also always a potential threat and has been a fact of life since the hijacking of the Achille Lauro in 1985. Terrorist plots have also been uncovered for attacks on other cruise ships in more recent years. Yet again, we can say that any such attach could lead to destruction that could also be most damaging to the marine environment.

There is one final ethical issue to briefly look at before we move on towards the end of this chapter.

Cruising and irresponsible consumerism

Increasingly, commentators are becoming concerned about what might be termed 'irresponsible consumerism' in relation to cruising. This appears to take many forms from so called 'booze cruises', due to discounted drinks packages, and the resulting fights, through to food waste. It also includes various types of marine wildlife-watching that are not examples of good practice, to adventure activities that are harmful to the environment and/or have a large carbon footprint. As tour operators slowly begin to embrace issues such as animal welfare and sustainability when planning their excursion programmes, it seems that some cruise operators have a way to go to catch up.

The cruise industry and responsible tourism

As we have noted before other sectors of the tourism industry have been proactive for some time in terms of sustainability challenges, usually as part of broader CSR (corporate social responsibility) policies. Tour operators have been particularly active, launching a number of voluntary initiatives, in destinations and in terms of raising awareness amongst their customers about a wide variety of issues. They have also tended to put money into local community-based projects, particularly those involving children or animals. The Tour Operators Initiative is one such scheme where industry invested in concrete actions that were not required by law and were voluntary.

Meanwhile, the hotel industry has a history of three decades of action to reduce its negative impacts on the environment, not least because this often also reduced its costs. Perhaps the most important hotel industry initiative was the International Hotels Environmental Initiative.

The airline industry, under pressure from the media and governments has, for some 20 years, been working with manufacturers to find ways of reducing its carbon footprint.

In all three cases I believe the main motive was probably the desire to show that each sector could put its own house in order through voluntary self-regulation in the hope that this would mean they would avoid government regulation which might be more onerous. And many companies have integrated their actions into their marketing communications with pages in their brochures and on their websites telling customers what they are doing. Hotels have put cards in guest rooms saying what they are doing to 'save the planet' and inviting guests to participate by switching off lights and not asking for their towels to be washed. One UK airline, for a while at least, put its carbon emissions data for its aircraft on the fuselage as a large graphic while at the time of writing one airline has an advertising campaign that claims it is a low emissions airline!

It is my contention that the cruise sector is a late arrival at this party! They were slow to see the potential challenges they might face about their activities and the impacts of cruise tourism on marine environments. However, they have now realised the risk they face if they are not seen to be acting more responsibly and more concerted and systematic action is being taken, not least because of high profile court cases in the USA and the fear of greater regulation if the sector does not 'put its own house in order' voluntarily.

As with other sectors of the tourism industry, the actions on sustainability has three dimensions, namely voluntary initiatives from professional bodies, voluntary action by individual corporations, and compliance with existing industry regulations. In the case of the former, the key body is the Cruise Lines International Association or CLIA. Its website sets out its position clearly under the heading '*Setting the course for sustainability*', it goes on to say: *'No industry has a stronger interest in protecting the oceans we sail and the destinations we visit than cruise. It is not simply our responsibility; operating sustainably is a business imperative'.* (www.cruising.org, 2019)

Its website says that the sector is committed to reducing the rate of carbon emissions across the industry fleet by 40% by 2030, citing a press release dated 18 December 2018. Apparently, this will be measured against a 2008 fleet baseline rather than from today. This press release goes on to talk about the use of cleaner liquefied natural gas or LNG fuel saying that by 2025 there could be 25 LNG powered vessels. However, based on current data, that is likely to represent less than 10% of all cruise ships! As always with the tourism industry, CLIA goes on to balance the issues around negative environmental impacts, understandably, with the positive economic benefits which it says are more than a million jobs worldwide and $134 billion of economic impact worldwide. A factsheet available through the website also offers further information including the following:

- *The cruise industry recycles 60% more waste per person than the average person does on land in the US.*
- *The cruise industry is building reefs, collecting critical ocean data, restoring fisheries and helping develop best practices for coastal communities.*
- *93% of the cruise industry has eliminated plastic drinking straws.*

(www.cruising.org, 2019)

This all seems very positive, but it all seems rather leisurely paced, modest in scope and ambition and lacking in holistic thinking.

Of course, like any business, cruise companies have to balance responsible business practice with commercial realities. Fortunately, though, in this respect, the cruise industry is in good shape. Their business is growing, margins are pretty good, and they seem to have plenty of money available to invest in new ships. Many lines appear to also pay only modest amounts of tax due to their particularly efficient tax arrangements. They, therefore, have the means to take effective action, in the short-term, to reduce their negative impacts on the oceans.

Let us now look at the voluntary actions of individual cruise brands and of parent companies such as the Carnival Corporation. Carnival publishes an annual sustainability report entitled *'Sustainability from ship to shore'* which covers all of the brands they own. This comprehensive 150-page report is glossy and covers a wide range of issues in an upbeat manner. I might suggest that this impressive document owes something to the outcomes of the court cases discussed earlier in this chapter and the pressure which Carnival has been under, particularly in the USA, in recent years to operate more responsibly and be more transparent.

The report covers 18 topics including environmental issues as well as community engagement and human resource management initiatives. It also includes a summary of performance and a set of goals under the title of, *'2020 Sustainability Goals'*. These goals include the following:

- *Continue to reduce waste generated by our shipboard operations by 5% by 2020 relative to our 2016 baseline*
- *Continue to improve water use efficiency of our shipboard operations by 5% by 2020 relative to our 2010 baseline*
- *Increase Advanced Wastewater Purification System coverage of our fleetwide capacity by 10 percentage points by 2020 relative to our 2014 baseline*

(www.carnivalsustainability.com, 2018)

The interesting points about these measurable targets are that they are very modest in scale and are based on baselines which are some time in the past, so they are low targets starting from a low base.

The same reference also shows a previous target for the reduction having been met in 2017 in terms of a 27.6% reduction since 2005. It then goes on to express their commitment to achieving the CLIA target of reducing emissions by 2030 by 40%. On the face of it this sounds good, but the 2030 target is for a 40% reduction against the baseline of 2008 according to the CLIA website. Therefore, some of the 40% has already been covered by the reduction of 27.6% achieved already from 2005 to 2017 and that might leave the reader thinking that a target of 40% by 2030 is very modest in reality. It could, in fact, presumably mean an actual reduction of less than 20% over the next decade or so which seems very limited given the scale of problems facing our oceans.

There are also goals around community engagement, guest and crew safety and security and diversity, for example, but they do not have measurable targets.

Furthermore, the Carnival sustainability report does not seem to explicitly address issues such as cruise ship operations in fragile marine environments or the impact cruise ships can have on wildlife habitats or the welfare of marine creatures.

The second largest player in the cruise market, Royal Caribbean, also produces an annual sustainability report. Based on the 2017 report which was entitled

'Seastainability', it is possible to identify several differences with the reporting undertaken by the Carnival Corporation. The main differences include:

- It is shorter at 54 pages and a little less of a glossy marketing type document; there are fewer photos and more detailed information.
- Its scope is broader, and it takes a more holistic approach towards sustainability.
- It is more explicit about some key issues such as plastic pollution.
- It says more about marine wildlife issues and it also talks about sustainable fisheries.
- It talks specifically about food sourcing in the context of sustainability.
- It mentions its partnerships with external stakeholders and its support for conservation projects.
- It talks about cruise destinations and conservation projects in them such as the Galapagos Islands.
- It gives details of its partnership with WWF and the fact that it has made donations totalling $1 million to WWF thanks to its customers.
- It makes the point that no more than 25% of its waste ends up in landfill, which suggests that the norm in the sector may be higher than that.
- It has a section towards the end which lists a large number of policies, targets and performance data.

Bonilla-Priego et al., in 2014, published an interesting study about the reporting of sustainability by cruise lines, as part of their corporate social responsibility activities. They suggested that *'cruising is a late adopter of reporting and companies disclose more management information than performance data which is typical in the early stages of development of CSR'*. They went on to say, *'companies that disclose information focus on soft, easy to mimic indicators.'* (Bonilla-Priego et al., 2014). Things have moved on somewhat since then but this evaluation of CSR reporting in the cruise industry still has some validity six years later.

In terms of individual cruise lines and the sector several points can be made about performance on sustainability issues.

First, they are improving but too slowly and their sustainability goals and targets are simply too low given the scale of the crisis faced by our oceans. Second, it seems likely that some of the improvement in their engagement with sustainability issues is the result of poor media coverage and high-profile court cases faced by some companies. Third, the sector is disadvantaged by having a large number of ageing ships that are not compatible with the environmental standards needed in today's world. Generally, the most recent and future ships will be a big improvement as they incorporate more responsible technologies, but they will represent the minority of ships afloat for a long time to come unless regulation requires them to be removed from service.

Lastly, most cruise lines do not seem to want to do too much to raise awareness amongst passengers of the potential negative impacts of cruising and the crisis facing our oceans. This was clear from the little survey I reported on in Chapter 1. Of course, no business wants to deter people from buying its products, but the industry could do much to encourage its customers to make a difference in terms of their own behaviour. This could include everything from food and water waste, to reduced use of plastics, to recycling, or to choosing shore excursions that have a lower carbon footprint, for example. I understand that this might not sit easily with the escapism, pampering and luxury which cruising is all about, but there is a way of giving customers a 'feelgood factor' by getting them to 'do the right thing'.

We need to recognise that there is a suite of regulations that cover the operation of cruise ships both national and international. Indeed, the industry claims to be one of the most regulated in the world. Regulations are set by a wide variety of organisations including: The International Maritime Organisation, International Labour Organisation and the World Health Organisation, as well as national agencies concerned with maritime safety, health, environmental safety and so on. And the regulations cover a wide variety of areas including

- The seaworthiness of vessels
- The safety of crews
- The health of crews and passengers
- Environmental standards
- Security.

Cruise ships are subject to inspections by national regulators when they enter ports in their territories. The country where the ship is registered, the so-called 'flag states', are also supposed to ensure ships registered in their country meet all requirements. However, it is well known that some of them are better at this than others which has given rise, rightly or wrongly to the idea that some are 'flags of convenience'! It is also worth pointing out that many of these regulations apply to all shipping not just cruise ships.

Well publicised court cases in recent years have, unfortunately, shown that companies do not always comply with the regulations, although this is often blamed on crew not following company policy. However, in such an apparently highly regulated and well managed industry it is difficult to see how this could happen.

Having now looked at what the cruise industry is doing voluntarily and the current regulatory environment in which it operates it is time to look at changes that could or should be made to the regulation and management of the cruise industry in future.

Potential developments in the management of the cruise industry

Perhaps the first thing to say is that, until recently, the marine environment was attracting insufficient attention from policy makers, the industry and society as a whole. However, as we noted in the first two chapters, this is no longer the case and the world is now well aware of the challenges facing the oceans, including those contributed to by the growing cruise industry. Furthermore, there is now a much larger body of accurate data available to inform public policy.

It would seem that the obvious place to start planning the future management of the cruise industry is with a review of existing regulation. In general, it would appear that the regulations need to be strengthened in recognition of the plight of our seas. Particularly strict regulations would appear to be needed around carbon emissions and pollution of the seas. New regulations are probably also needed to counter emerging issues such as plastic pollution as well as to tackle damage to marine habitats and harm to marine wildlife. More radical action may be needed to prevent old ships with particularly poor environmental standards to be removed from service. And, of course, any regulation is only as good as the effectiveness of enforcement so that also needs to be a priority. The industry needs to be given more ambitious targets because currently it is working with targets that are going to be a simple matter of 'too little too late'.

However, we also need to look much more than we do currently at where cruises go and their impact on specific places. There is a need for researchers to develop the concept of carrying capacity in terms of different types of cruise destination, so that management systems can be put in place that are 'fit for purpose' for any location because marine environments, as we saw in Chapter 2, are infinitely varied. Perhaps we need to go even further and develop 'no go' areas for cruise ships, which are either total exclusion zones or where cruise ships are only permitted to visit at only a few particular times and/or where there is a limit the number of ships that can visit. This already exists in a very limited way in Alaska, for example, but it is still rare, and this is a quota, not a ban. What I am suggesting is, perhaps, a ban on cruise ships in places that are not on normal shipping lanes, and have extraordinarily fragile environments, and are already in crisis. Antarctica is one such example of such a place which there is no reason for anyone, apart from researchers, to visit. There are plenty of other great pleasures for cruisers to explore, perhaps we just need to be gotten away from the idea that we should be able to visit anywhere if we have the money?

Finally, the future management of the cruise industry is not just about the impact of cruising on the open seas, it is also about their impact on their ports of call, both positive and negative. In recent years we have seen the emergence of a backlash against too much tourism in tourist 'honeypot' destinations, led by groups of citizens, under the banner of 'over-tourism'. We have seen protests and campaigns about the impact of too many cruise ships arriving in destinations from Juneau in Alaska to Venice in Europe.

The arrival, in a coastal resort or a small island, of potentially thousands of cruisers, who jump on buses and pack as much in as possible over just a few hours overwhelms transport systems and causes inconvenience for local people. It causes over-crowding which adversely affects the quality of life for both residents and traditional tourists who are spending days or even weeks in the destination. In addition, the wash from the engines of a cruise ship can be damaging to the foundations of buildings on fragile townscapes like in Venice or can cause harm to the shoreline, which could increase the likelihood of flooding. It seems that we need to develop and implement workable concepts of carrying capacity for cruise ships in relation to ports of call.

I recognise that the last two paragraphs run counter to two principles enshrined in international law and the ethos of our society, namely freedom of navigation on the oceans and the freedom to travel, respectively. However, the world is facing a fight for survival and having to meet challenges that are totally new to us so it may be that these treasured ideas have to be thought about again, particularly in relation to cruising, which while highly enjoyable and economically significant, is not essential to either human life or the health of our oceans.

Conclusions

Before I present my own conclusions to this chapter I would like to inform the reader that at the end of this chapter she or he will find one of two opinion pieces that have been kindly contributed to the book by two world renowned academics. This one is a perspective on responsible tourism and the cruise industry by Professor Harold Goodwin.

First, there is no doubt that the cruise sector is highly successful, having achieved rates of growth and profit margins that are the envy of most other sectors of the broader tourism industry.

However, with this success have come problems, as the cruise sector relies on the oceans, which are currently in crisis just as the industry is growing rapidly and is extending its reach to every part of the planet.

It is clear that the cruise industry has struggled to keep pace with the challenges it faces from its burgeoning growth at a time when the planet is struggling for survival, a struggle that is taking place as much at sea as on land.

For some time, the cruise sector was able to hide in the shadows of the airline industry which was seen as the villain in the tourism industry in relation to carbon emissions and carbon footprint. But in recent years questions have started to be asked about the emissions of cruise ships while others have focused on issues such as fuel pollution, air pollution, damage to wildlife habitats and labour practices. These have put the cruise industry at the heart of the debate over responsible tourism.

After a slow start and a period of being in denial the industry is starting to 'up its game' to meet the challenges but it is moving too slowly to keep pace with the development of the crisis. It needs to take a more holistic approach to sustainability and embrace the spirit of responsible business rather than merely paying lip service to it. This will not be easy in a business environment that is volatile and full of uncertainty, but the industry needs to act quickly or it risks being the subject of tougher government regulation or negative media coverage.

It is also being complicated by the fact that the cruise sector is at the centre of what is probably the 'hottest' current debate in tourism, namely 'over-tourism'. More and more destinations, or to be more accurate, the residents of destinations, are beginning to question the value of cruise ships which disgorge thousands of passengers for a few hours swamping local infrastructure, disrupting daily life and then moving on, having experienced virtually nothing of the authentic atmosphere of the place.

If the operators are not careful, cruising could become the new villain of the tourism industry, although its customer base may well stay loyal. But if the sector fails to meet the challenge of sustainability and its impacts on the oceans these customers may start to be viewed as self-indulgent hedonists whose holidays are literally 'costing the earth!'

References

Amelung, B and Lamers, M (2007) Estimating the greenhouse emissions from Antarctic tourism. *Tourism in Marine Environments*. **4**(2-3) 121-133.

Baker, DM (2016) The cruise industry: past, present and future. *Journal of Tourism Research* **127**, 141-153

Bonilla-Prego, MJ, Font, X, and Pacheco-Olivares, M del R (2014) Corporate sustainability reporting index and baseline data for the cruise industry. *Tourism Management* **44**, 149-160.

Carić, H and Mackelworth, P (2014) Cruise tourism environmental impacts: the perspective from the Adriatic Sea. *Ocean and Coastal Management* **102**(A) 350-363.

www.carnivalsustainability.com (2018) www.carnivalsustainability.com/download/files/2018-carnival-sustianability-full.pdf [15 Jan 2020].

www.cruising.org (2019) www.cruising.org/sustainability [12 Jan 2020].

Dragović, B, Tzannatos, E, Tselentis, V, Meštrović, R and Škurić, M (2018) Ship emissions and their externalities in cruise ports. *Transportation research Part D: Transport and Environment* **61**(B) 289-300.

www.euractiv.com (2018)

www.forbes.com (2019)

Klein, RA (2011) Responsible cruise tourism: issues of cruise tourism and sustainability. Journal of Hospitality and *Tourism Management* **18**(1) 107-116.

Lester, J-A, and Weeden, C (2004) Stakeholders, the natural environment and the future of Caribbean Cruise Tourism. *International Journal of Tourism Research* **6**(1) 39-50.

www.marineinsight.com (2019) www.marineinsight.com/environment/8-ways-in-which-cruise-ships-can-cause-marine-pollution [18 Oct 2019]

Poplowski, K, Setton, E, McEwen, B, Hrebenyk, D, Graham, M, and Keller, P (2011) Impact of cruise ship emissions in Victoria, BC, Canada. *Atmospheric Environment* **45**(4) 824-833.

Stewart, E and Draper, D (2006) Sustainable cruise tourism in Arctic Canada; an integrated coastal management approach. *Tourism in Marine Environments*. **3**(2), 77-88 .

www.transportenvironment.org (2019)

Wood, RE (2000) Caribbean cruise tourism: globalization at sea. *Annals of Tourism Research* **27**(2) 345-370.

3

Questions and exercises

1. Discuss the reasons why the number of cruise passengers fell in the middle of the twentieth century but then grew again dramatically from the 1990s to the present day.
2. In a group, undertake a survey of media stories about cruising that are related in any way to sustainability over the past twelve months. Then, prepare a report giving an overview of the stories and whether they portrayed the cruise industry in a positive or negative light. Your report should conclude with some thoughts on how you think the cruise industry could and should respond to the types of stories that you found.
3. Look at several cruise company brochures and advertisements. Based on what you see, to what extent do you think the marketing of cruising today is compatible with the development of more sustainable cruising? How could it be improved do you think?
4. Identify and discuss what you believe to be the three most important negative impacts which cruising can have on the marine environment.
5. Select a cruise company and examine its website, looking at any content that is related, in any way, to responsible cruising or sustainability. Then, write a brief report setting out your main findings and your conclusions about the extent to which the company is committed to responsible cruising and sustainability.
6. In a group design a short survey for regular cruise passengers which is designed to gather their views on the impacts of cruising on both the marine environment and the ports of call, both positive and negative. Present your results highlighting any of your findings which came as a surprise to your group.
7. Using examples, examine the factors that determine the point at which a cruise destination reaches it carrying capacity.

Opinion Piece: Responsible Tourism and the Cruise Industry

Dr Harold Goodwin, Responsible Tourism Partnership

The growth of the cruise industry raises significant challenges for the three pillars of the sustainability agenda: environmental, socio-cultural and economic. These problems compounded by issues of extra-territoriality and conflicts between local authorities and national governments. Venice and Barcelona suffer from overtourism, a problem exacerbated by the arrival of larger and larger cruise ships which discharge thousands of tourists daily into cities already crowded with tourists. Neither city has any control over the number of ships docking. By contrast, the Mayor of Dubrovnik was able to sign an agreement with the Cruise Lines International Association (CLIA) to cap the number of cruise ships to two per day carrying a total of 5,000 tourists.

The regulation of cruise lines is complex. Flag states control cruise vessels and crews. Owners generally register under flags of convenience to take advantage of more favourable regulation on taxation, labour conditions and insurance. Flag state registration is often obscured by emphasising where the line is headquartered. Local administrations rarely have powers to regulate cruise lines; national administrations can regulate in ports and territorial waters. Port states inspect and enforce compliance with applicable international and domestic laws and regulations. In the wake of the sinking of the Titanic in 1914, the International Convention for the Safety of Life at Sea (SOLAS) has regulated the construction and operation of passenger vessels.[1] Since 1973, MARPOL has provided a regulatory framework aimed at preventing and minimising pollution from ships - both accidental contamination and that from routine operations.[2] This international convention is supplemented by flag and port state rules and regulations, and territorial waters regulations. [3] Flag states have the primary responsibility for enforcing regulations. A US General Accounting Office report in 2000 revealed flag states' poor record in prosecutions.[4]

Responsibility

Businesses often claim that they are sustainable. The concept is used widely in policy statements and promotional material, but it is difficult to determine whether the ideas are operative or inoperative. Is it more than greenwashing? Responsible tourism and sustainable tourism are not the same thing. Sustainability is the abstract aspiration, the goal. Lacking any precise definition of the objective or of the timeframe for delivery, businesses cannot be held to account on their sustainability claims. The Responsible Tourism concept emerged in the late 1990s as companies sought to differentiate their efforts from those paying no more than lip service to sustainability. Sustainable tourism and sustainability are statements of intent. Responsible Tourism is about what a company or destination does to address specific issues which arise from its activities or, more altruistically, issues which it can address through the way it conducts its business or through philanthropy.

In 2002, at an official side event to the World Summit on Sustainable Development in Johannesburg, the Cape Town Declaration on Responsible Tourism in Destinations[5] was agreed. It remains the document which defines the Responsible Tourism agenda and the criteria for the family of Responsible Tourism Awards licensed by WTM, London.[6] Transparency in describing the issue and what the company is doing to tackle it, and reporting the impacts of the company's efforts are core requirements.

We can use the characteristics of Responsible Tourism from the 2002 Declaration[7] to identify the challenges which the cruise industry should be addressing. The first characteristic is minimising negative economic, environmental and social impacts; the triple line covers most of the sustainability challenges for the cruise lines.[8] There are parts of the Responsible Tourism agenda which are particularly challenging for cruise lines, for example, involving local people in decisions about how tourism takes place in their community. The impact of large groups' onshore excursions or walking out of the port and up Las Ramblas in Barcelona precludes the making of the 'meaningful connections' and 'greater understanding of local cultural, social and environmental issues' sought in Responsible Tourism. Since 2014 TUI has developed shore excursions which meet specific criteria for sustainability, demonstrating that it benefits local people and minimises environmental impact.[9]

Since 2017 TUI cruises has donated five euros per booked shore excursion, through its Green & Fair programme, to environmental protection projects.[10] Cruise lines do engage in philanthropy[11], they could do more. Since 2011 Crystal Cruises has had a 'You Care, We Care' programme; Holland America has a similar programme called 'Cruise with Purpose'. Carnival launched Fathom offering voluntourism cruises in 2016, though the brand was quietly discontinued in 2019. Fathom failed to find sufficient clients and there was criticism of its approach to voluntourism.[12] The cruise lines provide a way that enables the elderly, frail and disabled to travel.[13]

The cruise industry has been widely criticised. Klein has published two papers analysing how the cruise industry has defended itself, highlighting its concern for the environment through PR, funding NGOs, developing interlocking boards of directors and funding candidates and campaign in state elections. [14]

The Covid-19 Crisis

Cruise lines constitute around 1% of global shipping, a tiny but high-profile part of the sector. CLIA predicted that its members would have 278 ships operating in 2020 and that 19 more vessels would join the fleet. In 2002 there was no thought of including health and safety in the Responsible Tourism declaration; it was not as an issue. Cruise ships can be amplifiers of infectious diseases because of the close human proximity of semi-closed ship environments,[15] and disease brought aboard by passengers at embarkation or as they return from shore excursions can quickly spread on board. Outbreaks of norovirus and Legionnaires disease on cruise lines arguably attract more attention; international regulations require that all cases are reported. [16] Cruise lines operate as closed towns, and the blame rests with the cruise line responsible for the health and safety and welfare of a closed community.

Covid-19 on cruise ships attracted considerable media attention, 'Stuck at Sea, Cruise Ship's Sick Crew Must Keep Vessel Going'[17]; 'Coronavirus: Are cruise ships really "floating Petri dishes"?'[18], 'Cruise ship nightmare: After measles, norovirus outbreaks, why does anyone still set sail?'[19], 'Cruise ships have become floating dungeons where the unlucky have been confined – ideal incubators of infectious diseases.'[20] *The Economist* is not given to lurid headlines but it did express concern about the impact on the cruise business. Writing about Holland America Line's Zaandam: 'As of March 30 at least 193 had fallen ill with flu-like symptoms, several have tested positive for Covid-19, and four have died. Fearing contagion, the ports of call have kept her out; and so the Zaandam has been stuck at sea.' The article was headlined 'The coronavirus may sink the cruise-ship business. The industry has few friends and its main customers, the elderly, may shun it for good.'[21]

The U.S. Centre for Disease Control documented the impact of Covid-19 on cruise lines pointing out that: 'Cruise ships are often settings for outbreaks of infectious diseases because of their closed environment, contact between travelers from many countries, and crew transfers between ships.' More than 800 cases of laboratory-confirmed Covid-19 cases occurred during outbreaks on three cruise ship voyages. Transmission occurred across multiple voyages from ship to ship by crew members; both crew members and passengers were affected; 10 deaths associated with cruise ships had been reported by March 27 2020.[22]

The Covid-19 pandemic resulted in the industry shutting down. Cruise industry leaders were stung by criticism from journalists during the Covid-19 pandemic arising from front-page coverage of passengers trapped aboard ships. The vice-president of Princess Cruises, U.K. & Europe, Tony Roberts, argued that they were victims of a global pandemic.[23] Carnival Corporation CEO, Arnold Donald, argued that the cruise industry is better equipped to handle Covid-19 than society in general. 'We already do isolation, deep cleaning, hold medical records, do temperature scanning, use hand sanitisers and promote regular hand-washing. We also document a lot of passenger information. So, we're uniquely suited in some ways.'[24]

Carnival was much less sanguine in filing with the Securities and Exchange Commission in Washington on March 31:

> Covid-19 has had, and is expected to continue to have, a significant impact on our financial condition and operations, which impacts our ability to obtain acceptable financing to fund resulting reductions in cash from operations. The current, and uncertain future, impact of the Covid-19 outbreak, including its effect on the ability or desire of people to travel (including on cruises), is expected to continue to impact our results, operations, outlooks, plans, goals, growth, reputation, cash flows, liquidity, and stock price.[25]

In the Covid-19 crisis, Congress has been unwilling to provide financial aid for the airline and cruise industries. Eight senators wrote in an open letter: 'If we give the airline and cruise industries assistance without requiring them to be better environmental stewards, we would miss a major opportunity to combat climate change and ocean dumping.'[26] A coalition of environment groups wrote to Senate and House leaders:

> Providing U.S. taxpayer dollars to massive foreign cruise ship corporations that pollute our environment, take advantage of tax loopholes, and flag ships in foreign countries would allow the cruise industry to return to business as usual, which is unacceptable… This polluting and destructive foreign industry does not deserve a bailout from American taxpayers, especially not in the middle of a national health emergency.[27]

Carnival's shares lost 80% of their value, recovering 25% when Saudi Arabia's public investment fund took an 8.2% stake in the company.[28] Carnival and the cruise sector has been denied government assistance because, as a consequence of their registration in flag states, they are seen as foreign-owned; by rejecting US governance they are seen by many as irresponsible.

Environment

The shipping industry is improving the operating efficiency of vessels at construction with ducktails[29], using lighter materials and solar panels, and the IMO now has an Energy Efficient Design Index and a Greenhouse Gas Emissions Rating System which enables charterers to choose more efficient ships. Keeping hulls smoother, route and weather planning, increase efficiency. 'Slow steaming' uses up to 30% less fuel. Currently, available efficiency approaches can reduce emissions by 20% to 40%, and potentially by 30% to 55% by 2030.[30] In April 2018 the IMO declared its intention to reduce the total annual GHG emissions of shipping by at least 50% by 2050 compared to 2008.[31] Ships built now will still be cruising in 2050; cruise lines need to be challenged to build more and more efficient liners and to retrofit where possible. Slow steaming, weather and route planning (for example sailing with the tide) require only changes of itineraries and operating practices and can be implemented now.

The cruise lines continue to be fined for air pollution, grey water dumping, breaching low sulphur fuel regulations and dumping plastic and garbage overboard.[32] In ports and territorial waters, enforcement of regulations is possible. In international waters, enforcement is the responsibility of flag states. There are two particular difficulties: one of reporting and evidence the other is that flag states risk ships moving to register elsewhere. As Klein has argued, despite their population size 'the cruise industry operates largely in a system of voluntary arrangements in which they police themselves.'[33]

Increasing numbers of passenger vessels are visiting Antarctica. In the 2017-18 season, over 50,000 people visited Antarctica, and four cruise-only vessels were operating.[34] Vessels over 500 passengers are not permitted to land them in Antarctica, and fuel oil standards adopted within the International Maritime Organisation require ships to burn lighter-grade fuels while in the Antarctic Treaty Area (the sea south of 60 South latitude).[35]

Economic

One of the aspirations of Responsible Tourism is to generate 'greater economic benefits for local people and enhance the well-being of host communities, improve working conditions and access to the industry.'[36] There few if any examples of cruise tourism achieving this in practice. Labour conditions on cruise lines are often criticised in the press and by NGOs. Bolt and Lashley used participant observation and semi-structured interviews to explore the hospitality crew's work experience. Given that they were interviewing employees, it is not surprising that respondents reported that the value of earnings, relative to what they might earn at home, was adequate compensation for being away from home for an extended period in challenging working and living conditions with a very diverse group of colleagues. At sea and in port, crew work within a total institution where they need permission to go ashore; they work long hours often with no days off, and in a confined environment where privacy and space for relaxation are absent. While off-duty, there are limited activities available to them, and they live in cramped accommodation below decks. There are issues about free time, length of contracts, and insecurity of employment, as cruise lines hire according to passenger occupancy through recruitment agencies who may or may not charge the crew member a fee. Flags of convenience states impose minimal crew protection; cruise lines benefit from limited regulations on working conditions, pay levels, nationalities employed, and tax obligations. Given the low wages and working conditions in many developing countries, there is a ready pool of labour for whom cruise line work is attractive. The UK NGO War on Want and the International Transport Workers' Federation (ITF) ran a campaign against 'sweatships' in 2002. There are occasional exposes in mainstream newspapers; in 2012 *The Guardian* revealed the basic salary of cruise ship staff at 75p an hour and 'Cabin stewards whose attitude was ranked below 92% by customers will forfeit an entire bonus payment worth approximately 15% of their basic salary.'[37]

Large cruise ships are probably best conceptualised as resorts or destinations, 'spaces of containment and revenue capture.'[38] Cruise ships full of relatively wealthy visitors can be a seductive proposition for destinations. However, the cruise lines are in the stronger negotiating position; the ships are highly mobile, and the ownership is concentrated. The ports are immobile, in competition with each other, and once the infrastructure demanded by the cruise lines is in place, revenue is essential to repay the loans. The cruise line is more powerful than the destination, port authority or excursion provider, and they exploit their commercial and political power.

The profitability of cruise lines relies on onboard revenue and excursion sales. Shore excursions are important to passengers; between 50% and 80% purchase a shore excursion at each port. For the cruise line, the sales are lucrative, it is alleged that as little as 10% of the excursion price is paid to the provider in some destinations, although a 50:50 split is more common. The passenger will judge the quality of the excursion in the context of the price they paid and generally where a passenger is dissatisfied the cruise line refunds the passenger and charges the cost back to the excursion provider.[39]

The same companies that provide shore excursions also provide port lecture and port shopping programmes. Maps guide passengers to preferred shops which promise the best deals. The cruise line collects an annual promotion fee or a commission on sales. There is little evidence that visitors return as traditional tourists to see more of the country; the country has been ticked off on their bucket list. A study by the Research Department of the Central Bank of Belize in 2007 concluded that: 'Certain areas of the cruise industry have taken on an oligopolistic nature. Tour spending and non-tour visitor spending were heavily skewed in favor of the minority...' and 90% of the goods sold in the tourism village at the port were imported.[40] A 2007 study found cruise visitors spent less than half as much per day as land-based visitors (US$44 vs. US$96, and terrestrial visitors stayed an average of 6.8 days spending US$653). Cruise passengers accounted for 75% of arrivals to Belize, but only 10% of employment in the tourism industry.[41]

Socio-cultural

Cruise ship tourism, with large numbers of tourists arriving in relatively concentrated periods of time, undermines the character of the destination to the detriment of visitors and residents alike. The scale of cruise tourism brings massification with it in Barcelona, Belize, Orkney, Port Stanley and Venice, and overtourism results. Hosts or guests, locals or visitors, feel that there are too many visitors and that the quality of life in the area, or the quality of the experience has deteriorated unacceptably. Often both visitors and guests experience the deterioration concurrently.[42] In Belize, non-cruise tourists are advised to avoid Xunantunich on 'cruise day'. As tourists are worth more than day visitors arriving as a mass from cruise lines, destinations need to manage the risk that tourists will be discouraged, undermining the quality of the destination. Those concerns motivated the Mayor of Dubrovnik to cap cruise arrivals.

There is little chance that cruise tourism can realise the aspiration in the Cape Town Declaration that tourism should be 'culturally sensitive', engendering 'respect between tourists and hosts, and building local pride and confidence'. On the other hand some cruise passengers book the same cabin, on the same cruise vessel, because they wish to meet again with the same crew members that looked after them previously, once or many times. Tourists are creating more enjoyable experiences for themselves through more meaningful connections with local people – on a cruise vessel the local people are the crew.

Sustainable cruising is an aspiration, progress is slow

The Responsible Tourism agenda for cruise lines is long and broad. There have been cruise line winners in the Responsible Tourism Awards: Holland America Line, in 2008, for having reduced dockside emissions by 20%, developed and implemented an 'Avoiding Whale Strikes' training programme, increased recycling by 50% and introduced new scrubber technology; Royal Caribbean Cruises in 2009 for reducing emissions in their newest ships by 50% over ships built around 10 years earlier; TUI was recognised in 2018 for Communicating its Better World, Better Holidays strategy which included a 11.5% reduction in CO_2 emissions per passenger night in cruise operations and its effort to improve excursions for passengers and locals alike .

TUI, the world's largest integrated tourism group, is committed to creating Better Holidays and a Better World with targets linked to the UN Sustainable Development Goals. All its 18 cruise ships are now covered by ISO 14001 certification, at 101 kg CO_2 per passenger night, emissions are down 11.7% on 2015 and 6.5% on 2017. Freshwater consumption per guest night was reduced by 32% between 2017 and 2018 and waste water by 14%, sulphur emissions were reduced by 36.7% and nitrogen by 12%, waste is separated on board for disposal in port.[43] There is a strong case to be made for TUI as leaders in responsible cruising, but there is much more to be done. Activist shareholders at its AGM in February 2020 criticised TUI for not doing enough to reduce air pollution in its cruise businesses.[44]

Cruise lines present a particular challenge for Responsible Tourism. Whilst there is some good practice and some progress is being made, those cruise lines which are investing more in sustainability, taking responsibility to make cruising better, are competing with others who are not. The regulatory framework and enforcement are weak, there needs to be more transparent reporting and more media and consumer pressure.

Endnotes

1 http://www.imo.org/en/OurWork/Safety/Regulations/Pages/PassengerShips.aspx

2 http://www.imo.org/en/About/Conventions/ListOfConventions/Pages/International-Convention-for-the-Prevention-of-Pollution-from-Ships-(MARPOL).aspx

3 Genç, R., (2016). Sustainability on Cruise Ship Management. *International Journal of Social Science Studies*, **4**(6) 80.

4 United States General Accounting Office (2000) Progress Made to Reduce Marine Pollution by Cruise Ships, but Important Issues Remain 19-21 https://www.gao.gov/products/RCED-00-48

5 https://responsibletourismpartnership.org/cape-town-declaration-on-responsible-tourism/

6 https://responsibletourismpartnership.org/world-rt-awards/

7 https://responsibletourismpartnership.org/cape-town-declaration-on-responsible-tourism/

8 Elkington, J., (1997). *Cannibals with Forks: The triple bottom line of twenty-first century business.* Capstone.

9 https://www.tuigroup.com/en-en/responsibility/sus_business/destination

10 https://www.tuigroup.com/en-en/responsibility/sus_business/destination

11 For example: Cruise Industry Charitable Foundation, Princess Cruises Community Foundation, Holland America has fundraising initiatives at sea On Deck for a Cause, Royal Caribbean's Walk for Sea Turtles

12 https://www.telegraph.co.uk/travel/cruises/articles/cruise-voluntourism-more-harm-than-good/

13 https://www.cruise118.com/advice/cruises-suitable-disabled-people/

14 For a discussion of the politics of this criticism see Klein, R.A. and Sitter, K.C., (2016). Troubled seas: The politics of activism related to the cruise industry. *Tourism in Marine Environments*, **11**(2-3), 146-158, and Klein, R.A., (2007). The politics of environmental activism: A case study of the cruise industry and the environmental movement. *Sociological Research Online*, **12**(2), 17-29.

15 Hill, C.D., (2019). Cruise ship travel. In *Travel Medicine* (pp. 377-382). Elsevier.

16 https://www.cdc.gov/nceh/vsp/surv/gilist.htm

17 https://www.bloomberg.com/news/articles/2020-03-30/sick-crew-keeps-cruise-ship-running-risking-spread-of-virus

18 https://www.bbc.co.uk/news/world-asia-51470603

19 https://www.nbcnews.com/think/opinion/cruise-ship-nightmare-after-measles-norovirus-outbreaks-why-does-anyone-ncna1043516

20 https://www.theguardian.com/commentisfree/2020/apr/14/cruise-ships-coronavirus-passengers-future

21 https://www.economist.com/business/2020/03/31/the-coronavirus-may-sink-the-cruise-ship-business

22 https://www.cdc.gov/mmwr/volumes/69/wr/mm6912e3.htm accessed 18-04-2020

23 https://www.travelweekly.co.uk/articles/368021/cruise-bosses-slam-appalling-and-objectionable-coverage

24 https://www.travelweekly.co.uk/articles/368055/carnival-corp-boss-denies-sector-was-slow-to-react-to-coronavirus

25 Carnival Corporation Form 8-K filing with the US Securities and Exchange Commission, 01/04/2020, https://www.carnivalcorp.us/financial-information/sec-filings/all?mobile=1&items_per_page=10&order=field_nir_sec_date_filed&sort=asc&page=387

26 https://www.markey.senate.gov/news/press-releases/whitehouse-markey-senate-democrats-request-requirements-for-airlines-and-cruise-lines-to-address-pollution-in-financial-assistance-package

27 https://skift.com/2020/03/23/why-lawmakers-are-so-uneasy-about-giving-the-cruise-industry-a-bailout/

28 https://skift.com/2020/04/06/carnival-sells-8-percent-stake-to-saudis-after-dire-financial-warning/ accessed 2020 04 19.

29 https://gcaptain.com/part-design-efficient-ship/

30 Reducing a ship's operating speed. Hawken, P. ed., (2017). *Drawdown: The most comprehensive plan ever proposed to reverse global warming*. Penguin 140-141

31 http://www.imo.org/en/OurWork/Environment/PollutionPrevention/AirPollution/Pages/GHG-Emissions.aspx

32 Cases listed on http://www.cruisejunkie.com/envirofines.html 2015-2020 accessed 19/04/2020. Klein, R.A., (2011). Responsible cruise tourism: Issues of cruise tourism and sustainability. *Journal of Hospitality and Tourism Management*, **18**(1), 107-116.

33 Klein, R.A. (2009). *Paradise Lost at Sea*, Fernwood, Halifax.

34 https://www.cruiseindustrynews.com/cruise-news/18920-antarctica-tourism-numbers-surge.html

35 https://iaato.org/en_GB/frequently-asked-questions#are-all-antarctic-tour-operators-members

36 https://responsibletourismpartnership.org/cape-town-declaration-on-responsible-tourism/

37 https://waronwant.org/sweatships and https://corpwatch.org/article/sweatships-sea-carnival-cruises-pay-below-international-standards, https://www.theguardian.com/business/2012/apr/29/cruise-firm-performance-bonuses-tips

38 Weaver, A., (2005). Spaces of containment and revenue capture:'super-sized'cruise ships as mobile tourism enclaves. *Tourism Geographies*, **7**(2), 165-184.

39 Klein, R.A., (2011). Responsible cruise tourism: Issues of cruise tourism and sustainability. *Journal of Hospitality and Tourism Management*, **18**(1), 107-116.

40 Novelo, A. Q. et al. (2007) Assessing the Direct Economic Impact of Cruise Tourism on the Belizean Economy, Research Department Central Bank of Belize, http://www.centralbank.org.bz/docs/rsh_4.5_conferences-working-papers/direct-economic-impact-of-cruise-tourism.pdf

41 CESD (2006). Cruise Tourism in Belize: Perceptions of Economic, Social & Environmental Impact https://www.responsibletravel.org/docs/Cruise%20Tourism%20in%20Belize%20-%20Full%20Report.pdf

42 Goodwin, H. (2017). The Challenge of Overtourism, Responsible Tourism Partnership https://haroldgoodwin.info/pubs/RTP'WP4Overtourism01'2017.pdf

43 TUI Group (2019) Sustainability Report Better Holidays, Better World.

44 https://www.tradewindsnews.com/regulation/cruise-emissions-in-focus-as-shareholders-blast-tui-s-record/2-1-754116 accessed 19/04/2020.

4 Wildlife–Watching in Marine Environments

If one looks at destination guide books from as late as the 1990s, very few would have had much to say about watching marine wildlife, except perhaps a few sentences about scuba-diving and the opportunity it gave the tourist to see interesting and beautiful fish as an added bonus for participating in this leisure activity. Otherwise, the tourist may have been recommended to visit a certain place because of the opportunities to fish for huge specimens that could be displayed as trophies and photographed to impress the folks back home. Yet, a few years later, marine wildlife-watching has become a major selling point for many coastal destinations around the world, from Australia to California, Sri Lanka to Alaska, South Africa to Scotland.

Interestingly, it is also an activity, out of all of the ways in which tourism and the marine environment interacts, that has attracted perhaps the most attention by tourism researchers. In this chapter we will look at how and why this change occurred together with a look at the impacts of the rise of marine wildlife-watching in its many forms.

However, first we need to be clear what we are talking about by looking at several typologies of marine wildlife-watching.

Typologies of marine wildlife-watching

There are several ways of looking at marine wildlife-watching and some of these are illustrated below in diagrammatic form, starting with an attempt to segment the marine wildlife-watching market in Figure 4.1.

On the right-hand side of the diagram it is reasonable to assume that the availability of wildlife-watching experiences will not play a role in the choice of destination whereas in the case of the left-hand side these opportunities will have been a major factor in the choice in the selection of the vacation destination.

In 2010 Catlin and Jones published a longitudinal study of whale shark watching in Ningaloo Marine Park in Western Australia that compared the market in 1995 and 2005. They said their findings *showed a shift in the industry to the mainstream from the periphery. In comparison with the past, shark tourism there now attracts more generalist tourists. There is now a greater distribution of age groups, less skilled individuals, a higher tolerance of crowding, and a larger focus on the non-wildlife components of the experience.* (Catlin and Jones, 2010)

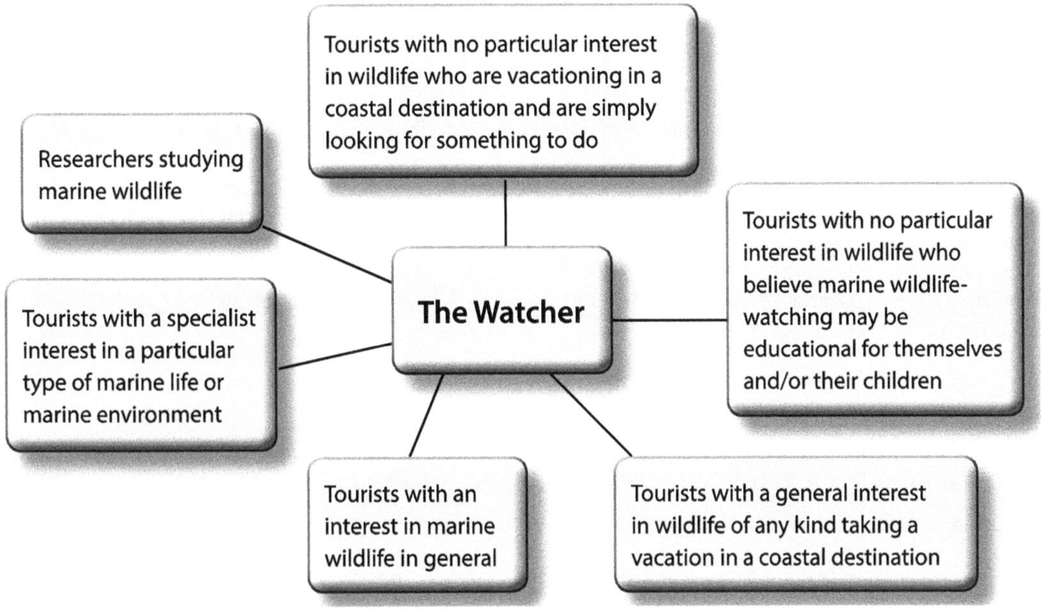

Figure 4.1: Who are the marine wildlife-watchers

Figure 4.2 focuses upon the types of wildlife that are being watched; these are more diverse than one might imagine.

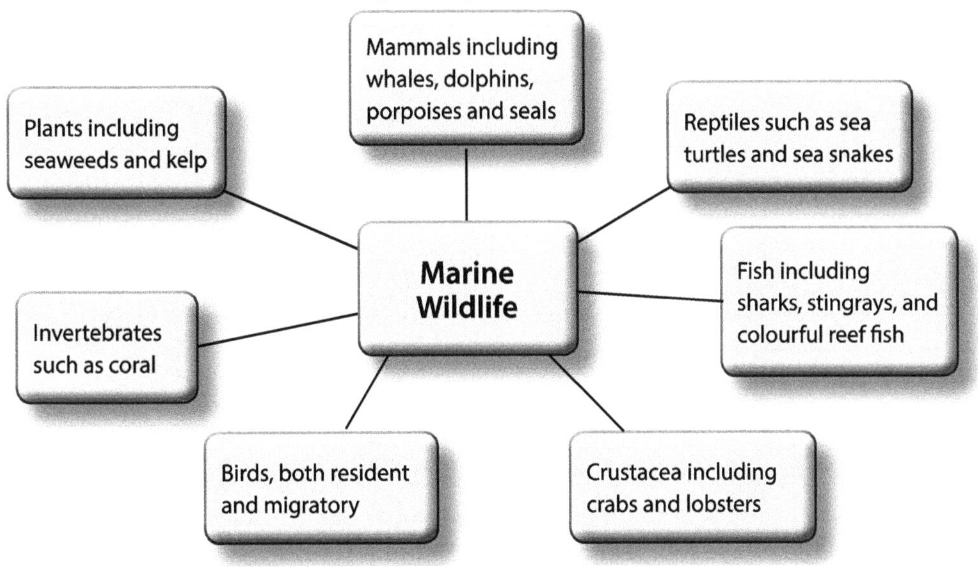

Figure 4.2: What marine wildlife is being watched?

Of course, it is important to recognise that as we noted in Figure 4.1, some tourists are interested in one particular type of marine wildlife while others find all marine wildlife of interest.

When considering the impact of wildlife-watching one issue that needs to be considered is the degree of interaction that takes place between the wildlife and the watcher. This is presented in Figure 4.3 in the form of a continuum. Worryingly, but perhaps not surprisingly in the era of experiential tourism, it is the extreme form of interaction shown at the far right that is probably growing fastest. It is seen by some tourists as a form of co-creation involving the voluntary and even enthusiastic participation of the marine creature.

The tourist sits passively in a boat and observes the wildlife	The tourist is enclosed in a submersible observing the wildlife	The tourist is on a boat and is involved in feeding the wildlife in some way	The tourist is in the water but enclosed in a cage to separate them from the wildlife	The tourist is in the water swimming amongst the wildlife	The tourist is actually interacting with the wildlife such as feeding, touching or riding

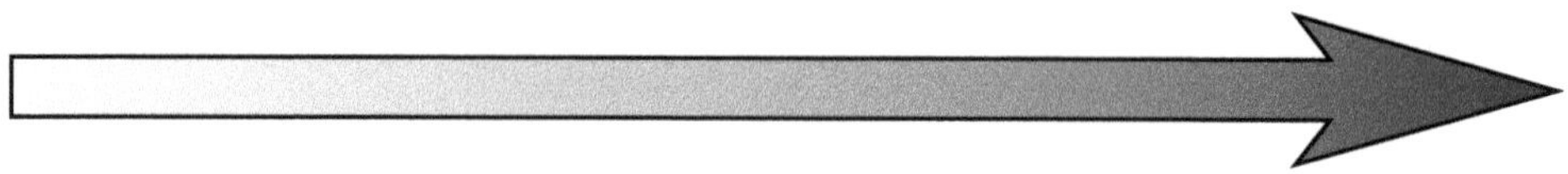

Figure 4.3: What is the level of interaction between the watcher and the marine wildlife?

In Figure 4.4 we look at the diversity of marine wildlife-watching based on the duration of the experience, again in the form of a continuum. It is clear that the duration of the experience varies significantly but as yet there appears to be a well-developed body of evidence to show if the actual duration has any real effect on the impact of wildlife-watching on the marine environment.

Tourists spend a few minutes watching wildlife from the shore, such as a seal on a rock or a passing dolphin	Tourists spend around 30 to 60 minutes in the water interacting with marine wildlife including feeding, touching or swimming amongst marine creatures	Tourists participate in a wildlife cruise which may last one from one hour to one day	Tourists join an overnight marine wildlife cruise	Tourists spend their whole vacation of up to two weeks looking at marine wildlife

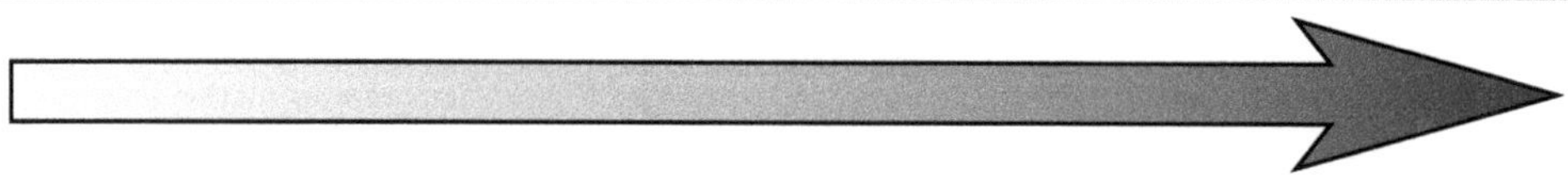

Figure 4.4: What is the duration of the wildlife-watching experience?

Finally, Figure 4.5 looks at where the wildlife watching takes place in our oceans. Unfortunately, it is impossible to find global data to show which parts of the marine environment are used most for wildlife-watching, statistically speaking.

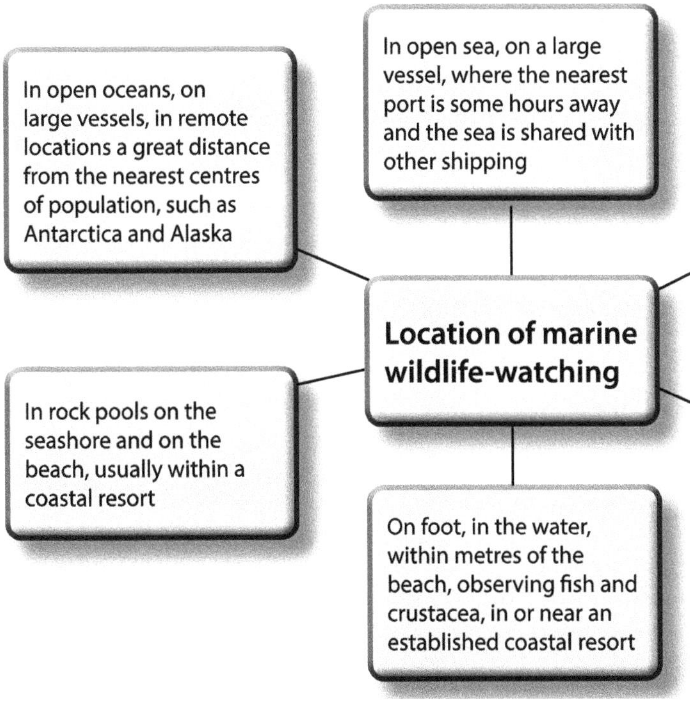

Figure 4.5: Where does marine wildlife-watching take place?

Figure 4.5 is not comprehensive but it does clearly illustrate the broad range of the locations and situations in which marine wildlife-watching takes place, from looking for baby crabs in a rockpool a few metres from the promenade to looking at whales, on a large cruise ship, fighting its way through the waves in Antarctica.

Having now tried to explore some potential typologies of marine wildlife-watching we will now take a brief look at its evolution over recent years.

The development of marine wildlife-watching

I believe that the roots of marine wildlife-watching probably date back to the 1950s and 1960s when safaris began to develop as tourist experiences, primarily in Africa. Until then hunting had been the main motivation for tourists to 'engage' with the wildlife of, and parts of Asia. Animal heads, skins and photographs of trophies were the main souvenirs of such trips. I believe that there are several reasons why wildlife-watching has replaced hunting as the main tourist motivation for most visits to Africa, as well as being a major factor in visits to other countries where hunting was prevalent, including India and Sri Lanka, for example.

- First, while hunting still continues to attract tourists to many parts of the world, it is no longer fashionable to the same degree and often now takes place clandestinely and most hunters like to have a low profile unlike their flamboyant predecessors who wanted to show off their kills. When today's hunters take boastful photos of their 'trophies' they tend to find themselves the subject of media criticism and attacks on social media.

- Second, as former colonies became independent after World War Two, they did not usually want to see a continuation of hunting of their wildlife by foreigners, who had exploited and decimated their wildlife during the colonial era.
- Third, the governments of countries which were home to lions, tigers, elephants and so on began to realise that there was more money to be made from conserving wildlife for tourists to look at year after year, rather than the short-term gain from hunting, which would eventually end when the last animals were shot. As a result, they established reserves where tourists could view wildlife safely.
- Fourth, in the main generating markets for tourism at the time in the 1950s and 1960s, Europe and North America, the 'new' medium of television was beginning to make people even more aware of the wildlife that lived in Africa and Asia. This, in turn, stimulated interest in visiting these places, particularly as the television programmes made everyone aware that wildlife was under threat from man and might not always be here for us to see. This perhaps injected a little urgency into people's desires to go and see the wildlife while they still could!
- Finally, the burgeoning new tour operation sector which developed after World War Two was able to start packaging wildlife-watching experiences and make them more affordable to destinations such as Kenya.

Over time, wildlife-watching tourism has developed new dimensions as well as spreading to more and more destinations. For some it is the main purpose of their vacation, while for others it is a major factor in their choice of destination. Watching wildlife on land is now a truly global phenomenon, stretching from the Amazon and Galapagos Islands in South America to the unique biodiversity phenomenon of Madagascar, from the savannah plains of Africa to the Arctic tundra, from the outback of Australia to the national parks of Sri Lanka.

In some ways, the rise of marine wildlife tourism mirrors what happened on land, but with subtle differences. The issue of whale hunting attracted media attention, which in turned raised interest about whales. As whale hunting bans came into place, coastal communities that had been involved in whaling needed to find new ways to earn a living. Thus, whale-watching was born, and people began to see that there might be more money and a more sustainable income from watching whales rather than hunting them.

At the same time, there were other factors at play which were increasing tourist interest in particular species of marine wildlife and increasing the desire of tourists to see them in 'the wild'. Three examples will suffice to illustrate this as follows:

- Media coverage of the life of sea turtles, the huge distances they cover to lay their eggs that exhaust them and the high mortality rate of their young, together with the impact of hunters and poachers on their numbers, simultaneously raised concerns for their welfare and created a desire to see them in their natural habitats.
- Nature programmes told people that dolphins were very intelligent while the TV show, *Flipper* in the 1960s, reinforced this idea. Meanwhile, the tricks performed by trained dolphins held captive in zoos and marine parks were presenting them to the public as creatures that were 'entertaining'!
- Movies like *Jaws* painted a picture of sharks that was, at once, both fascinating and scary.

Just like with land-based wildlife, there was no shortage of entrepreneurs as well as NGOs and charities willing and able to give tourists an opportunity to see marine creatures in their natural environment.

However, again in parallel with what happened on land, interest in wildlife was not about all wildlife but was highly selective. Big creatures like whales, cute creatures such as dolphins and seals, heroic creatures like sea turtles, and scary creatures like sharks were all fascinating for tourists, as were coral reefs with their colonies of colourful fish. However, there was generally little interest in most marine wildlife, from kelp to crustacea, tuna to sea birds, apart from a few people obsessed with a particular species of marine life. This is an issue that will be taken up further later in this chapter.

In the first instance, as we noted earlier, marine wildlife-watching helped former whaling communities to find new ways of earning a living. It also began to help fishing communities to find new uses for their boats as the depletion of fishing stocks began to make fishing less economically viable. Many of these settlements had never been tourist destinations, but they now began to attract tourists due to their ability to offer marine wildlife experiences. I have personally observed this phenomenon in places as diverse as Husavik on the north coast of Iceland, Western Australia, California and Sri Lanka. In other words, marine wildlife-watching has been responsible for transforming whaling and fishing communities into tourist destinations with all that entails.

In recent years we have seen a subtle shift too from wildlife-watching to wildlife interaction, moving from purely passive observation to active interaction with marine creatures. This includes everything from swimming with dolphins in many locations, to touching stingrays at Hamelin Bay in western Australia, to being in a cage surrounded by sharks in South Africa. This raises major ethical issues around tourist behaviour and the balance between what tourists want to do and what it is appropriate for them to do.

Sadly, the rise of social media and the phenomenon of the 'selfie' has encouraged this trend, for a picture of you swimming with a dolphin or a photo of the teeth of a shark taken through the bars of a shark cage is a lot more exciting than a photo of you sitting on a whale-watching boat with dozens of other tourists!

The market and its motivations

Let us continue with this focus on tourists and motivations as we take a closer look at the market for marine wildlife-watching and the motivations of the tourists who participate in this market. The first thing to say is that there is no reliable data available for the market for marine wildlife-watching as a whole because of problems of both definitions of what it comprises, together with the difficulties of measuring any tourism market. The same is true to a large extent of land-based wildlife tourism too, in reality. However, some research has been done specifically on the market for whale-watching, which is, perhaps, the most high-profile type of marine wildlife-watching, although even then reliable up-to-date statistics are difficult to find.

In 2001, Erich Hoyt produced a report on behalf of the International Fund for Animal Welfare. Interestingly most of the research to measure the scale of whale-watching appears to have been undertaken by animal welfare organisations. The same organisation published a report in 2009 which updated the work of Hoyt.

Table 4.1 presents the findings of both groups, illustrating the growth of whale-watching, worldwide, between 1991 and 2008.

Table 4.1: The growth of whale-watching 1991-2008

Year	Total whale-watching trips	Total expenditure (US dollars)
1991	4,000,000	Unknown
1994	5,400,000	500,000,000
1998	9,000,000	1,050,000,000
2008	13,000,000	2,100,000,000

Source: Hoyt, 2001 and later sources

This data shows that between 1991 and 2008 whale-watching grew at a rapid rate particularly during the 1990s, with the growth rates slowing after 2000. Unfortunately, it has not proved possible to find more up-to-date and reliable data for the size of the global market.

However, while the growth rate may have slowed down worldwide, there has been substantial growth in certain parts of the world, where changing migration patterns now bring whales close enough to shore for boats to be able to take tourists out to see them. The IFAW report in 2008 highlighted particularly rapid growth in whale-watching was being seen in Asia, the Caribbean, Central America and South America. It is likely that the growth seen over the past three decades has been a function of both supply side developments, with more and more destinations offering whale-watching, together with growing public interest in whales stimulated by the travel and news media.

Figure 4.1 shows that the marine wildlife-watching market is fragmented and contains many different distinctive segments. Sadly, there is no real research on the market and segmentation of the market on a global scale to help us understand the motivations and decision-making processes of marine wildlife watchers. Nevertheless, it seems reasonable to suggest that each of the watchers will have their own personal motivators and that the motivators will also vary between the different types of wildlife-watching shown in Figure 4.1. Any list of generic motivators though, I believe, would be likely to include words such as learning and education, awareness-raising, fun, and so on. But there might also be ones that were unconscious or which people might be unwilling to admit to such as entertainment or status enhancement or just finding something to do to pass time! Some specific activities such as shark cage diving might also elicit replies such as getting an adrenalin rush.

However, there have been some interesting studies of particular countries or the markets for watching specific species. For example, in 2019, Geldenhuys et al. published the results of their research on the segmentation of the market for whale-watching and shark cage diving in South Africa. Unfortunately, their results did not distinguish sufficiently between the two types of marine watching activity as the two are very different, with one being passive and above water and the other being more active and being immersed in water.

Their results do seem to have been in line with those of other researchers who have studied marine wildlife tourism. They said that the average such tourist, at least in South Africa, was female (57%), with an average age of thirty-eight, with relatively high salaries and quite well educated. Many were new to marine wildlife-watching which they suggested might mean that these were activities that might be seen as once-in-a-lifetime or 'bucket list' activities (Geldenhuys et al., 2019). They split their sample into six segments which they labelled '*thrill* seekers', '*risk takers*' '*thalasophiles*', '*consorts*', '*experience seekers*' and '*adventure junkies*'. (Geldenhuys et al., 2019). They then went on to look at these six segments in terms of variables such as gender, language, education, and desire to participate again in future in similar activities.

There seems to be a clear need for more research to be conducted globally for this market, preferably using a standard methodology so that results can be compared between destinations and segments.

The supply side – destinations and operators

We have already noted that more and more coastal communities have sought to attract marine wildlife-watchers as whaling has been abandoned and traditional maritime activities including fishing have declined. Many of these places have been new to tourism, seeing it as a lifeline, a way of regenerating their hometowns. This means that many of the first operators were pioneers, who were more likely to be ex-fishermen than owners of existing tourism businesses. Alternatively, those boat owners who had traditionally run angling trips or short sightseeing trips in their boats began to diversify into marine wildlife-watching when they saw the potential income generation opportunity that it represented.

Marine wildlife-watching requires only limited infrastructure, such as a small boat, which does not need to be purpose-built necessarily, but it does require the ability to operate the boat safely and the knowledge to give the best chance of finding the wildlife and to provide an informative commentary for the tourists.

The best data we have on the growth of destinations and operators involved in marine wildlife tourism relates largely to whale-watching. It is, therefore, important to recognise that this is not a complete picture and to acknowledge that most countries or territories with a coastline will have at least one or two of the types of marine wildlife-watching that were identified in Figure 4.1. Nevertheless, if we focus just on whale-watching the growth in destinations and operators is easy to see. If we start with destinations, according to Hoyt, in 2001:

- Whale-watching was taking place in 87 different countries and overseas territories or dependencies.
- Twenty-two countries had entered the market, offering commercial whale-watching between 1995 and 2001, just six years, including St Lucia, Namibia, Oman, Taiwan and Fiji.
- The USA, Canada and the Canary Islands were each hosting at least one million whale watching trips per annum.
- Some communities, which were not traditional tourist destinations, had been economically transformed be whale-watching. These included: Kaikoura in New Zealand; Provincetown MA in the USA; Andenes in Norway; Husavik in Iceland; Hermanus in South Africa; Monkey Mia in Australia; Guerrero Negro in Mexico, and; Ogasawara in Japan.

Source: Hoyt, 2001.

By 2008 the IFAW was suggesting that whale-watching was now offered commercially in 119 countries and territories. In the same report they suggested that there were now some 3,300 whale-watching operators worldwide who employed some 13,200 people between them. For many coastal destinations marine wildlife-watching is now one of their major attractions for tourists. Yet, often, it seems there is no real strategic approach to its management. In 2013 Bentz et al noted that there was '*no island region-wide strategy for wildlife-watching tourism in the Azores*'. (Bentz et al., 2013)

The destinations and operators have continued to innovate in order to attract new marine wildlife-watching business.

First, some destinations have created artificial reefs by deliberately sinking ships, for example, to attract those who want to see reef fish, in particular. This is not only expensive but controversial; it can be seen as either an aide to conservation or an intrusion into an established ecosystem.

Second, new ways are constantly being found for people to explore the marine environment from vessels such as flat-bottomed boats and submersibles to 'underwater walks' using helmets. Each has the potential to harm the marine environment through either noise pollution, fuel spillage or damage to the seabed.

The challenges facing marine wildlife tourist destinations and operators include:

- The fact that wildlife tourism in any sea is vulnerable to changes in nature that are beyond its control. This may include everything from changes in migration routes and breeding patterns to the effects of rises in sea level and water temperature.
- The seasonality of marine wildlife-watching, not only in terms of when tourists take their vacations, but also the seasonality of wildlife migration, for example. The two may not coincide with each other which can create opportunities to extend the tourist season, of course, but can also mean there are few people there to see the migrating wildlife!
- Growing competition from new destinations and from new operators who see potential in this market. This can put downward pressure on both prices and margins.
- Increasing concerns over the ethics of marine wildlife-watching and the operating standards of some operators, which is giving the sector a bad name.
- Prices for marine wildlife-watching experiences are generally not low, reflecting the operating costs of the boats and the fact that seasons may be short, meaning as much income as possible has to be earned over a short period.

Marine wildlife – the good, the bad and the ugly

Before we go any further into this subject, I would like to make a point, I believe is important. Levels of interest in marine wildlife vary dramatically between one species and another. In reality, most people are only interested in a few species, as we have already seen from the fact that most of what has been said to date in this chapter has been about whales and sharks and also, to a lesser degree, seals, sea turtles and dolphins. This is a real issue when we are considering the impacts of tourism on the marine environment, because these environments are ecosystems in which every species has a role to play and no one species is any more important than any other. If we ignore this simple fact our efforts to try to minimise the negative impacts on our oceans will be doomed to failure.

So, what kind of marine creatures are of most interest to tourists? I divide them into six categories as follows:

- The 'monsters', creatures whose appeal lies in their sheer size, such as whales.
- The 'cute' such as penguins, these can also be seen sometimes as 'humorous' or amusing in their habits or the way they move.
- The 'almost human' who it is believed are intelligent enough to be well above the normal level of sea creatures, like dolphins.
- The 'vulnerable heroes or heroines' who struggle against the odds to survive and reproduce such as female sea turtles.
- The 'scary' and 'wicked' that are feared for the threat they pose to human beings such as sharks, but also because in some cases it is believed they actually target humans.
- The 'freaks', those creatures which stand out as particularly odd-looking such as stonefish.

Some marine species, however, can exist in more than one category. Sea horses, for instance, are clearly 'freakish' in appearance but they are also usually viewed as 'cute'. Some people view stingrays as scary while others see them as cute, as can be seen at Hamelin Bay in Western Australia where people go in the water to touch them.

Within categories, there can be hierarchies of public interest, in effect, so that in the scary category the shark is at the top of the hierarchy but there are other creatures lower down including sea snakes and electric eels, for example.

In recent years we have seen a combining of wildlife-watching and adventure tourism in two ways. First, tourists are increasing deliberately seeking out aggressive shark species such as Great White Sharks in South Africa, Australia and Florida, for example. Dobson in 2006 wrote about this phenomenon, so it is not new (Dobson, 2006).

Second, in places such as South Africa, shark interactions where tourists are immersed in the water protected by cages have become a popular adrenaline experience. The market for shark cage tourism has been studied by several authors including Geldenhuys in a paper published in 2019.

It may seem like a minor point but when we are talking about public attitudes towards, and support for, marine conservation, attitudes towards particular species are particularly important. We need to show people that every element in the marine ecosystem is important and that they are totally inter-dependent.

The impact of marine wildlife-watching on marine environments

It is now time for us to look at the impacts that marine wildlife-watching has on marine environments. Some of these points will be taken further when we move on in the next section to look at the issue of responsible and irresponsible wildlife-watching. I am aware of the potential overlap with the contents of Chapters 5 and 6 at this point so will try to avoid duplication by sticking to the impacts of watching only. However, it is important to recognise that marine activities, such as diving, which we will look at in the next chapter, while sporting activities, are also, at least partly, motivated by a desire to view marine wildlife.

Figure 4.6 illustrates some of the main impacts of marine wildlife-watching on marine environments, although it is not a comprehensive overview covering every possible impact, by any means. The reader will quickly notice that the impacts are almost universally negative for the marine environment, although some would argue that watching raises awareness, which may lead to interest in wildlife conservation. We will look at this idea later in the chapter. Others would argue that one should also take into account the positive economic impacts of wildlife watching for coastal communities. While this is a valid point the main focus of this book is on the impact which tourism has on the marine environment rather than the communities that live around the fringes of our oceans, important though they are.

I would like to make four brief points about the contents of Figure 4.6.

- First, I believe that much of the harm done to marine wildlife is generally caused by a lack of knowledge on the part of tourists, rather than any actual intent to cause harm on the part of the tourist.
- Second, while it is clear that some wildlife tourism operators are aware of the potential harm their activities can cause and choose to overlook them in the pursuit of profit, others may be unaware of the potential consequences of their operations.

- Third, both tourists and operators need to be better informed and that implies we need much more empirical evidence of the harm which wildlife-watching can cause, and evidence that comes from particular locations, so its relevance is readily apparent to both tourists and operators.
- Fourth, a number of the issues covered in the diagram are not only harmful to the marine ecosystem, they actually pose a risk to the tourists themselves, such as swimming with large sea mammals and trying to stroke a seal!

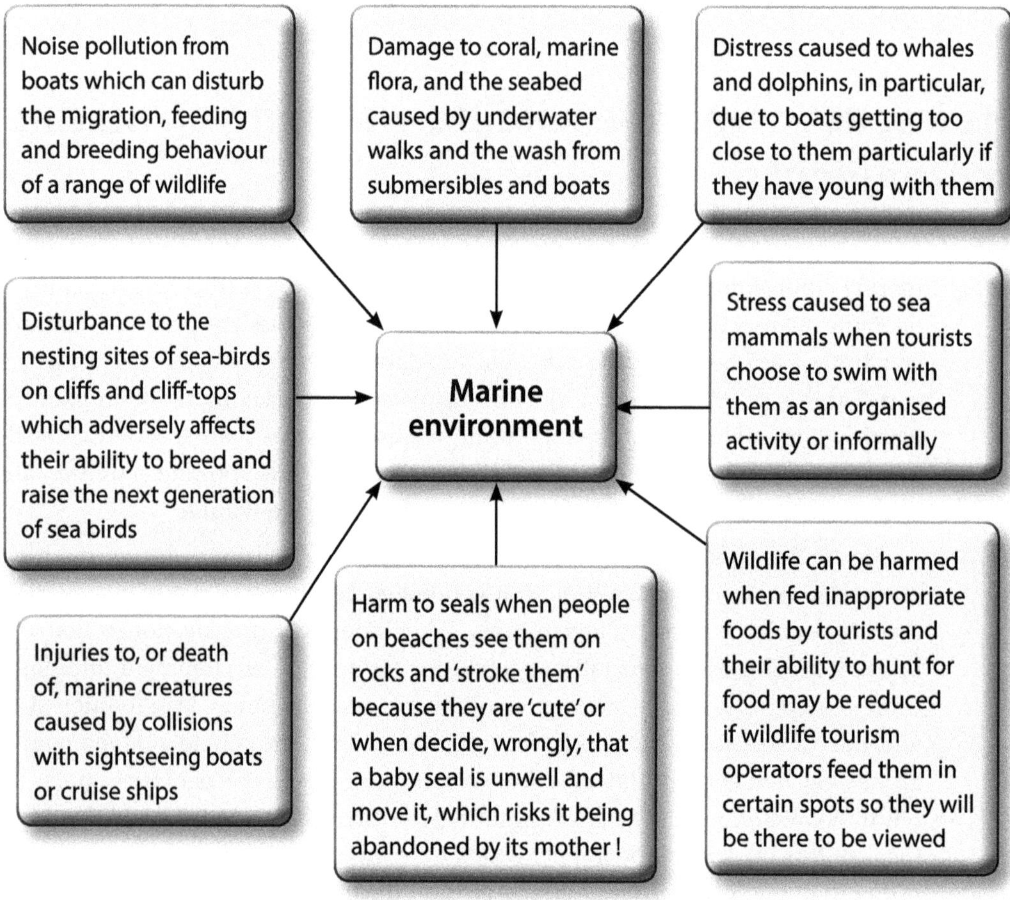

Figure 4.6: The impacts of wildlife-watching on marine environments

Responsible and irresponsible marine wildlife-watching

While most tourism literature talks about sustainable tourism, authors such as Harold Goodwin have preferred to use the term 'responsible tourism'. I also like this term because it implies that everyone can do something to move us towards more sustainable tourism by taking responsibility for their own behaviour and activities and acting more responsibly, whether that be tourists, industry, local communities or governmental bodies. It implies more of a bottom-up approach, perhaps, than sustainable tourism, which is a term that has become, sadly, more associated with top-down policymaking and pontificating at governmental and supra-governmental body level, with too many words and not enough action.

It is a matter of opinion where the line is drawn that separates responsible tourism from irresponsible tourism and, in the absence of regulation or internationally recognised benchmarks for responsible tourism, it is a line that individual tourists, operators and destinations have to draw for themselves. Knowing where to draw the line also needs us to have confidence in our understanding of the impacts of tourism activities and that confidence may be lacking at the moment and for good reason, as our knowledge is more limited than we would wish.

However, being responsible tourists or operators requires pre-requisites that go beyond awareness and knowledge. Perhaps the most important of these pre-requisites is the willingness on the part of tourists or operators to make certain sacrifices and deny themselves certain things for the good of the marine environment. This might mean not swimming with dolphins, even though you can, for a tourist and not feeding wildlife to attract them to viewing points, for an operator. One involves a denial of pleasure on the part of the tourist, while the other means risking a reduction in income and customer satisfaction for the operator.

In the meantime, let us look at five examples of marine wildlife-watching which are increasingly being seen as irresponsible. They are as follows:

- **Whale-watching** which is conducted in a way that is harmful to the welfare of the whales. This can involve approaching at too high a speed or even providing food which disrupts the normal feeding behaviour of the whales. Other boats cross the path of the whales rather than approaching them slowly from the side which causes them stress. Likewise, some boats split up the pod by going between whales, which again causes stress to the creatures. Those interested in this topic can obtain useful information at www.responsibletravel.com.
- **Insensitive observation onshore of sea turtles** when they are laying their eggs or when the hatchlings are making their way to the sea. Tourists often disturb both processes by getting too close and making too much noise or taking photos with a flash. And when the hatchlings head for the sea they are either confused by the bright floodlights at some watching sites while the light also makes it easier for the predators to spot and attack the baby sea turtles. Sometimes hotels or resorts with beaches used by sea turtles for laying their eggs may try to make tourists feel better by making a small donation to a conservation project for each tourist who watches the spectacle. But this does not help if the watching is conducted inappropriately and some conservation projects are better than others!

- **The hunting and feeding habits of marine wildlife**, whether that be sea eagles, fish or sharks, can be altered dramatically if operators feed them at the watching spots to guarantee viewings for tourists. Burgin and Hardiman noted that *'human-mediated feeding often accompanies wildlife-watching in nature....Outcomes include habituation, behaviour change and dietary impacts'*. (Burgin and Hardiman, 2015)
- While the subject is still being debated in respect of swimming with dolphins specifically, in general, there seems no doubt that tourists' **swimming with marine wildlife** is a pretty one-sided activity, giving the tourist a memorable experience but simply disturbing the life of the creatures. There seems little reason, normally, therefore why this activity should be undertaken by tourists. Seeing people to try to ride dolphins shows a singular lack of both respect and understanding. Many tour operators are already starting to remove elephant rides from their programmes, so perhaps it is time to either end or control the practice of swimming with wildlife.
- Traditionally tourists have also had the opportunity to observe marine wildlife in captivity in theme parks and zoos around the world. This keeping wildlife in captivity was often sold to tourists as being motivated by the purposes of research and conservation. This idea has now been challenged, not least because some of the creatures were trained to perform in unnatural ways to entertain visitors. This subject will be discussed in a little more detail towards the end of the chapter.

In fairness, it has to be said that there are many examples of good practice in marine wildlife-watching and most people would probably prefer to see it made more responsible rather than being banned. Many believe the answer is regulation or voluntary codes of conduct, which we will explore later in the chapter.

Instead, at this point we will look at the potential value of wildlife-watching to support marine conservation

Marine wildlife-watching and marine conservation – a symbiotic relationship?

Many people believe that there is a potentially symbiotic relationship between wildlife-watching at sea or from the seashore and marine conservation. The argument goes that watching raises awareness and may inspire some tourists to become committed to marine conservation and seek to get involved. Unfortunately, there seems little hard evidence to support this argument at the moment but that is not to say it is not true for some tourists. Indeed, some researchers have found that watching wildlife does increase awareness of the issues and create some desire to get involved in marine conservation.

Zeppel and Muloin in 2008, for example, stated that their study in Australia *found that tourist learning during mediated encounters with marine wildlife contributes to pro-environmental attitudes and improved on-site behaviour.* Some respondents indicated that the experience had given them an *intention to engage in conservation actions that benefit marine species.* (Zeppel and Muloin, 2008). Unfortunately, there is little empirical evidence that I have been able to find to show that this engagement with conservation as a result of a wildlife-watching trip actually happens.

However, there are several factors that will influence if this is likely to happen including:

- The quality of the wildlife-watching experience enjoyed by the tourist and the extent to which the operator conveyed messages to their customers about responsible watching and marine conservation
- The opportunities for tourists to get involved in marine conservation, which are presented to them while the memory of the trip is still fresh in their minds. This can be via sponsoring a project, or becoming involved in an organisation, or taking part in a voluntary activity such as a beach clean.
- The attitudes of the tourist and how open their mind is to messages about marine conservation, rather than simply seeing the watching experience as a fun activity.

In 2016, Eagle et al. looked at the way on which social marketing might be used to reduce plastic pollution in the oceans using the effects of plastic on sea turtles, which are a popular target species with marine wildlife watchers. (Eagle et al., 2016)

We talked in the title of this section of a potentially symbiotic relationship which suggests a two-way relationship, of course. So how can marine conservation enhance marine wildlife-watching? The obvious answer is that by conserving marine environments the future of wildlife-watching at sea will be made more sustainable. But there is another less obvious answer. If tourists can become involved in marine conservation and it is put at the heart of marine wildlife-watching, then their watching trips should be enhanced through a 'feelgood factor' that they know that their enjoyment is not at the expense of the wildlife.

This 'feelgood' factor seems to be supported by Andersen and Miller who, working with whale-watching tourists in the San Juan Islands in Washington, USA, stated that '*marine environmental education is a powerful tool for shaping human conduct and for enhancing quality of life*'. (Andersen and Miller, 2006)

Apart from the tourists, though, this will only happen if this idea of a symbiotic relationship is enthusiastically embraced not only by operators but also by the public sector bodies and voluntary organisations that work in the field of marine conservation.

Codes of conduct or regulation?

This question goes to the heart of every debate about sustainability in any sector of tourism. Traditionally, the tourism industry has preferred voluntary codes of conduct as part of its objective of showing that regulation is not needed, as industry can behave responsibly through self-regulation. It argues that regulation is heavy-handed, bureaucratic, expensive and hard to enforce. In general, this argument has tended to be accepted by governments in most places. And in some cases, this self-regulation has proved effective, as we have seen in the hotel industry with self-regulation and promotion of good practice through the International Hotels Initiative.

While codes of conduct are a popular way of seeking to manage marine wildlife-watching they seem to work in some locations but not in others. Duprey et al. in 2008 noted that, as part of a code of conduct, the introduction of a voluntary 'rest period' had been effective in reducing vessel traffic around dolphins in Kaikoura, New Zealand from 2.63 interactions per hour to 1.46. However, they also noted that there were also regulations in place alongside this to protect the marine wildlife. (Duprey et al., 2008) On the other hand, Waayers et al. in 2008 observed non-compliance behaviour by tourists to a voluntary code of conduct designed to protect nesting sea turtles in the Ningaloo Marine Park in Western Australia. (Waayers et al., 2008)

However, in the tour operation sector in recent years, we have started to see some industry players coming around to the idea of regulation for a simple reason. Some tour operators have gone beyond what consumers and governments expect in terms of responsible tourism and have taken action to minimise their negative impacts and maximise their positive impacts. This has inevitably cost them money. Now, some complain that other tour operators are undercutting them by not engaging with responsible tourism to the point where they are in danger of losing ground to these competitors. Some of them are arguing, therefore, that some form of basic regulation may be required to ensure some form of 'level playing field' in the tour operation market.

Where codes of conduct are introduced in tourism, they may be either focused on tourists or operators or both. In marine wildlife-watching the emphasis to date has clearly been on codes of conduct rather than regulation. A search of Google at the time of writing which found some 40 articles about codes of conduct or regulations for marine wildlife tourism revealed 36 articles that focused on codes of conduct and just four which mentioned specific regulations or rules.

However, a number of countries and tourist destinations do have rules governing marine wildlife-watching. In 2016, the UK government saw fit to remind boatowners of the rules relating to 'protected marine animals', those deemed to require special protection because of their importance or their rarity or both. They said in a press release:

The Marine Management Organisation is reminding boat owners and operators of registered passenger and charter vessels about the rules and best practice relating to marine wildlife. As the warmer weather reaches English shores it brings with it increasing opportunities to observe dolphins, porpoises and whales....It is essential that the health and well-being of the animals is considered at all times....Endangered species are protected by wildlife legislation. ...It is an offence to intentionally and/or recklessly disturb these animals, with offences of disturbance carrying a maximum penalty of six months imprisonment and/or an unlimited fine. (www.gov.uk, 2016).

The same press release goes on to offer basic advice and request that operators follow a voluntary code developed under the Wildlife Safe or WiSe scheme.

This UK government press release raises two interesting points namely:

- Like similar regulations in other countries it focuses just on the protection of specific species rather than on the marine ecosystem as a whole.
- It recognises that not all interactions between vessels and wildlife at sea involve commercial wildlife-watching vessels. For example, in Cornwall in the UK, a wide variety of marine wildlife is often seen from ferries such as that which plies between Penzance and the Isles of Scilly. And we saw in Chapter 3 that cruise ships often provide an opportunity for their passengers to view marine creatures. It is, therefore, essential that the operators of such vessels are also aware of good practice in relation to wildlife-watching at sea.

There remain challenges with both codes of conduct and regulation. With the former the problem is how to get operators to conform to a code of conduct when there is no compulsion to do so. Not complying may be more profitable and customers may be happy to see a code broken if it increases their chances of seeing marine creatures. As far as regulation is concerned the problem is one of enforcement which may be less effective than one would hope due to a lack of resources or corruption, for example. And operators who do behave responsibly already complain that everyone suffers through regulation due to the irresponsible behaviour of a minority of operators.

Buultjens et al. stated that the government of Sri Lanka had recognised that whale watching in the country had grown rapidly in recent years and had begun to regulate the activity. However, they also saw that bad operator behaviour remained a concern, there were still concerns over whale welfare and the long term sustainability of the whale watching industry was in doubt. (Buultjens et al., 2016) Interestingly, research undertaken by my students in Sri Lanka in 2016-2018 seemed to bear out these findings.

The problem for tourists who want to be responsible and 'do the right thing' is that their situation is difficult. While many operators may claim to be responsible or 'green' there is little independent third-party verification of such claims so it is difficult for the customer to know the truth.

However, there also remains the problem of tourists who do not necessarily want to 'do the right thing' and put their own desires ahead of the interests of marine wildlife. Acevedo-Gutierrez et al. in the Kaikoura Peninsula in New Zealand stated on the basis of a study of 362 respondents that *posted signs were ineffective in increasing compliance with regulations.* (Acevedo-Gutierrez et al., 2011)

Watching marine wildlife in captivity

There is a long tradition of people seeing marine wildlife, and land-based wildlife for that matter, in zoos. In the days before most people could afford to travel outside their own country on vacation this was the only chance they had of seeing creatures such as sharks, seals, sea lions, penguins or tropical fish. Then, emerging coastal destinations began to realise that tourists were also interested in seeing marine creatures, such as dolphins and even whales, performing tricks and putting on shows for the visitors. This very much mirrored what already happened in the circuses that visited most towns and cities in many countries around the world, where monkeys in particular would often be dressed up and lions and elephants would perform for the amusement of humans.

However, over time, circuses began to be criticised for making animals perform and today, in several countries, circuses are not permitted to have wildlife in their shows. Even zoos faced opposition because they were seen as keeping animals captive, in unnatural surroundings, for the amusement of the public. They had to, and still have to, defend themselves by emphasising that they are places of research, conservation and education rather than entertainment, but many people are far from convinced by this argument.

Undoubtedly, the most controversial issue relating to marine mammals being kept in captivity relates to shows involving dolphins and whales. There are a number of objections to these shows including the following:

- The idea of using quite intelligent mammals to perform tricks.
- The ways in which the creatures are trained, which some allege involves a degree of cruelty.
- The suggestion that there is no benefit for research or conservation arising from having sea mammals performing for the crowd.
- The fact that the creatures have been traditionally bred in captivity especially for these shows.

The opposition to keeping and breeding of whales in captivity increased in the wake of the movie, *Free Willy* in 1993, a 'tearjerker' about a captive killer whale.

This was then followed up with three sequels in 1995, 1997 and 2010. In a strange way the growing popularity of recordings of whale sounds or 'whale music' as background music to aid relaxation in recent years served, perhaps, to make whales seem special and too important to be used simply to entertain people

Much of the criticism has been focused on the Sea World® Theme Park in Orlando, Florida, in the USA including numerous stories in the print media as well as television documentaries. Under pressure, Sea World® announced in 2016 that it would stop its orca breeding programme. However, in autumn 2019, its website was still advertising the following experiences:

- A comedy show featuring a sea lion.
- The opportunity to swim alongside dolphins, rays, tropical fish and otters.
- Dining in a restaurant from which visitors can see pilot whales.

However, the company has modified its approach to sea mammal performances, and they are no longer prominently displayed on the website as Sea World seeks to promote itself strongly as a fun day out with rides as well as wildlife and as a centre for conservation.

At the same time, the Loro Parque attraction in Tenerife in Spain is still promoting its marine mammal show. Its website stated in October 2019 at the beginning of a section entitled '*attractions for all the family*' that: *'At Loro Parque, you can watch the Dolphins as they perform right before your eyes and see the Orcas with their top-class performance'*. (www.loroparquetenerife.com, 2020). And despite the negative media coverage of such spectacles I feel that it is important to point out that in 2019 this park was ranked first of36 'things to do' in Puerto de la Cruz in Tenerife.

On the other hand, the tourism industry has been responding to growing public concerns over the keeping of marine mammals in captivity and their use in shows. In 2014 Virgin said it would only work with attraction suppliers which did not take orcas from the wild, while in 2017 they went further saying they would add no further captive wildlife attractions to their programmes. In 2019, TripAdvisor announced that they would no longer be promoting, or selling tickets to, attractions that featured whale or dolphin shows although it appears that reviews of such attractions would still be permitted.

In the same year, the government of Canada also took action imposing a ban on shows involving dolphins and whales together with the breeding of these mammals in captivity as well as a ban on keeping them in captivity at all. These changes may partly be due to a perception that attitudes towards keeping marine wildlife in captivity have changes. Ten years ago, Draheim et al. found in Bayanhibe in the Dominican Republic that tourists *had a high level of concern for dolphin welfare and would rather see wild than captive dolphins.* (Draheim et al., 2010)

These developments may herald the beginning of the end of keeping sea mammals in captivity and/or using them to perform in shows. But some will say it does not go far enough for other marine wildlife is still being kept in captivity in artificial surroundings. Others will argue that it is important for people who cannot see these marine creatures in their natural environment to be able to do so in zoos and that this is important for education and can contribute to conservation. In any event, there is no doubt that wildlife in captivity and even performing wildlife is still, rightly or wrongly, popular with large parts of the population of the world. Ong, for example, has written about this issue in a paper which looked at the 'cuteifying' spaces and staging marine animal performances for the entertainment of the middle classes at a theme park in the Pearl River Delta area of China. (Ong, 2017)

Lessons from wildlife-watching and wildlife conservation on land

There are some interesting lessons to be learned from the experience of wildlife-watching on land for those concerned with marine wildlife-watching. And as we have just seen above there are many parallels between the two in relation to the keeping of creatures in captivity in zoos. But in this section, we will focus upon wildlife-watching 'in the wild'.

The first lesson perhaps, to which we will return in a later chapter, is the need to create places where wildlife can be protected. From the late nineteenth century this has usually involved the creation of protected areas like national parks. These areas aim to conserve nature by limiting development and many encourage visitors while others restrict visitor numbers, and a few have certain areas where visitors are banned. Some national parks, though, have become so popular with visitors that their over-riding mission of conservation is being threatened by the sheer volume of visitors they receive.

However, in terms of wildlife conservation on land, in Africa in particular, we have seen the development of 'reserves', places where wildlife was protected but also located or even re-located to make it more accessible to tourists. Often, the income from visitors was crucial to the funding of the conservation work.

As we will see later in the book, we have seen an increasing trend worldwide towards the designation of marine national parks and marine nature reserves. The experiences of such parks and reserves on land could be very useful to those managing the marine versions, in everything from funding to communication strategies, public participation in decision-making to identifying key performance indicators.

There is always a fundamental dilemma with such protected places that their designation draws attention to a place and in doing so makes people want to visit it which, in turn, puts increased pressure on the environment that can undermine the mission of the protected area.

Land-based wildlife conservation has also created an issue about the rights of people as well as wildlife, and which should take priority. In many national parks, in areas that rely on the income from wildlife tourism, the priority has often been the wildlife, as they appear to have a higher monetary value than the human population. This has led, particularly in Africa, to instances where people have been moved so wildlife can be introduced to areas where they did not live before. This raises some fundamental ethical questions which are not so likely to arise with marine wildlife conservation given the seas themselves have no resident population. However, most of our seas do have populations that live around their fringes and large numbers of people who make their living from the oceans, so questions about the priorities between wildlife and people are bound to arise at some point.

Land-based wildlife tourism has also faced challenges around interactive experiences, perhaps most recently in relation to the riding of elephants by tourists. A number of tour operators have now stopped contracting suppliers for excursions that involve riding elephants. Intrepid Tours was a pioneer in this respect, stopping offering such tours in 2014 and others have followed suit, possibly influenced by the lobbying of pressure groups such as People for the Ethical Treatment of Animals.

Another issue seen on land, which has now been seen at sea, is the practice of operators providing food to attract wildlife to the spots where tourists come to view wildlife, thus disturbing the normal hunting and feeding behaviour of wild animals. Even some respected wildlife sanctuaries have used feeding times as events to attract tourists so they can use the resulting income to further their conservation work. More worrying though is the practice of commercial operators providing food to guarantee sightings to increase customer satisfaction and boost their profits. In the longer term such practices cause massive harm to animal behaviour and disturb the delicate balance of the ecosystem. The practice is so widespread that there are examples everywhere, although some are extreme, such as the feeding of wild bears in Finland which I was told about during a visit a couple of years ago. Hotels and resorts have also started doing the same now to give their guests something else to look at, whether the animal be a deer or an elephant, tropical birds or a large water monitor (as I have seen for myself)!

And, finally and controversially, there is the debate over wildlife sanctuaries that claim to be rescuing and saving injured or orphaned or rare wildlife. This raises moral and ethical issues about the extent to which humans should intervene in nature and evolution. If the creature has been injured or orphaned due to the actions of human's intervention, this may be fine but if not, then perhaps nature needs to take its course? However, most people would probably say that we should intervene in nature to save species which are threatened with extinction, often as a result of habitat destruction by man. However, there are cases apparently where such sanctuaries have been set up simply to make money with the animals not being well looked after and no real conservation work undertaken.

As we have seen there are above parallels between wildlife-watching on land and at sea and there are lessons to be learned from each other.

The impact of climate change on marine wildlife-watching

Just before we begin to draw conclusions from this chapter let us take a brief look at climate change and what it might mean for marine wildlife-watching. I think there may be several consequences as follows:

- As the temperature of the waters increase, we will see some species disappearing from certain locations and others arriving in their place. The warming of the seas and the rise in sea levels are likely to lead to changes in breeding locations and migration routes which could create new marine wildlife destinations while others lose business as they are no longer visited by the species which tourists want to see.
- The warming of the seas and the rise in sea levels probably damage and destroy a number of marine wildlife habitats and reduce the wildlife to be seen, including everything from coral reefs to mangrove swamps to kelp forests.
- Sea level rise as a result of global warming will also threaten the future survival of many low-lying coastal communities that act as starting points for marine wildlife-watching trips, as they begin to feel the effects of more extreme weather leading to regular flooding.
- Attempts to reduce the pace of global warming are likely to put pressure on marine wildlife tourism operators to stop using polluting forms of transport and make use of boats which have a much lower carbon footprint or are even carbon neutral.
- Unless air travel makes huge strides in reducing its carbon footprint through technological innovation, it seems likely that, at some point, limits will need to be placed on air travel, particularly long-haul flights. This would have a large negative impact on many marine wildlife-watching destinations, from Australia to California, Iceland to South Africa.

At the same time, properly managed and controlled marine wildlife tourism might be able to make a small contribution to the battle against climate change if the tourists who participate in it could be converted into activists who would change their consumer behaviour and persuade others to do the same based on what they learned from their time watching marine wildlife.

Conclusions

We have seen that marine wildlife-watching has grown steadily over the past three decades in terms of volume, although it is perhaps fair to say that hard data about this growth are not easy to find. However, at the same time, it has also spread geographically, and is continuing to spread, to the point where there are few corners of the world which are not yet receiving visits from marine wildlife tourists.

In the early part of the chapter we saw that the activity of wildlife-watching at sea is almost infinitely varied, involving different types of tourists looking at different types of wildlife in a range of different types of location.

It has been noted that in some ways the development of marine wildlife tourism has mirrored that of land-based wildlife tourism albeit generally somewhat later. This mirroring is clearly illustrated, too, in the transition that took place from hunting to wildlife-watching, starting with big game in Africa and later encompassing whale-watching. Of course, in both cases hunting has not completely stopped although in relation to whaling the scale of hunting is now minute.

I then went on to look at the fact that interest in marine wildlife does not extend equally to all species that live in the oceans. I suggested that people only seemed to be interested in species that were viewed as big monsters or cute or humorous, or vulnerable but heroic, or scary and aggressive, or freaky, just plain odd in appearance. This is problematic in terms of marine conservation because the marine environment is a complicated ecosystem in which every element or species relies on other species for their existence. Therefore the species people enjoy watching can only survive or flourish if every element in their ecosystem is in a healthy state, whether that be the populations of fish they feed upon, the marine vegetation amongst which they deposit their eggs, or the habitat in which they live.

If marine wildlife tourism is going to help play a role in marine conservation it is vital that those who operate tours of the seas in search of wildlife make clear to tourists that the health of marine environments is what matters, not just whether you can see a whale today.

The impacts of marine wildlife tourism are the same in principle as those for any type of tourism, being divided between environmental, economic and social. However, as far as the ocean and its welfare is concerned the impacts are almost always negative in terms of environmental impacts. The social and economic benefits which are often used on land as a counterargument to negative environmental impacts are obviously irrelevant to the ocean itself. They may be significant to the communities that live on the fringes of the ocean and those who make their living on the sea, but they are of no value whatsoever to the oceans themselves.

This situation is made rather worse by the fact that much marine wildlife-watching appears to be carried out in a rather irresponsible manner, due to either a lack of knowledge or a ruthless quest for profits. However, there are many examples of good practice which need to be adopted industry-wide either voluntarily or as a result of government regulation. Indeed, this theme of regulation versus voluntary codes of conduct remains one of the most important ongoing debates in marine wildlife tourism. It is hard to believe, based on experience to date, that voluntary industry action alone will make marine tourism sustainable in the foreseeable future.

Experience from what has happened on land raises questions about whether simply designating marine national parks or marine nature reserves will on their own solve the challenges posed by marine wildlife-watching and ensure the sustainability of marine environments.

All of this is set against a background of climate change and global warming, something which threatens the future health of our oceans and the survival of the communities that live on their fringes. If this process is not slowed or preferably stopped or even reversed, then the future of marine wildlife-watching looks grim, but that will be as nothing compared to the impact on all tourism and all life on earth.

On a more positive note if we can harness the interest people undoubtedly have in the oceans and marine wildlife marine tourism could yet play a valuable role in helping us create a more sustainable world.

References

Acevedo-Gutierrez, A, Acevedo, L, Belonovich, O and Boren, L (2011) How effective are posted signs to regulate tourism? An example with New Zealand fur seals. *Tourism in Marine Environments*, **7**(1) 39-41.

Andersen, MS and Miller, ML (2006) Onboard marine environmental education: whale watching in the San Juan Islands, Washington. *Tourism in Marine Environments*. **2**(2) 111-118.

Bentz, J, Dearden, P and Calado, H (2013) Strategies for marine wildlife tourism in small island: the case of the Azores. *Journal of Coastal Research* **65**(sp1), 874-879.

Burgin, S and Hardiman, N (2015) Effects of non-consumptive wildlife-oriented tourism on marine species and prospects for their sustainable management. *Journal of Environmental Management* **151**, 210-220.

Buultjens, J, Ratnayke, I and Gnanapala, A (2016) Whale watching in Sri Lanka: perceptions of sustainability. *Tourism Management Perspectives* **18**, 125-133.

Catlin, J and Jones, R (2010) Whale shark tourism at Ningaloo Marine Park; A longitudinal study of wildlife tourism. *Tourism Management*. **31**(3), 386-394.

Dobson, J (2006) Sharks, wildlife tourism and state regulation. *Tourism in Marine Environments* **3**(1) 15-23.

Draheim, M, Bonnelly, I, Bloom, T, Rose, N, and Parsons, ECM (2010) Tourist attitudes towards marine mammal tourism: an example from the Dominican Republic. *Tourism in Marine Environments* **6**(4) 175-183.

Duprey, NMT, Weir, J and Würsig, B (2008) Effectiveness of a voluntary code of conduct in reducing vessel traffic around dolphins. *Ocean and Coastal Management* **51**(8-9), 632-637 .

Eagle, L, Hamann, M, and Low, DR (2016) The role of social marketing, marine turtles and sustainable tourism in reducing plastic pollution. *Marine Pollution Bulletin* **107**(1) 324-332.

Geldenhuys, L, Van der Merve, P and Saaymon, M (2019) Determining the market for marine wildlife tourism in South Africa. *African Journal of Hospitality, Tourism and Leisure* **8**(4) 1-18.

www.gov.uk (2016) www.gov.uk/government/news/boat-operators-reminded-of-rules-on-protected-marine-animals [18 Oct 2019]

Hoyt, E (2001) Whale Watching 2001: *Worldwide Tourism Numbers, Expenditures, and Expanding Socio-Economic Benefits*. International Fund for Animal Welfare. Yarmouth Port, MA

www.loroparquetenerife.com (2020) [2 Jan 2020]

Ong, C-E (2017) 'Cuteifying' spaces and staging marine animals for Chinese middle-class consumption. *Tourism Geographies* **19**(2) 188-207.

Waayers, D, Newsome, D and Lee D (2008) Research note. Observations of non-compliance behaviour by tourists to a voluntary code of conduct: a pilot study of turtle tourism in the Exmouth region, Western Australia. *Journal of Ecotourism* **5**(3) 211-222.

Zeppel, H and Muloin, S (2008) Conservation and education benefits of interpretation on marine wildlife tours. *Tourism in Marine Environments* **5**(2-3) 215-227.

4

Questions and exercises

1. In a group, develop a survey of people who have participated in any kind of marine wildlife-watching experience. As a result of your findings, try to place them within one of the six broad types presented in Figure 4.1. How easily do they fit into these types? If you were to develop Figure 4.1 further which types or categories would you use?
2. Discuss the reasons why the market for marine wildlife has grown so rapidly over the past 30 years. Compare and contrast the development of marine wildlife tourism with the development of wildlife tourism on land.
3. Select a tourist destination which is well known for offering marine wildlife-watching opportunities. Having gathered as much relevant data as possible, create a short management plan suggesting how marine tourism in the destination should be developed in the future.
4. Select an example of a Code of Conduct for marine wildlife-watching. Analyse it highlighting what you consider to be its strengths and weaknesses. To what extent do you think it will help make wildlife-watching more responsible?
5. Examine the websites of whale-watching operators in three different countries. For each website, record everything that is related in any way to sustainability or responsible whale watching. Then compare the three sites and produce a presentation highlighting the differences and similarities between them. Conclude by highlighting anything in your results that surprised or worried you.
6. Using examples, critically evaluate the statement that 'only regulation can ensure that marine wildlife tourism can be sustainable'.

5 Leisure Activities in Marine Environments

In recent years we have seen a veritable 'explosion' in the use of the sea for a growing variety of marine leisure activities. This is now a truly global phenomenon that can be seen from Iceland to South Africa, Australia to Florida. Activities in the sea have always been part of coastal tourism, from paddling and swimming to sailing and diving to angling and boat trips. However, in the past decade or two we have seen the invention of new activities and the developments of variations on traditional marine activities. We now have coasteering, wild swimming, paddle-boarding, RIB and banana boats and sea kayaking, all giving tourists further opportunities to get pleasure from the marine environment.

Many of these activities also reflect a change in marine leisure with an increase in active rather than passive activities and an increase in adventure activities. This has, inevitably, increased the risk level of sea-based leisure activities in some ways. Several of the more adventurous new activities also involve travelling further from land or to less developed areas of coast, increasing the risk further. Interestingly, it appears that many tourists become 'hooked' on some sea-based leisure activities once they have experienced them on vacation. From that point onwards their desire to continue to participate in an activity will often influence their choice of vacation destination. The innovative developments we have seen in terms of sea-based leisure activities have led to a huge increase in participation in marine leisure activities. This has been stimulated by, and reflected in, the investment made in equipment and infrastructure for such activities by governments, commercial operators, and hotels and resorts.

However, some of these activities represent a risk to the marine environment too in terms of their impacts. There are great variations in the nature and scale of the impacts of these different leisure activities on the marine environment, and we will explore these in this chapter looking at the activities one by one. It is clear that in general leisure activities do not have a positive impact on the ocean – they do not make it a better or healthier place usually. Therefore, it is a matter of just how negative their impacts are from serious or severe to minor or maybe even neutral, and this might be the best we can hope for realistically.

Before that, though, I would like to offer several typologies of marine leisure activities that highlight different aspects of their nature and impacts. These will be presented in diagrammatic form and are of necessity, tentative and subjective. The characteristics of marine leisure activities outlined in Figure 5.1 show that there are considerable variations between different types of activities based on a range of variables including how much equipment is required, how weather dependent they are and how much personal energy the participant needs to expend. Some of the factors shown in this diagram also determine how expensive it is to take part in these activities.

Other important characteristics of marine leisure activities include whether they can be undertaken individually and/or in groups and whether they can be undertaken competitively, pitting one participant against another.

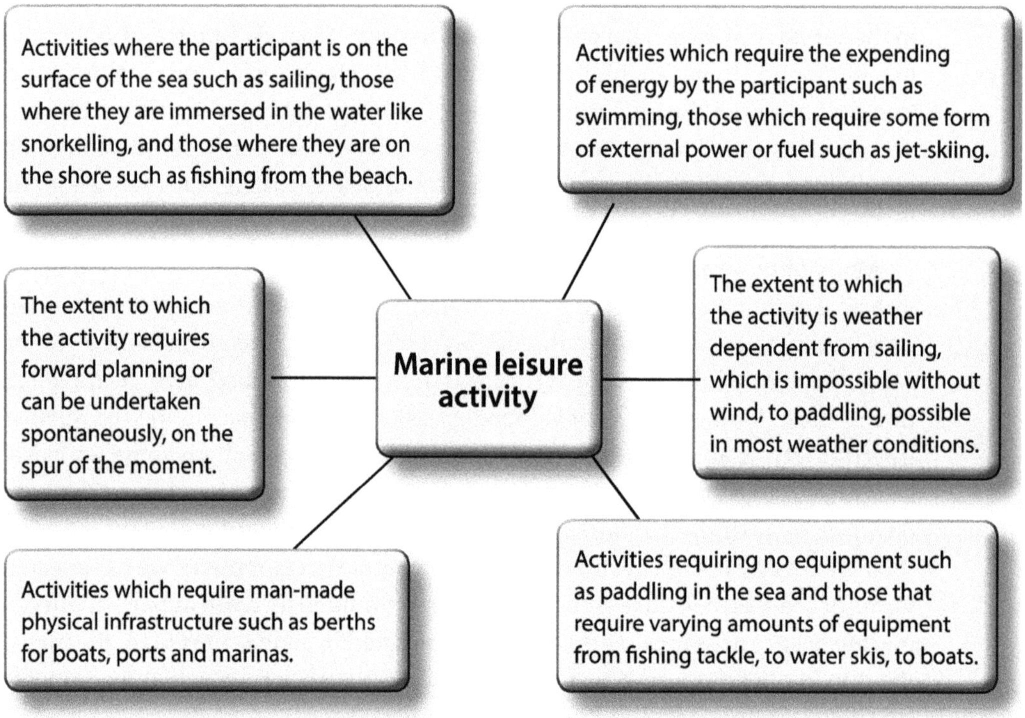

Figure 5.1: Types of activity based on their characteristics

Figure 5.2 looks at marine leisure activities in terms of what requirements they have of the participants. It shows that some require no particular skills and only minimal levels of fitness, while others require significant levels of skills and knowledge and/or physical fitness. It seems that some activities require both knowledge and skills, while others require a relatively high level of personal fitness, and others require little fitness, knowledge or particular skills.

Most will have potential risks and safety procedures that participants will need to observe, although these vary greatly between activities. Those swimming from a beach need to be aware of the risk of rip tides and that they should not swim if a red flag is flying. Sailing enthusiasts, however, have a myriad of risks to worry about, many established safety procedures to be aware of, and probably responsibilities laid down by law which they must follow.

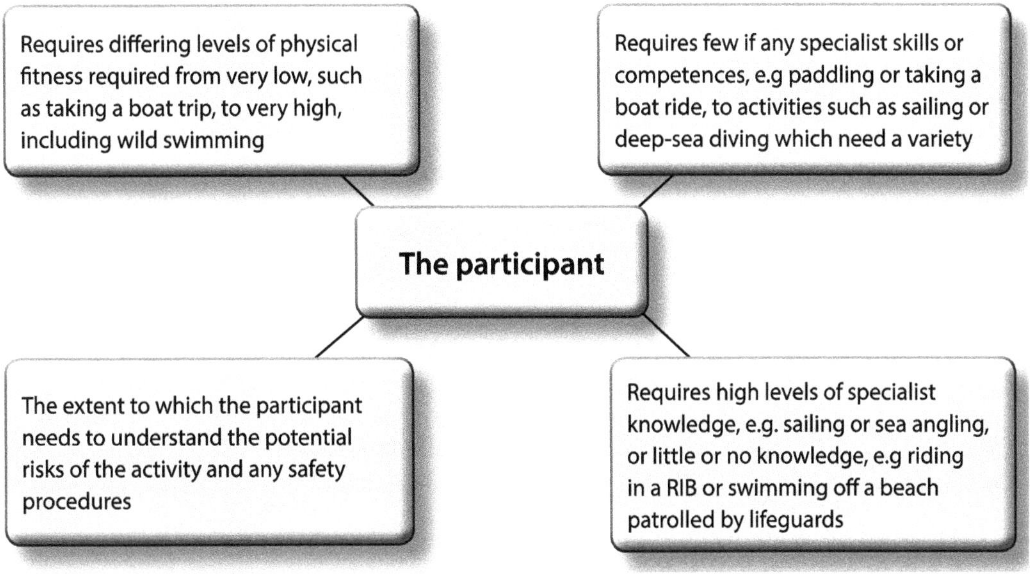

Figure 5.2: Types of activity based on their requirements of participants

Figure 5.3 looks at different marine leisure activities in terms of their potential impacts on the marine environment. It shows a wide range of factors relating to the impacts of marine leisure activities, some of which will be explored further when we look at different types of activity later in the chapter. I acknowledge that some activities such as swimming may be almost wholly neutral in their impacts on the marine environment but it cannot think of a single activity that has a positive impact on the ocean environment, although many bring economic benefits to operators and tourist destinations, as well as pleasure for the participants.

■ Paddling and swimming

In many ways, the most natural of leisure activities in the sea are paddling, swimming and wild or adventure swimming. Almost everyone who has visited the coast for leisure has paddled at some time even if it was just a matter of walking out into a few centimetres of water. Most have also swum in the sea at some point even if only for a few minutes close to shore. However, in recent years we have seen the rise of wild swimming or adventure swimming where the pleasure comes from going beyond the norm, swimming in remote or challenging locations, away from other swimmers.

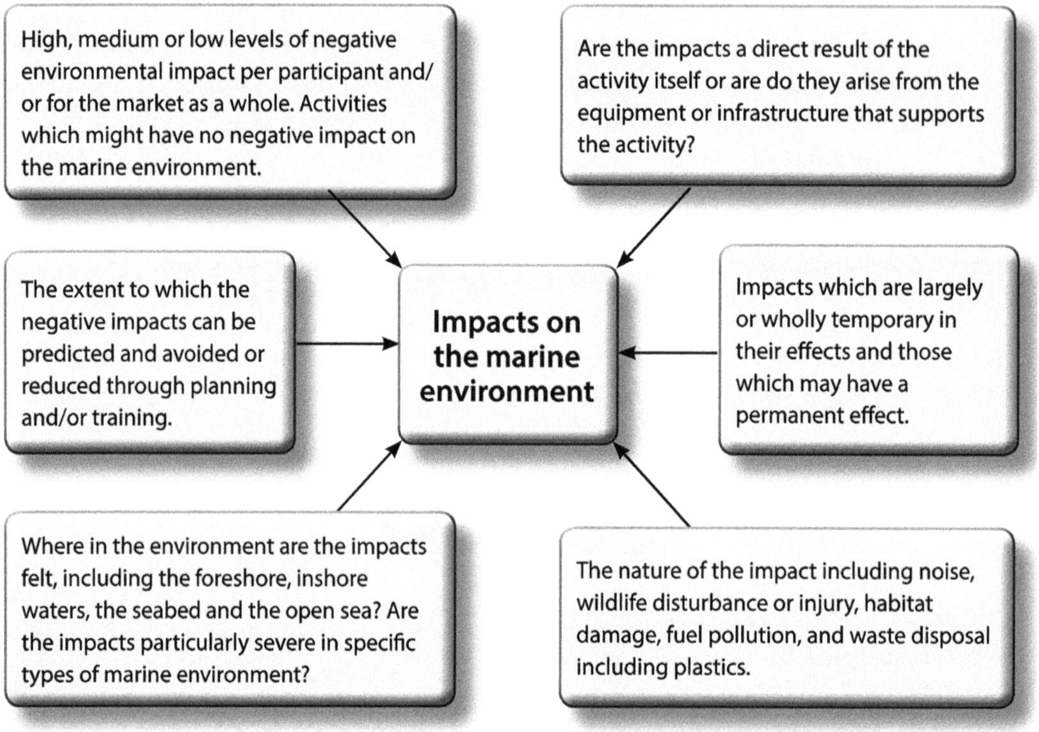

Figure 5.3: Types of activity based upon their likely impact on the marine environment

As one would probably imagine paddling and swimming has only a very small impact on the marine environment in most cases, despite the scale of the activity worldwide. Indeed, if one looks into the subject there seems far more concern about the effects of swimming in the oceans on human health, with a belief that these are increasingly negative due to pollution.

However, there are two impacts worth noting. First, paddling can disturb wildlife habitats for those creatures that live on the shoreline. Second, and perhaps more significant, is the impact of millions of people entering the sea covered in suntan lotions, which are full of chemicals, some of which will inevitably be washed off into the sea

■ Coasteering

Coasteering is a word that dates back to the 1970s. It is an interesting activity that brings land-based and sea-based activity together, as it involves exploring an area of the intertidal zone using a range of techniques involving everything from swimming to rock climbing, surfing to diving in sea caves. It is highly active and requires a certain level of fitness. It has proved particularly popular in the UK in areas such as Pembrokeshire and Wales.

The impact of coasteering on the ocean will largely depend on how much of the time the participant spends in the water and the extent to which they need the use of boats with engines powered by fossil fuels. Wildlife disturbance is likely to be very low normally and there could be more negative impacts on rocky cliffs through the dislodging of rocks or the use of artificial climbing gear, potentially, than on the marine environment.

Surfing

While the roots of surfing go back centuries, when it was a key part of Polynesian culture, its history as a global leisure activity can be traced back to Hawaii and the late nineteenth and early twentieth century. Today, it is a worldwide phenomenon with its own culture, styles of fashion and lifestyle, served by globalised suppliers providing equipment and clothing. There are a number of major tournaments and a significant number of professional surfers. Surfing has become an integral part of the popular image of several places including much of the coast of California and specific locations such as Bondi Beach in Australia. It has also become a major attraction factor for visitors to places as diverse as Cornwall in the UK, the west coast of Portugal, Hawaii, Indonesia, Chile and South Africa. It has also been further developed by the invention of offshoot activities, inspired by surfing, such as windsurfing and paddle-boarding. This is an almost exact mirror image of what we have seen with winter sports and the development of snowboarding, for example.

On the face of it surfing might appear the epitome of a low impact marine leisure activity, just a person and a board. However, as surfers themselves have acknowledged, this is not the case. A surfing website, www.stillstoked.com, published an excellent article about the impact of surfing on the environment and the rest of this section is based on information from this article.

In this article they suggested there are now around 20 million regular surfers and predicted that the surfing industry would reach over $US 10 billion by 2024. They then went on to tell readers that:

- *Surfboard blanks (foam cores) are made of polyurethane.... It does not break down in landfill. 400,000 new toxic surfboards are produced each year.*
- *Surfboard resins, especially the most common polyester resins, are extremely toxic and non-biodegradable.*
- *Fins are made from the same toxic materials.*
- *Wetsuits are made from neoprene which is, essentially a toxic rubber made from petroleum.*
- *Leashes are more of the same: petroleum-based and non-biodegradable.*
- *Wax is also made from 70% petroleum and contains the chemicals benzene and toluene.*

- *Sunscreen is loaded with toxic chemicals which have a devastating impact on ocean life. Sunscreen contains the chemicals oxybenzone and octinoxate which decrease coral defences against bleaching, damaging their DNA. The same chemicals that keep you from getting too much colour actually strip the colour from the coral which prevents their development.*

  Source: www.stillstoked.com, 2019

This article is impressive in that it takes a holistic view, talking about the impacts of surfing on the ocean but also on land in terms of pollution and toxic waste. It then goes on to give information about alternative products that were less harmful for the environment.

The article is just one example of how the surfing community appears to be trying to minimise its negative impacts on the oceans and go further in terms of conserving the marine environments in general and becoming involved in action on broader environmental issues. In the USA the Surfrider Foundation, is '*dedicated to the protection and enjoyment of the world's ocean waves and beaches through a powerful activist network*' (www.stillstoked.com, 2019). Meanwhile, there are initiatives such as 'Take 3 for the sea', which encourages surfers to pick up three pieces of rubbish as they finish each surf session (www.stillstoked.com). And in the UK, there are 'Surfers against Sewage', a group of surfers who campaign against sewage pollution on beaches and in the oceans. This is part enlightened self-interest, but it also reflects an interest in broader environmental quality and conservation.

In the surfing world it seems that responsible surfing is something being driven by surfers themselves and activists, rather than it being led by professional bodies or industry. Frankly, they seem to be ahead of some other marine leisure activities in terms of responding to the negative environmental impacts of their chosen activity.

Diving

From the point of view of our oceans diving is a step or two beyond surfing because much of it involves going below the surface of the water, albeit only a very short distance in the case of snorkelling, compared to swimming and surfing where most activity is focused on the surface of the water. It also differs from these other activities in that interacting with the marine environment, especially flora and fauna, is, perhaps, the major motivator for participation.

There are three main types of diving which tourists may participate in:

- **Deep-sea diving.** While a single definition of this does not exist, it is generally considered to involve dives below the scuba-diving limit of around a depth of 40 metres. It offers the diver the opportunity to explore marine environments not accessible to participants in other types of diving. However, it requires specialist equipment and special training

- **Scuba-diving.** This allows the diver to explore beyond the limits of snorkelling by utilising an underwater breathing supply allowing them to breathe underwater. Training is still required but it allows the diver to explore marine habitats close up, for a period of time.
- **Snorkelling.** This is a way of seeing a small area of the ocean from the surface using a mask and a breathing tube. It requires no special training and nothing beyond the mask and tube, so it is an easy activity to engage in casually while on vacation

Source: www.oysterdiving.com, 2019

For snorkellers and scuba divers, coral reefs are a particular attraction, while wrecks are a popular destination for deep-sea divers, and for scuba divers when they are in relatively shallow water. Controversially, some coastal resorts have created artificial reefs by dropping man-made structures into the sea or deliberately sinking ships or boats. While these are an artificial intrusion into the ocean environment, they can become successful wildlife habitats in their own right.

It is impossible to measure the market for the different types of diving because while some people participate regularly at home and on vacation, others try it just once while they are on vacation. However, various figures for the number of 'regular' divers are found online in the order of six million for scuba-divers and twenty million for snorkelling. This data is attributed to the Sport and Fitness Association, although I have not found the original source.

However, the Diving Equipment and Marketing Association (DEMA) has conducted and published some research from the market in the USA. They put the number of active scuba divers in the country at between 2.7 and 3.5 million and the number of active snorkellers at around 11 million (www.dema.org, 2019).

In terms of diver motivations and behaviour, Bentz et al. in 2016 distinguished between those people with a general interest in diving and those with a specialist interest who had more experience and knowledge. (Bentz et al., 2016) In terms of diver motivations It has been found that *novice divers* (in Brazil) *preferred megafauna while experienced divers liked the 'cryptic' creatures.* (Giglio et al., 2015) Both papers suggested that such differences in types of divers and their motivations needed to be taken into account when seeking to manage diving sites effectively.

For many tourist destinations around the world, diving has become an important market, and many now promote themselves as diving destinations. If you look at tour operator brochures and traveller posts on social media, it is clear that the desire to participate in some form of diving has become a motivator for millions of tourists to visit particular locations.

The cruise industry can also contribute to the challenges caused by diving. It has been estimated that of the three million cruise passengers who visit Cozumel in Mexico each year, around 90% express a desire to take part in some form of marine leisure activity, including scuba diving. (Mota and Frausto, 2014)

In terms of the potentially destructive impact of diving on the ocean environment most attention has focused on coral reefs, which attract divers because of the beauty of the reefs and the colourful and varied wildlife for whom the reef is home. They are fragile and easily damaged, and even the diving industry has recognised the harm that irresponsible divers can do. This concern around the impacts of diving on coral reefs is not a recent phenomenon. More than 25 years ago, Hawkins and Roberts were warning that '*unless the pace of tourist development in the northern Red Sea is reduced soon the carrying capacity of coral reefs seem sure to be exceeded with widespread reef degradation the likely result*'. (Hawkins and Roberts, 1994) Their concerns seem to have been well founded for a few years later, in a study of Eilat on the Red Sea, researchers were suggesting that '*current rates of recreational diving on some reefs at Eilat are unsustainable*'. (Zakai and Chadwick-Furman, 2002)

There is a particular risk in terms of those who are not regular divers. but who decide that their vacation is a perfect opportunity to snorkel or scuba dive for the first time. They can cause great damage quickly, due to sheer ignorance, in their eagerness to see as much as possible in as short a time as possible. Of course, divers can have other impacts, including disruption of wildlife because of their presence, and damage to the seabed and marine flora when they walk on the seabed. Occasionally they also dispose of bits of equipment which causes pollution of the marine environment. Diving from boats can also cause damage when the anchor is dropped onto the seabed.

In 2018, the Reef Resilience Network placed an interesting report on its website that identified several interesting points in relation to the impact of snorkelling and scuba diving on marine environments. It noted that:

- *A number of studies have documented impacts to coral reefs from divers and snorkellers. They can be caused by fin kicks, pushing or holding coral, dragging gear, and kneeling/standing on corals.*
- *Scuba divers typically have more impacts on corals than snorkellers, particularly divers wearing gloves and photographers with equipment. This is because snorkellers mainly float above the corals on the surface of the water.*
- *Studies suggest that inexperienced divers cause greater damage than more experienced divers.*
- *A recent study showed that the nationality of visitors can have a greater impact of incidences of reef damage than other factors such as dive experience and ability. This*

suggests that user attitudes, perceptions and beliefs are just as important as the actual impacts that recreational activities can have on reefs. It also suggests that in some places, management strategies may need to be targeted to specific groups.

- *Trampling of corals is also common on shallow, near shore reef flats and has led to extensive damage in areas with high levels of human use.*
- *The incidences of coral damage from diver contact inspired the concept of 'diver carrying capacity'. While this approach has been applied to numerous reefs throughout the tropics some conservation organisations suggest assessing carrying capacity may have limited practical value. Quantifying carrying capacity can be challenging it varies widely based on the ecological conditions of a reef, the potential resilience of a reef, and visitor behaviour.*

Source: www.reefresilience.org, 2019

This article went on, interestingly, to identify causes of damage to reefs from other marine leisure activities, such as boating and cruise ships.

The rapid growth of recreational diving in recent years means that a lot of often irreversible damage has already been done, but the Professional Association of Diving Instructors (PADI) is not ignoring the problem and is trying to encourage dive schools to promote responsible diving, not least because the livelihoods of their members depend on the marine environment! On their website they say that PADI '*has demonstrated a longstanding commitment to environmental conservation through its course offerings…to protect and preserve our ocean planet*' (www.padi.com, 2019). Their website focuses a lot on responsible diving with examples of their activities, including the positive impacts which divers can have through litter clearing initiatives.

However, the problem is that some PADI schools may be tempted to put profit ahead of responsible diving, while there are many non-PADI operators who are unlikely to be too concerned about responsible diving. And, for the tourist keen to have a go at snorkelling, for the first time for just a couple of hours, responsible diving is unlikely to figure in their thinking.

Again, the issue of codes of conduct arise here and many voluntary codes exist at the level of both professional associations and individual destinations from the UK to Bali, the Azores to Thailand. However, these are often very vague or minimalist and there is no mechanism for ensuring that all divers follow them.

An interesting article from divein.com recognised the impact which divers can have on coral reefs. Although the site is a pro-diving site, they admitted that '*Coral reef diving is an all-time favourite to most scuba divers. But like it or not, scuba divers are causing substantial damage to the world's coral reefs*' (www.divein.com, 2019). The article went on to explain how diving can damage reefs and suggested a simple voluntary code of behaviour as follows:

Do's

- *Study about marine life that you might see while diving.*
- *Educate yourself about the maximum limit of interacting with corals.*
- *Make sure you and your buddy are aware of the impact on coral reefs.*
- *Firmly obey guidelines of safe diving.*
- *Take respectful attitudes for coral-friendly diving.*

Don't's

- *Don't touch any of the corals.*
- *Don't let your body come in to contact with corals.*
- *Don't let your diving equipment hang loose. Secure all equipment so nothing bumps into the corals.*
- *Don't pollute the water with anything.*
- *Don't break off or take any pieces of the corals.*

(www.divein.com, 2019)

The problem specifically relating to coral reefs is that they are already under threat from the impact of rising water temperatures, diving damage is just speeding their decline. Furthermore, it has been estimated that coral reef tourism is worth a staggering $36 billion so moves to control it may not be popular with destination marketers and operators. (Spalding et al., 2017)

A number of studies in recent years have sought to identify the carrying capacity of dive sites in a number of locations including:

- The Isabel Islands National Park in Mexico (Ríos-Jara et al., 2013)
- Mabul Island in Malaysia (Zhang et al., 2016)
- Koh Chang Marine Park, Thailand (Roman et al., 2007)
- The Saba Marine Park, Netherlands Antilles (Hawkins et al., 2005)

At the same time, some researchers have been investigating whether artificial reefs are a potential solution to the damage that divers do. In 2007 Stolk et al. presented a critical review of the literature on this topic. (Stolk et al., 2007). The opinions of divers about artificial reefs were reported by Shani et al. in 2012 in a paper which suggested that artificial reefs might be a way of resolving the ecological concerns around natural reef diving and the desire of divers to pursue their sport. (Shani et al., 2012) However, as we will see later in the book the construction of artificial reefs can bring their own environmental problems.

Let us now turn our attention from activities where humans largely rely on their own energy to those marine leisure activities which involve the use of vessels of one sort or another.

Sea kayaking

Kayaking on lakes and in rivers has been around for a long time as a leisure activity but in recent years it has also become popular to kayak in the open sea. It is seen as a good way of exploring out-of-the-way, remote locations in a way which has minimum negative impact on the marine environment. It can be seen as part of the growth of interest in what might be termed 'adventure tourism' and popular destinations include Iceland, Alaska and even the rugged western coasts of the British Isles.

This is generally true that virtually no leisure activity is neutral in terms of its impact on the marine environment. An EU funded project, the UK Marine SACS Project, which ran from 1996 to 2001, charted the impacts of sea kayaking and canoeing on marine environments, and suggested that there were observable impacts, when launching the boats, in terms of most types of inshore waters and estuaries, as well as impacts upon seals and dolphins. Otherwise they regarded the impacts as 'minimal' in relation to noise disturbance, disturbance to wildlife, and erosion and turbidity. (www.ukmarinesac.org.uk, 2019).

Sailing

Again, it would appear that sailing should have a very low impact on the marine environment given it is based on wind power rather than motors requiring fossil fuels. Even then, though, many sailing boats also have an auxiliary motor for those days when there is insufficient wind to propel the vessel. However, even pure sailing vessels, without an engine, can generate negative impacts on the marine environment, some of which were identified as part of the previously referred to SACS project conducted in the UK between 1996 and 2001. According to that study there were a range of potential impacts with the scale of impacts varying between different types of marine environment.

These impacts included the following:

- The launching of boats from shores or other 'natural' locations, not designed and built specifically for launching boats, may cause damage to the foreshore environment from trampling and the dragging of boats.
- Sewage discharge from sailing boats can affect the marine environment, particularly in estuaries where there is 'low flushing'; in these locations it can also contribute towards a reduction in oxygen levels in the water.
- Wildlife living in the sediment and benthic communities may be exposed to contamination from anti-fouling paints when maintenance takes place on the water or on mud or sand flats at low tide. Anti-fouling paints can contribute to concentrations of heavy metals in particular locations, although this had not been quantified at the time of this study.

- Litter and waste discarded by boat users can contribute to marine pollution albeit on a very limited scale.
- Seals appear to be prone to being disturbed by the presence even of sailing vessels.
- Disturbance to marine birdlife from the loud noises emitted when sails are set.
- Localised seabed impacts may occur from either anchor drag or boat grounding.

 Source: Based upon www.ukmarinesac.org.uk

Some of these impacts also relate to vessels with engines, of course. As we will discuss in more detail later in the chapter, some of the impacts from boating come, not from the vessels themselves, but from the infrastructure required by boaters such as marinas.

Powered boats

Most boating is undertaken in vessels with some form of artificial propulsion in the form of some type of engine. The impact of these boats is numerous. According to the pressure group, Sailors for the Sea, some of the main issues are as follows:

- Oil and fuel spills from boats can cause great damage to marine waters. There are ways of reducing the risk of this happening and there are actions that can be taken to minimise the impact if it does happen.
- The marine environment can be damaged by the 'blackwater' sewage discharge from marine 'heads' or toilets.
- Emissions from engines can be reduced by using cleaner types of engine and reducing fuel consumption.
- Leakages of oil and toxic chemicals can occur from bilges that are not regularly inspected and maintained.
- The use of toxic cleaning products on board the vessel can cause dangerous and damaging chemicals to enter the marine environment.
- Many owners use copper-based paints for painting the bottoms of their vessels and this can harm marine wildlife.
- Inappropriate ways of storing the boat when not in use can harm the marine environment.
- Waste disposed of at sea harms both marine habitats and wildlife.
- 'Greywater' from onboard sinks or showers and produced by boat cleaning, together with the release of detergents into the water is a major polluter of the marine environment.
- Poor practice when anchoring a boat causes damage to marine ecosystems, particularly coral reefs, sea grasses and shellfish beds.

- Boats getting too close to marine wildlife can have a significant harmful impact on marine creatures including seabirds, seals, dolphins, and whales.
- Owners whose boats are not moored at a marina or in a harbour will need to tow their boat to the launching place, thus adding significantly to the carbon footprint of their boating trip.

Source: Based upon material on www.sailorsforthesea.org, 2019

It is also well documented that noise pollution from powered boats has a negative impact on marine creatures.

Sailors for the Sea and other groups within the boating community have been very active in trying to spread good environmental practice although, as ever, we can never be sure of many people receive and respond positively to the good advice they provide. At the same time, there have been some positive developments such as the introduction of electric powered boats which have less harmful impacts on the marine environment in terms of both fuel spillage and emissions, although the use and disposal of batteries may still be of some concern with some boats.

The added problem with boating is the fact that it is an activity that is often focused on small geographical areas, which means that its impacts are highly concentrated in space, most obviously around the marinas and harbours. The slicks of oil and floating litter and debris one finds in most such places are clearly unwelcome additions to the marine environment.

However, I would now like to turn my attention to three specific types of powered craft which I believe are particularly problematic in relation to the marine environment namely RIBS/Zodiacs and jet-skis.

RIBS

RIBS or Rigid Inflatable Boats, of which Zodiac are perhaps the best- known brand, have become increasingly popular in recent years. They have powerful engines and are relatively cheap, light and fast, and are used both by private enthusiasts but also by wildlife-watching operators and cruise ships. Their main advantage is their ability to get into remote or shallow places where normal boats are not able to go. Virtually unsinkable, it is claimed, and highly agile, they are increasingly used by operators offering whale-watching or dolphin-watching cruises when the spotting sites are not too far from the shore. The fact that RIBs can get pretty well anywhere due to their flexibility and shallow draught means that virtually no wildlife habitat is out of reach for them, no matter how fragile it might be.

At the same time, they are frequently used by cruise ships to access ports of call where the cruise ship is too large to come alongside the berths or where there are no cruise ship berths, just small harbours. This is particularly the case with cruises to so-called adventure destinations, such as Antarctica and Greenland. However, they are more frequently used by cruise ships for excursions where the boat ride is part of the experience rather than just a means of transport. Indeed, they are part of the adrenalin rush side of the tourism industry, which is not surprising given that most can reach speeds of 30-40 knots and some can even go as fast as 70 knots.

However, the reader will not be surprised to learn that RIBs are very controversial craft for use as wildlife-watching vessels because of their speed and noise, which are both alleged to have a substantial and harmful impact on the very wildlife they are used to spot, namely whales, dolphins and seals. Some of their other potential impacts, such as fuel spillage and emissions, can be reduced or eliminated by replacing hydrocarbon fuels with electric power. Likewise, there are environmentally friendlier ways of looking after the hulls than anti-fouling paints which contain chemicals that are harmful to the marine environment.

But the speed, and resulting disturbance of the water, while it is the real problem, reflects the attributes of RIBs that are the very reasons wildlife watchers might choose an operator who has RIBs against those with more conventional vessels. Not only does the speed allow them to get to the viewing points more quickly and spend more time viewing, but the thrill of the ride is as much an attraction for some people as the wildlife.

Of course, there are voluntary codes of conduct in some destinations and examples of good practice in the cruise sector but, in general, there is a general consensus that RIBs remain a significant threat to the well-being of marine wildlife.

Jet skis

Most beach users will be familiar with jet-skis, or personal watercraft as they are more correctly known. Originally developed by Kawasaki in the 1970s they are now found in the inshore waters of coastal resorts across the world. They usually carry from 1 to 3 people, have powerful engines, and can travel of speeds of over 40 knots.

Most people who do not like them emphasise the risk they pose to human beings rather than to wildlife or marine environments. There have been many accidents in which swimmers and jet-skiers themselves have been killed or injured. Because they operate close to shore, they often find themselves in conflict with other leisure users, particularly swimmers, surfers and shore anglers. For others, their main objection to them is one of noise. Even the government of the Emirate of Abu Dhabi talks on its website about the noise pollution caused by jet skis (www.dot.gov.abudhabi, 2019).

Because they operate in inshore waters, their main impacts will be wash from their propellers, which will affect the fish, seagrasses, and other marine creatures that live in these waters. In some waters their operation may also affect mammals such as seals and dolphins.

The UK SACS project, that has been quoted several times, identified negative impacts from jet-skis including significant levels of wildlife disturbance as well as some damage to the beach environment during launching, and some erosion and turbidity around shallow inlets and sandbanks (www.ukmarinesac.org.uk, 2019).

While some areas have voluntary codes of conduct for jet-skis, they remain controversial with many local communities trying to get them banned. Such campaigns have been seen in places as diverse as Brighton in the UK and New South Wales in Australia. Sometimes such bans have been reversed by governments, as was the case is 2014 in Negril in Jamaica, to be replaced by regulation. At times the lifting of bans or rejections of bans may be motivated by the idea that such bans mean lost income for tourist destinations.

Boat-assisted leisure activities

Up until now we have focused upon boat-based activities where the craft itself was at the heart of the experience and using it was the main purpose of the leisure activity. However, it is important to recognise that there are also activities where the boat is a facilitator rather than the focus of the activity. These activities include parasailing, paragliding, water-skiing and the so-called 'banana boats', inflatable boats that look like bananas and are towed behind motor-boats. Indeed, all of these activities rely on boats with engines to tow the craft to which the tourist is attached! Again, there are safety concerns with each of these activities and potential conflicts with other inshore waters leisure users.

However, our main concern with them relates to the fact that they also have a potential impact they have on marine wildlife and environments as the towing craft are usually powered by hydrocarbon fuels and they are fast and powerful and create strong washes that will affect wildlife, sea beds and beaches. Meanwhile, the launching of these various craft from the shore may also harm the marine environment as equipment is dragged across sand and over the seabed or through the water. Yet again, some codes of conduct are in place but, as ever, the issue is one of the levels and monitoring of compliance.

Before we leave the topic of boats it is important to recognise the harm that all kinds of vessels can do to the marine environment through the damage done by their anchors to the seabed and marine ecosystem. For example, an interesting study in the marine reserve of Cap Creus in Spain looked at the impact of anchors on a particular type of marine flora. (Lloret et al., 2008)

Sea angling

In parallel to commercial fishing is the leisure activity of sea angling. This takes several forms including:

- Fishing conducted by rod and line from the shoreline whether that be a beach or rocks. This is the least expensive option and is widely practiced around the world, often by local residents or day visitors rather than by tourists.
- Fishing conducted by rod and line from a boat, usually one chartered from a commercial operator with relevant expertise, or one which is privately owned or rented. This is often a group-based activity rather than something undertaken individually.
- Sport fishing – fishing with rod and line for 'trophy' or 'specimen', usually large fish such as tuna, marlin or shark. This usually involves a chartered boat and a crew who are specialists in this field. Most participants will be tourists, and many may have flown halfway across the world to try to catch a 'trophy' fish. In some ways this is the equivalent of the old big-game hunting on land.

The scale of sea angling can be gauged from data published in 2013 for England by DEFRA that suggested that in 2012 four million days were spent sea angling with just under 75% of these being spent fishing from the shore and a little over 25% being from boats. The researchers estimated that there were some 884,000 regular sea anglers in England and their activities generated £1.23 billion per annum and supported some 23,600 jobs (Armstrong et al., 2013).

Sea angling produces four types of potential impacts on the marine environment, namely:

- The impacts of the **boats** that sea anglers use. These are largely the same as those associated with powered boats and have already been adequately discussed above. As they may try several fishing spots during a trip, the risk of seabed damage from anchor drag can be significant.
- The marine environment can also be damaged by **discarded equipment** such as fishing lines and lead weights, although sea fishing associations try to encourage their members not to discard equipment irresponsibly.
- The impact on **fish stocks**. Over-fishing can denude stocks but, this is not a common occurrence and regulations are often in place to reduce this risk.
- The impact on **marine wildlife**. This can occur when a fish that is on the line escapes but may still have the hook in its mouth or may even remain tied to a length of line which may have metal pieces on it. Discarded hooks and fishing lines may be ingested by marine creatures, such as seals, and cause them harm.

It should be noted that there are often regulations in place to help reduce some of the negative impacts of sea angling on the marine environment. For example, there may be restrictions on angling during breeding seasons and regulations covering when a fish is too small to be taken away. As ever, though, there remains the issue of to what extent such regulations are enforced. There are also voluntary codes of conduct, which vary in their effectiveness from place to place.

In fairness, it has to be said that many anglers have embraced the idea of more responsible angling and have embraced the idea of 'catch and release' whereby any fish caught is returned to the sea alive rather than being killed and taken away for food or as a 'trophy'. The survey of sea-angling in England in 2012, referred to earlier, estimated that 75% of the fish caught by shore anglers were returned to the sea alive, including some defined as under-sized fish by the regulations. This compared to a figure of around 50% for fish caught from boats. (Armstrong et al., 2013).

For many non-anglers perhaps the most controversial type of fishing is so called 'sport fishing' where anglers go in search of specimen fish from species that are often endangered such as sharks and bluefish tuna, to give but two examples. While catch and release is becoming more popular in this type of fishing, there still seem to be too many anglers taking too many pictures of themselves next to huge fish hanging from hooks on harbour quays or hoisted high on the masts of charter boats. This is an activity largely indulged in by affluent tourists rather than local people and often tourists from countries thousands of kilometres away. It is not about food but rather the acclaim that comes from one's peers for catching a mighty creature of the deep. Zoning has been used in some places to restrict where recreational fishing can take place to protect fish stocks, such as in Ningaloo Marine Park in Western Australia. However, not all sea anglers will comply with such zoning and some may not even be aware of it. (Smallwood and Beckley, 2012)

To end this section, and put sea angling into perspective, it is almost certainly correct and probably appropriate to say that far less harm is done to the marine environment and marine wildlife by sea angling than is caused by the commercial fishing industry. This is something we will explore a little in Chapter 7. On the other hand, unlike sea angling, the main purpose of commercial fishing is to provide food for people rather than giving pleasure to, and enriching the leisure time, of tourists, day visitors and the residents of coastal resorts.

The infrastructure for marine leisure activities

Most marine leisure activities require infrastructure and we will now look, briefly, of the impact of this infrastructure on the marine environment. Many sea-based leisure activities involve boats and therefore there is often a need for harbours or marinas where boats can be kept safely when not in use and where maintenance work can be undertaken. These are also the places where the services and retail outlets required by boat users can be found, such as dry docks and ships' chandlers who provide the equipment used on boats, from clothing to GPS devices, life vests to ropes.

Most harbours used by leisure boaters or commercial operators are long established and any disturbance caused to the marine environment from their construction was over long ago. However, on the other hand, every month new marinas are being developed all over the world, sometimes with limited environmental impact assessments before their construction begins. Many tourist destinations believe that possessing a marina is essential if they are to attract the lucrative marine tourism market. Several authors have studied the impact of marina construction on marine environments. These seem to fall into two types.

First, there are the impacts from the construction process itself which continue for as long as the construction is ongoing. Apart from pollution from waste disposal and accidental fuel spillage, the most significant such impact appears to be related to noise and the vibration which occurs from machinery. This creates what is termed 'acoustic trauma' for marine life and can cause harm or even death to marine creatures as well as disrupting feeding and breeding. Of course, in addition there may be damage to wildlife habitats caused by changes to the seabed caused by the construction works.

The second set of impacts are around the longer-term impacts of the construction of a new marina. These include coastal erosion which can be difficult to predict accurately. Klein and Zviely published an interesting study on this topic in 2000 looking at the example of the Herzliya marina in Israel.

Both marinas and established harbours often require dredging as well as dredging of the channels by which boats enter and leave them. This dredging can have a dramatic impact on marine habitats particularly as it happens periodically rather than being a one-off event.

On the other side of the argument, mooring boats in marinas and harbours avoids the damage caused when boats are launched from the shore, including trampling by people and the effects of boats being dragged across sand and mudflats, causing harm to the marine life there.

At the same time, marina's and harbours carry their own ongoing risks to the marine environment including accidental fuel spills and accumulations of waste disposed of by boat owners. Given that many boats are serviced in marinas and harbours, they can also cause pollution from the chemicals and materials used in the maintenance and cleaning of boats if these are not used responsibly.

In Chapter 6 we will look at the issue of infrastructure again in the broader context of the impacts of tourism on the ocean fringe, from coastal tourism.

Conclusions

We will now try and pull together the main threads that have run through this chapter. The first point to make is that the use of our oceans for leisure activities has increased dramatically in recent decades. As with tourism as a whole, it is now a truly global phenomenon. Furthermore, in addition to the more traditional activities such as swimming, sailing and angling we have seen the invention of new activities in recent years, including coasteering, parasailing, and jet skiing, for example.

We have seen that marine leisure activities are diverse; some require only the energy of the participant while others require some form of propulsion, usually involving some kind of engine. Certain activities take place on the surface of the water while others involve immersion in the water. There are activities which require the participant to have specific knowledge or skills and others that do not.

Not all of the participants in these activities are tourists but most are either staying visitors or day visitors who do not live in coastal communities.

Perhaps most importantly, though, we have seen that all activities have impacts on the marine environment to some extent, although the exact impacts vary substantially between the different activities. As is usually the case with tourism these impacts are almost universally negative and harmful and their impact is magnified by the fact that they are often concentrated in relatively small geographical areas, normally on the oceans fringe or in inshore waters. In general, it also seems that the impacts are greater when the activity requires specific equipment of one type or another or when transportation by motorised vessel is an integral part of the activity.

However, we have also seen that in respect of many of these leisure activities efforts are being made to reduce their impacts and make them more sustainable. In some cases, these efforts are being led by professional bodies such as PADI in relation to diving. Interestingly, in other areas such as surfing and sea angling, the drive for change is being led by the participants themselves, working voluntarily as activists to raise awareness and encourage changes in behaviour. In general, it seems that those involved in marine leisure activities as both consumers and suppliers seem to be more engaged with the idea of sustainability than appeared to be the case with the cruise sector when we looked at that in Chapter 3.

However, as with the cruise sector and marine wildlife-watching, there is still a debate to be had over the respective merits of voluntary codes of conduct versus regulation. Both exist in the realm of marine leisure activities but, again, there seems a greater reliance on voluntary action by industry and consumers rather than on compulsion through government regulation. We need to ask if that is right when, despite many examples of good practice, the negative impacts of leisure activities on marine environments appear to be getting worse due to the continued growth in demand.

I believe that two final points need to be made before the end of this chapter.

First, marine leisure activities do not only have negative impacts on the marine environment, they can also harm those that participate in them. Risk is an ever-present reality in a number of marine activities including the danger of being drowned while swimming in unfamiliar waters, suffering 'the bends' while diving or falling overboard from a boat. One of the many risks, and one which always attracts media headlines, is when sharks attack people as they swim or surf. In the past few years a number of such news stories have come from the Indian Ocean island of La Reunion. Between 2011 and 2016, there were reported to be 493 shark attacks across the world of which 43 proved fatal. Of these attacks, 19 were in La Reunion but of these 7 were fatal. In other words, over this five-year period, 16% of the total of fatal shark attacks were in La Reunion (www.deeperblue.com,2019). Many of these attacks have been attributed to bull sharks. There has been considerable debate over why this should be the case, particularly after there were two further fatal attacks in 2019. Some have suggested that the growth of shark attacks is a result of more sharks being attracted to La Reunion due to human activities which have an impact on the marine environment such as agricultural run-off and over-fishing.

Second, we saw in Chapter 2 that our oceans are in crisis, facing challenges from many different causes. While tourism impacts may not be the most important factor in this situation they clearly contribute to the crisis. And as we have noted before, tourism is less important to sustaining human life than fishing or farming, for example. As a purely discretionary rather than an essential activity, maybe marine tourism needs to be significantly controlled in order to help protect our oceans?

In Chapter 6 we will focus on the place where most interactions between tourists and the sea take place, namely the ocean fringe, the beaches, resorts, and inshore waters which are at the heart of coastal tourism.

■ References

Armstrong, M, Brown, A, Hargreaves, J, Hyder, K, Pilgrim-Morrison, S, Munday, M, Proctor, S, Roberts, A, and Williamson, K (2013) *Sea Angling 2012 - a Survey of Recreational Sea Angling Activity and Economic Value in England.* Department for Environment, Food and Rural Affairs. London.

Bentz,J, Lopes, F, Calado, H, and Dearden, P (2016) Managing marine wildlife tourism activities: analysis of motivations and specialization levels of divers and whale watchers. *Tourism Management Perspectives* **18**, 74-83.

www.dema.org (2019) www.dema.org.share/download.aspx?id=7811B097-8882-4707-A160-F999B49614B6 [14 Nov 2019]

www.divein.com (2019) www.divein.com/articles/a-scuba-divers-impact-on-a-coral-reef [12 Dec 2019]

www.dot.gov.abudhabi (2019) www.dot.gov.abudhabi/pwc/en/info/Marine_Environment [12 Nov 2019]

Giglio, VJ, Luiz, OJ and Schiavetti, A (2015) Marine life preferences and perceptions among recreational divers in Brazilian coral reefs. *Tourism Management* **51**, 49-57.

Hawkins, JP, Roberts, CM, Kooistra, D, Buchan, K and White, S (2005) Sustainability of scuba diving tourism on coral reefs of Saba. *Coastal Management* **33**(4) 373-387.

Klein, M & Zviely, D (2001) The environmental impact of marine development on adjacent beaches: a case study of Herzliya Marina, Israel. *Applied Geography* **21**, 145-156.

Lloret, J, Zaragoza, N, Caballero, D and Riera, V (2008) Impacts of recreational boating on the marine environment of Cap de Creus (Mediterranean Sea) *Ocean and Coastal Management* **51**(11) 749-754.

Mota, L and Frausto, O (2014) The use of scuba diving tourism for marine protected area management. *International Journal of Social, Behavioural, Economic, Business and Industrial Engineering* **8**(10) 3358-3363.

www.oysterdiving.com (2019) www.oysterdiving.com/blog/whats-the-difference-between-deep-sea-diving-snorkellingand-scuba-diving [30 Oct 2019]

www.padi.com (2019) www.padi.com [11 Nov 2019]

www.reefresilience.org (2019) www.reefresilience.org/stressors/local-stressors/coral-reefs-tourism-and-recreation-impacts [13 Nov 2019]

Ríos-Jara, E, Galván-Villa, CM, Rodriguez-Zaragoza, FA, López-Uriarte, E, and Mũnoz-Fernández, VT (2013) The tourism carrying capacity of underwater trails in Isabel Island National Park, Mexico. *Environmental Management* **52**, 335-347.

Roman, GSJ, Dearden, P and Rollins, R (2007) Application of zoning and 'limits to acceptable change' to manage snorkelling tourism. *Environmental Management* **39**, 819-830.

www.sailorsforthesea.org (2019) [21 Dec 2019]

Shani, A, Polak, O and Shashar, N (2012) Artificial reefs and mass marine ecotourism. *Tourism Geographies* **14**(3) 361-382.

Smallwood, CB and Beckley, LE (2012) Spatial distribution and zoning compliance of recreational fishing in Ningaloo Marine Park, north-western Australia, *Fisheries Research* **125-126**, 40-50.

Spalding, M, Burke, L, Wood, SA, Ashpole, J, Hutchison, PZ, J, and Ermgassen (2017) Mapping the global value and distribution of coral reef tourism. *Marine Policy* **82**, 104-113.

www.stillstoked.com (2019) www.stillstoked.com/feature/safety-environmental-impact-eco-friendly-surf-products [19 Oct 2019]

Stolk, P, Markwell, K and Jenkins, JM (2007) Artificial reefs as recreational scuba diving resources; a critical review of research. *Journal of Sustainable Tourism* **15**(4) 331-350.

www.ukmarinesac.org.uk (2019) www.ukmarinesac.org.uk/activities/recreation.rOb_02_2.htm [27 Nov 2019]

Zakai, D, and Chadwick-Furman, NE (2002) Impacts of intensive recreational diving on the reef corals at Eilat, northern Red Sea. *Biological Conservation* **105**(2) 179-187.

Zhang, L-Y, Chung, S-S, and Qiu, J-W (2016) Ecological carrying capacity assessment of diving site: a case study of Mabul Island, Malaysia. Journal of *Environmental Management* **183**(1) 253-259.

Questions and exercises

1. Discuss how sea-based leisure activities and the use of the sea for pleasure by tourists, have developed over the past 100 years. To what extent can we predict how they might develop in the future and identify the factors that will influence their development?
2. Using examples, discuss what you consider to be the three most important negative impacts which marine leisure activities can have on the marine environment.
3. Compare and contrast the approaches to responsible marine activities being taken in relation to surfing, diving and sea angling. In each case who are the key players involved, what actions are they taking and what do you believe are their motivations? Which approach do you find the most impressive and why?
4. Select a set of regulations from a country of your choice that are concerned with the impacts of one or more marine leisure activities. How effective do you feel these regulations will be in reducing negative impacts on the marine environment and why do you think this? Comment on how easy or difficult these regulations would be to enforce?
5. Compare and contrast the various impacts on the marine environment that may arise from shore-based sea angling and boat-based sea angling.
6. Discuss the likely impact of scuba diving and snorkelling on the marine environment, using research data. Which do you think is likely to produce the most harmful impacts?
7. Working in groups, examine how marine leisure activities affect different types of marine environment in different ways. Try to identify the key factors that determine the type and scale and type of impacts in the different types of marine environment. For the purposes of this exercise you will probably need to develop your own typology of marine environments. The contents of Chapter 2 may be of assistance in this task.

6 Coastal Tourism and the Ocean Fringe

This focus of this book is on the marine environment, but one cannot understand the impact of tourism on the marine environment without looking at the ocean fringe, the interface between the land and the ocean. In this chapter we will concentrate on how things that happen on land in relation to tourism impact on the marine environment. However, it is also important to note that this relationship is two-way and that tourism on land is affected by the ocean in terms of coastal erosion, for example, as well as being impacted by changes in the temperature of sea water and rising sea levels.

When we discuss the ocean fringe in this chapter we are going to be focused upon the area of land and sea which encompasses the following;

- The intertidal zone which is *the seashore which is covered during high tide and exposed during low tide, revealing a unique biome* (www.biologydictionary.net, 2019). This is where beach-based activities such as swimming, surfing, shore fishing and parasailing are to be found.
- Inshore waters, the area of sea close to shore, perhaps extending a few kilometres beyond the shoreline. This is the zone in which marine activities such as sailing and boat fishing take place.
- Coastal settlements along the shoreline including resorts, hotels, restaurants and transport infrastructure which is located on or close to the shoreline.

In other words, we are focusing on that zone which represents the interface between the natural marine environment and human activity. Since the earliest days of modern tourism this interface between land and sea has been at the heart of the tourist experience. An interesting overview of the historic evolution of coastal tourism and its impacts was offered by Gormsen in 1997.

It all began, in Europe at least, with the early sea bathers, cautiously immersing themselves in the sea but staying only a few metres from the beach. It included people sun-bathing on the beach when the tide went out and looking in the rock pools left by the receding tide to spot minute examples of marine wildlife. And, it involved walking along the beach or walking to the cliff tops to admire the views of the wild ocean.

Over time infrastructure grew up to accommodate those who loved vacationing by the sea, and this included hotels, restaurants, amusement parks and attractive promenades.

Until the twentieth century this form of tourism was primarily limited to Europe and parts of the USA. From the mid twentieth century, coastal resort tourism was at the forefront of the rise of mass tourism which started in Europe, but such tourism is now to be found across the world, from Mozambique to Rio de Janeiro, Bondi Beach in Sydney to the Spanish Costas. At first, most such mass tourism was based in resorts which evolved from existing fishing villages such as Benidorm in Spain. Infrastructure such as hotels grew to meet growing demand over time, until tourism came to be the dominant or even the only real industry in the place. At the same time, new and more sophisticated leisure activities were developed to exploit beaches and the sea, including parasailing, jet-skiing, sailing, and sea angling from boats. These sometimes required specialist infrastructure which had an impact on the marine environment, most notably marinas. In recent decades we have seen an increase in marine-based adventure sports and more active forms of marine leisure such as scuba-diving, wild swimming and coasteering.

As governments saw the potential economic benefits of tourism, more and more of them have sought to gain a share of this lucrative business as a matter of urgency. Instead of waiting for organic change and individual entrepreneurs to transform sleepy fishing villages into lively tourist destinations, they have taken it upon themselves to create purpose-built destinations, such as those developed on the coast of Languedoc in France in the 1970s. It is therefore interesting to note that 20 years ago the French government commissioned a major report looking at the impacts of tourism, leisure and sport on marine flora and fauna (Bellan and Bellan-Santini, 2001)

Elsewhere existing settlements best known as commercial ports were transformed very quickly into major tourist destinations through huge investment. Dubai is perhaps the best example of this, but there are examples in many parts of the world including Tunisia and Egypt. Fifteen years ago, Shaalan stated that '*many beach resorts are now in operation and there are still hundreds to be constructed. However, previous tourism development in Egypt has resulted in a series of negative environmental impacts.*' (Shaalan, 2005) In existing coastal small-scale tourist destinations we have seen the construction of major new resort complexes, which are much larger than traditional coastal destination hotels. They also tend to offer marine leisure activities that have potentially negative impacts on the marine environment. Some purpose-developed destinations and resort complexes were better designed than others and it is clear that some of them were developed in ways that took little or no account of their potential impact on the marine environment.

And as if that was not enough we have also seen, in the past 25 years, the development of a number of new airports, to meet the demands of growing tourist numbers, that have had impacts on marine environments, usually due to land being 'reclaimed' from the ocean to create the site of the airport. Examples of this include the new international airport built for Hong Kong on Chep Lak Kok island and Kansai International Airport in Japan. Tourism destinations and resort complexes have also at times have also expanded by 'reclaiming' land from the sea. The most dramatic example of this to date is undoubtedly The Palm Jumeirah development in Dubai, on which there are large hotels, apartments, restaurants and marina berths. Other versions of the same phenomenon are now to be found in both Qatar and Bahrain. Perhaps the most amazing example of an artificial island developed for the purposes of tourism is Thilafushi in the Maldives. This island was created for one single purpose, namely as a dump for garbage, most of which is due to tourism!

So, we have seen that our use of the ocean fringe by tourism increasingly involves larger-scale and more radical projects and that this is being seen in more and more countries as destinations and resort operators seek to compete with those destinations and operators that are seen as success stories.

In the rest of this chapter we will look at how tourism on the ocean fringe impacts on the marine environment.

The main interactions between coastal tourism, the ocean fringe, and the marine environment

There are four main ways in which tourism in and around the ocean fringe impacts the marine environment and these are illustrated in Figure 6.1

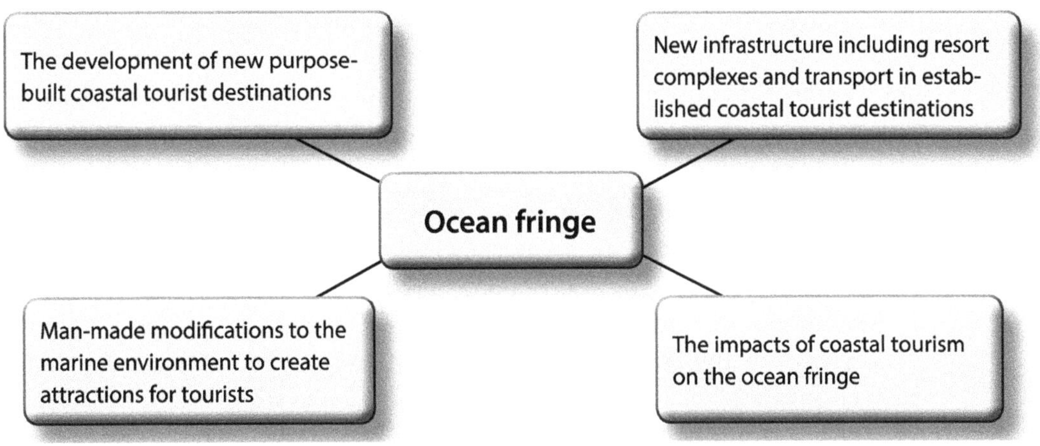

Figure 6.1: The key interactions between coastal tourism, the ocean fringe, and the marine environment

The development of new purpose-built coastal tourist destinations

Brand new purpose-built destinations are not easy to define as few of them are absolutely brand new; there is often something there already, whether it be a few fishermen's houses, small villages or a modest port. The point about places in this category is that there tends to be no existing tourism and when these places are earmarked for conversion into purpose-built tourist destinations, the scale of development tends to be enormous, swamping what was there before. Furthermore, the pace of development is much faster than anything they have seen before. Most such developments have taken place in countries which wish to cash in on the growth of tourism and have highly centralised and powerful national governments that are able to guide development to where they want it to go or intervene directly in the market as players in their own right.

The creation of new destinations is not new, but it is the scale and the pace of development which is relatively new. Technological developments have made this possible, not least in the 'reclamation' of land from the sea, for the construction of tourism infrastructure whether than be a resort complex or an airport.

Not surprisingly, such projects have major impacts on the marine environment, both during their construction and when they are up and running. The construction phase is likely to result in substantial disturbance to wildlife and their habitats but the ongoing impacts once the destination is completed and is functioning are more difficult to predict. The impacts during construction are obviously magnified many times over if the project involves land 'reclamation' from the sea and the creation of artificial islands as we will see later in the chapter. Indeed, it appears that such islands have become increasingly popular in recent years.

The impact once the destination is developed probably depend on how well planned the destination was and how accurately potential impacts were predicted and taken account of by the designers. Over time, one would hope that the ongoing impacts of new destinations on the marine environment lessen as our knowledge of the impacts and the techniques available to minimise them improve.

These purpose-built destinations can seem rather sterile to visitors when compared to places that grew organically and adapted to tourism over a period of time. Nevertheless, they do seem popular with tourists and in the past decades we have seen a number of countries, including Tunisia, Egypt, the UAE, and Qatar, using this approach to quickly enter the tourism market. I remember being taken on an official visit to the Gulf of Aqaba in Jordan to see an area that had been identified as a future tourist destination based on a master plan that was spread out for us on a table in an office.

Nor is this era of developing new mega-sized purpose-built tourist destinations over apparently. In 2016 the government of Indonesia announced a plan to develop no fewer than ten brand new tourist destinations as part of its 'Wonderful Indonesia' programme. These were spread across the country and were being targeted at lucrative markets within Asia, such as South Korea, as part of the government's ambition that Indonesia should welcome 20 million tourists by 2019. (www.en.tempo.co, 2019). And Indonesia is not alone. In 2019 the *Phnom Penh Post* announced ambitious plans to develop the Bokor National Park into an 'historical tourist destination' based on a master plan, entitled, '2035 Bokor City Master Plan. (www.phnompemhpost.com, 2019)

Such projects, while wholly understandable in terms of the economic benefits which tourism brings to poorer countries in particular, raise questions about whether such mega projects are appropriate in an era when air travel and the idea of continuous economic growth are under scrutiny because of climate change. They also raise questions in terms of the issue of community involvement in tourism planning, something which is considered important in most texts about sustainable tourism. One wonders how much say local communities have when such master plans are developed and whether they reflect the wishes of local communities as well as central governments.

The development of new infrastructure in established coastal tourist destinations

Once established, any tourist destination needs to get ahead of competition from existing destinations as well as anticipating and countering potential future competitors. This usually means developing new infrastructure in the form of attractions, accommodation or transport. This not a new phenomenon but the scale and diversity of the projects is new. Again, these developments can have significant impacts on the marine environment, both during construction and when they are operating. Existing destinations normally see investment when there is a need to:

- Meet growing demand
- Stimulate a growth in demand
- Rejuvenate or re-position the destination to make it more resilient and enhance its sustainability

This investment can be either from the public or private sector and often a combination of the two. However, whoever is making the investment, the construction and perhaps usually also the operation, of the infrastructure is likely to have significant impacts on the marine environment. In Figure 6.2 I endeavour to identify the main types of infrastructure developments that might have an impact on the marine environment.

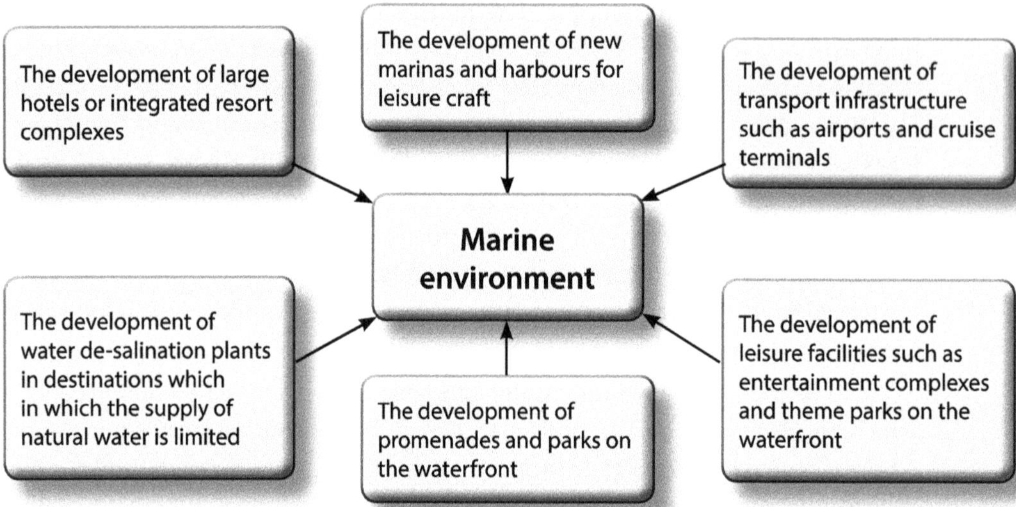

Figure 6.2: Types of infrastructure development that can have an impact on the marine environment

Before we look at the types of development highlighted in Figure 6.2, perhaps, we need to begin by saying that negative impacts from these types of infrastructure development are not necessarily inevitable. Furthermore, their scale will depend on the quality of the environmental impact assessment that is undertaken and the mitigation measures which are put in place. It is highly likely that the type of development which has the greatest impact on the marine environment is the development of new hotels and integrated resort complexes, by which we mean complexes that resemble mini-destinations because they incorporate accommodation, a number of bars and restaurants, leisure and entertainment facilities and potentially casinos and conference and exhibition venues. The development of such units has substantial and continuous impacts because:

- There are far more of them than there are airports or marinas, for example.
- As more and more places seek to become tourist destinations new hotels and resort complexes are opening, literally every day, all over the world.
- They are usually sited as close to the beach and the sea as possible so their interactions with the ocean are greater than if they were located a little further inland away from the shoreline.
- They want to be located in the most stunning location with the best views, and such places are also often those with fragile marine environments. This is often seen in relation to coats fringed by mangrove swamps, for example, which play a vital role in combatting the impacts of climate change.
- To satisfy their clientele they tend to organise water sports activities for guests, including those with relatively high impacts on the marine environment such as snorkelling and scuba-diving, and jet-skiing.

- Some units are deliberately located on beaches which are also important wildlife habitats, such as those where turtles come ashore to lay their eggs and from which the baby turtles launch themselves into the ocean for the first time. This is then used by hotels and resorts as an attractor for consumers and some even provide lighting, so it is easy for guests to see what is happening, but which also makes it easy for predators to spot and attack the baby turtles!
- Once a hotel or resort arrives in a previously pristine area that received no tourists before, other entrepreneurs see an opportunity and before long more units are built, and the place begins to become a true tourist destination.
- Unlike some types of infrastructure development where the impacts may decline dramatically once construction is completed, hotels and resorts continue to have impacts whether that be through the impact of water sports, plastic pollution and waste disposal or just noise pollution from guests and loud music.

Of course, none of this is inevitable, if there is a strong land-use planning system in place with strictly enforced building controls in place. However, many of the countries where tourism is growing fastest are less affluent countries without such systems in place for one or more of the following reasons:

- They may lack the personnel with the technical skills to manage a sophisticated land-use planning system.
- Even where regulations are in place to control development there may be a lack of resources to allow them to be effectively implemented.
- Central government has the power to impose development projects on local communities without consulting them.

Even where planning systems and regulations governing development do exist, they can be undermined by two factors, namely:

- Public sector decision-makers will often be swayed by the short-term economic benefits of new hotel and resort developments, which is not surprising when a country is relatively poor.
- Corruption, which tends to be common in poorer countries with a small rich elite, who tend to control the levers of power and are attracted to the tourism industry by the potential financial rewards and the prestige that can come with hotel or resort ownership.

We will return to the issue of planning and tourism management in more detail in Chapter 10.

Like cruise ships, resorts seem to be getting bigger, so it is reasonable to assume that their impacts are also growing in parallel. Resorts with more than 500 guest rooms are increasingly common and this means that there could be more than 1,000 guests on site at any one time in addition to the staff. The impacts can easily be imagined if 200 of these guests go scuba-diving, 50 take out jet skis and 100 have a barbecue on the beach on a single day, all concentrated in a small area of sea and beach in front of the hotel. This spatial concentration of impacts is one of the problems associated with large hotels and resort complexes. We also need to acknowledge that large resort complexes and hotels have a carbon footprint that means they are contributing towards global warming and thus the problems faced by our oceans that were outlined in Chapter 2. They also consume resources which may be in short supply in the area such as water, not for drinking or cooking alone, but also for luxuries such as swimming pools and golf courses.

I would suggest that the development of some other types of infrastructure, whether it be marinas, airports, or new promenades, tend to have most of their direct impacts on the marine environment primarily during the construction phase. However, fuel spills and waste disposal can occur once the marina is operating and the marine environment will continue to be affected by noise pollution and air pollution from the airport once it is open. At the same time marina construction can have unintended consequences, creating coastal erosion along other parts of the coast (Klein and Zviely, 2001)

Before we conclude this chapter I would like to focus on one other type of infrastructure project, namely water desalination plants. Many of these have been developed to support residential communities so they can survive in harsh, arid environments, but they have also been developed to allow tourism to exist in what are in effect deserts. This process clearly has an impact on marine environments.

An interesting article by Folk on the EcoMENA website in 2019 highlighted issues around de-salination such as:

- Brine that is left behind after de-salination may be put straight back into the ocean in large concentrations which can be deadly for marine life.
- When water is taken into the plant and processed to produce clean water, small fish, crabs, fish eggs and plankton can be sucked in too and killed.

(www.ecomena.org, 2019)

There are, at least, partial solutions to both problems but these can be more expensive and are not always adopted. In 2010, Roberts et al. published a study of desalination on the marine environment drawing on previous published studies. Their conclusions included the following:

- *'Discharge scenarios can lead to substantial increases in salinity and temperature, and the accumulation of metals, hydrocarbons, and toxic anti-fouling compounds in receiving waters.*
- *A clear consensus across many of the reviewed articles is that discharge site selection is the primary factor that determines the extent of ecological impacts of desalination plants.*

- *The greatest environmental and ecological impacts have occurred around older multi-stage flash (MSF) plants discharging to water bodies with little flushing.'*
  (Roberts, Johnston and Knott, 2010)

To conclude, let us note that the impacts of any infrastructure development on a marine environment depends on a number of factors. A crucial factor is the nature of the environment and how far it is vulnerable or resilient. Every one is different and this needs to be borne in mind when thinking about the impacts upon them.

The impacts of coastal tourism on the ocean fringe

Leaving aside new infrastructure projects, the everyday operation of the tourism industry and the activities of tourists create their own impacts on the marine environment. To be fair these need to be separated out from two other impacts:

- Impacts caused by local residents in the destination even though many of these may only be in there because they are employed in the tourism industry.
- Impacts caused by pollution which has its source inland, but which enters the sea via rivers that flow through the coastal tourist destination.

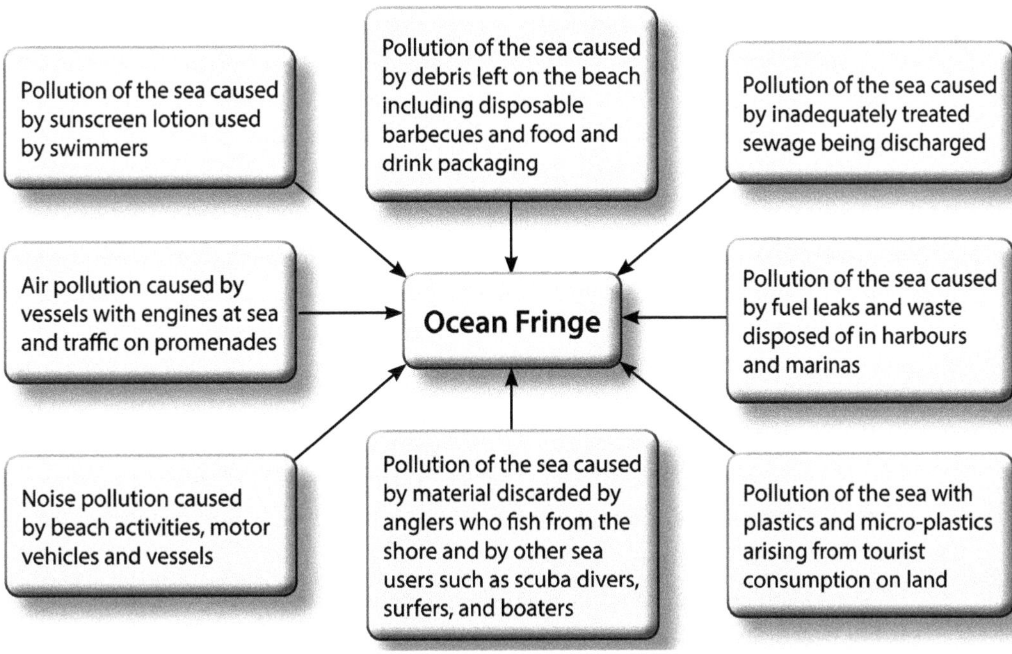

Figure 6.3: Impacts on the marine environment that arise from tourism activity in the ocean fringe

Leaving this aside, one of the management challenges faced by coastal tourist destinations is how to cope with fluctuations in population size. In the quiet season there may be 10,000 people in the destination, but this may swell to over 100,000 at any one time during the peak season. The problem is, therefore, to decide whether to plan infrastructure for 10,000 people, 100,000 or some number in between.

Managing the daily environmental impacts of tourism is usually a public sector responsibility but in many destinations in so-called 'developing countries' it is clear that the public sector agencies lack the resources to discharge this responsibility effectively. Figure 6.3 identifies the range of impacts on the marine environment that arise from tourism activity in the ocean fringe primarily concerning the built-up area of the coastal tourist destinations, beaches and the intertidal zone. It focuses specifically on those effects which are directly attributable to tourist activity and the tourism industry.

Let us now look at the impacts highlighted in Figure 6.3 in rather more detail, beginning with the effects of sunscreen products on our oceans.

Because of the risk of skin cancer, we are constantly being advised to apply sunscreen lotions to our bodies regularly and in generous quantities. Yet there is clear evidence that such products have the potential to cause significant harm to our marine environment. *The Guardian* newspaper in the UK reported on a study of this subject in 2015 and stated that a '*UV filtering chemical is killing off baby coral around tourist resorts, particularly in the Caribbean and Hawaii.... Oxybenzone, a UV -filtering chemical compound found in 3,500 brands of sunscreen worldwide, can be fatal to baby corals and damaging to adults in large concentration ...Oxybenzone alters coral DNA, makes coral more susceptible to potentially fatal bleaching and acts as an endocrine disruptor, causing baby coral to encase itself in its own skeleton and die, according to the findings ...Between 6,000 and 14,000 tonnes of sunscreen lotion winds up in coral reef areas each year.*' (The Guardian, 21 October 2015).

The leader of that research project, Dr Craig Downs, published an article on the same subject on the 'Marine Safe' website in 2016. He said, '*I first came to understand the significance of sunscreen lotion in marine pollution during an investigation into declining coral reefs in the US Virgin Islands. A local resident complained ...about an oily iridescent sheen on the surface of the water that lingered after the mass of tourists had gone home; it was supposedly caused sunscreen washing off the swimmers. Swimmer pollution threatens coral reefs across the world, from the Gulf of Aqaba in the Red Sea to Playa Hermosa in Costa Rica.... An ''organic'' certification doesn't mean a sunscreen is safe for the environment. Several plant-based oils can be toxic to reef organisms.* (www.marinesafe.org, 2016).

The marine environment is also being harmed by the debris we leave behind on the beach, even if we never go into the sea. Anything left on a beach that is covered by the sea at high tide is likely to be swept into the ocean, so it contributes to the pollution of the sea. Further to this, such debris can also prove harmful to those marine creatures that make their home on the beach. Beach debris comes in many shapes and forms including drinks bottles, sandwich wrapping, empty sunscreen bottles and insect repellent, newspapers and magazines and discarded lighters and cigarette butts. Unfortunately, much of the waste people leave behind on the beach is made of plastic or includes micro-plastics. A recent addition to the harmful things left on beaches is the disposable barbecue. All this debris will, at some point, find its way into the sea and into the food chain of marine creatures. And, as Dyson noted in 2010, debris can harm tourism as well as the environment for *'more remote sites become more popular and subsequently more degraded due to pollution, which degrades the quality of the experience'.* (Dyson, 2010)

There are also issues around the impact of those who use the sea for their leisure activities. Sea anglers who fish from the shore can discard hooks or lengths of line, which can become swallowed by marine creatures or entangled around seabirds. Scuba divers, snorkellers and surfers sometimes also discard items of kit which can either cause pollution or harm wildlife. Just a short walk around the many marinas and harbours in coastal tourist destinations quickly reveals the pollution that can come from badly maintained or irresponsibly operated boats including oil slicks from fuel spills and rubbish dumped overboard.

Perhaps, the most serious pollution caused by tourism in tourist destinations is simply the rubbish that comes from shoreside businesses that meet the needs of tourists, particularly plastics and micro-plastics. It has been estimated that as many as eight million individual bits of plastic enter our oceans every day! The micro-plastics can be the result of plastic waste dumped in the sea breaking up, but some are also contained in personal hygiene products. So, they can be flushed into the sea in a tourist destination if a hotel guest just uses a particular shower gel or shampoo. Micro-plastics can be catastrophic for wildlife which swallows them, but they also enter the food chain and can end up being eaten by humans too!

In any destination, tourists create a lot of sewage and in some, this is untreated or inadequately treated before being discharged into the sea. Even where adequate treatment facilities exist, sewage can enter the ocean as a result of storms or heavy rainfall. An article on the US-based Sciencing.com website in 2018 talked about the hazards under-treated sewage could pose for the marine environment. They mentioned biohazards that could deplete parts of the sea of oxygen with devastating results in terms of fish stocks. Drug traces in sewage can lead to accelerated insect reproduction. Nutrient hazards were discussed, including eutrophication in the ecosystem. They suggested that around 50% of wastewater flowing into the Mediterranean Sea contained untreated sewage. (www.sciencing.com, 2018).

Noise pollution is also an issue as the noise from people and boat engines, together with vibration from the latter, can disturb wildlife, which is a particular concern at times when they are breeding.

Finally, air pollution caused by the cars of tourists and the emissions from boat engines also ultimately impact on the marine environment. For example, nitric oxide from vehicle exhausts can stimulate the growth of algae which can lead to depleted oxygen levels that threaten marine wildlife.

However, despite this rather depressing list of negative impacts, there are some positive signs. In west Cornwall in the UK where I live, for example, there are several voluntary initiatives by local residents who are trying to tackle some of the negative impacts of tourism on the ocean fringe, including:

- Activist groups like 'Surfers against Sewage' who campaign on a range of issues, not just sewage, and try to encourage more responsible behaviour by tourists and locals alike.
- Regular beach cleaning by local resident volunteers such as those organised by the excellent Friends of Portheras Cove, the nearest such scheme to my home.
- A voluntary campaign in the main local coastal town, Penzance, which aims to create a 'Plastic Free Penzance'.

Unfortunately, in many places the public sector seems to be lagging behind when it comes to tackling the negative impacts of tourism on the ocean fringe.

Man-made modifications to the marine environment to create attractions for tourists

Tourists choose to visit a place because they like what it offers but given the great competition that exists in the destination sector every destination is constantly searching for ways of gaining competitive advantage. One of the methods used is to modify the natural environment of the ocean to make it even more attractive for tourists. This can take quite dramatic forms including the creation of:

- Artificial islands, often to provide a location for new hotels.
- Artificial reefs to make the destination more attractive to scuba divers and snorkellers.
- Artificial beaches where there is either no natural beach or the beach is not attractive because of its colour, texture or the materials of which it is composed.

Each of these interventions in the natural marine environment obviously has an impact on that environment, some of these impacts can be widespread and/or long-term in their impacts.

Artificial islands

It is important to recognise that not all artificial islands have been constructed to serve the tourism industry. Some have been developed to provide homes where over-crowding or a lack of usable land is a problem. Others have been developed for strategic and commercial reasons such as those developed by China in the South China Sea. However, some have certainly been developed to attract tourists, such as the island created in the 1980s to provide a site for Tokyo Disneyland.

Given the scale of the project, not surprisingly, most of the attention on artificial islands, in the context of tourism, has focused on the three Palm Islands in Dubai, in the UAE. A Master's thesis by Bayyinah Salahuddin in 2006, which has been much quoted, was one of the earliest attempts to evaluate the impact of this artificial island development on the marine environment. While acknowledging that the project created new marine habitats, he suggested that its construction had also *'buried and asphyxiated wildlife, increased turbidity, and changed the alongshore sediment transport'*. He acknowledged that the developer had put some actions in place to mitigate the harmful effects but stated that the process has been likened to *'cutting down a rainforest to build an apartment block, planting two trees and claiming that they not only replenished the rainforest, but made it better'*. (Salahuddin, 2006).

Since then, other sources have claimed that the construction of the Palm Islands caused several negative impacts for the marine environment, including:

- Destruction of seagrasses.
- Increased shoreline erosion which has also included beach erosion.
- Obstructing and changing water currents leading to increases in pollution in some parts of the Persian Gulf where the pollution is no longer 'flushed out'.
- The destruction of turtle nesting sites

While it seems clear that the construction of these artificial islands caused negative impacts on the marine environment it may be some time before we know if these are long-term or irreversible. However, there is also the issue of when small existing islands are converted into tourist resorts. In the Maldives by 2008, more than 80 uninhabited islands had been converted into tourist resorts. One author claimed that *'the consumptive leisure lifestyle of the tourists has been harmful* [in terms of] *sewage, garbage and waste pollution'*. (Domroes, 2008)

6

Artificial reefs

Tourist destinations around the world have constructed artificial reefs in recent decades to attract tourists. They are very controversial, with some seeing them as unwarranted human interventions in the marine environment, while others believe they can enhance the marine ecosystem by helping conserve marine species. Those who are in favour of them argue that they can help regenerate corals by providing a surface that baby corals can attach themselves to and as they mature, they become a habit for a range of marine species. If they can be used to divert divers and fishermen away from more fragile natural ecosystems, then they can help conserve these other places or so the argument goes. And they undoubtedly have a positive impact on local economies.

There are also strong arguments ranged against them. One of these is based on the objects or materials that are introduced into the sea to stimulate the development of a reef. Sometimes boats are deliberately sunk to create a reef; in Florida a number of decommissioned naval vessels have met this fate. Even old aircraft have been used to create artificial reefs. Of course, the risk is that these may cause pollution through everything from leaking fuel to toxic material or bits of debris that break off from them. Some reefs have been constructed from concrete, which again might raise concerns about what could leach out into the ocean over time. Perhaps the biggest concern though about artificial reefs is the issue of the integrity of the marine environment and its authenticity. We view debris being thrown into the ocean and pollution of the sea from plastics as increasingly unacceptable, yet it is appears to be fine to drop concrete structures or even an aircraft into the ocean if it will attract divers and may provide a new habitat for marine wildlife.

However, there are examples of responsibly constructed artificial reefs, some of them guided by local regulations. In Australia, for example, such regulations exist, and a reef can only be constructed if a permit has been obtained from a government agency. Consideration of permit requests involves a range of factors including the site and layout, the materials and the resources of the applicant.

Artificial beaches

Many destinations have invested in creating artificial beaches, either wholly or in part because:

- They have no beach, or their beach is unattractive to tourists because of its colour or texture.
- They are losing their existing beach over time due to coastal erosion, sand mining or storm damage. The latter is becoming more common due to global warming, and many claims that the coastal defences built to protect communities from these storms are causing erosion and making the problem worse.

Even relatively recently developed destinations, such as Hoi An in Vietnam, seem to already be suffering from beach erosion. (Thinh et al., 2019). This issue was discussed in a book entitled *The Last Beach* by Orrin Pilkey and Andrew Cooper, published in 2014. According to *The Guardian* newspaper, Pilkey said in an interview that, '*Most natural sand beaches are disappearing, due partly to rising sea levels and increased storm action, but also due to massive erosion caused by the human development of the shore… Ugly sea walls have removed beaches altogether … Many of the world's most famous beaches are now ecologically dead and dependent for their survival on being replenished with sand and gravel. The death knell has sounded for large stretches of beaches along densely developed shorelines like those in Florida, Spain's Costa del Sol, Australia's Gold Coast and Brazil's Rio de Janeiro.*' (The Guardian, 15 December 2014).

This apparent increase in the erosion of beaches for whatever reason has created demand for beach restoration services or what is called 'beach nourishment'. A number of companies offer such a service; two of them had the following case studies on their websites in autumn 2019:

- 'Artiproje', a Turkish company, mentioned projects in Gocek and at the D-Hotel Maris, both are in Turkey (www.artiproje.com.tr, 2019)

- 'Coastal Science', a US company detailed projects at Nag's Head in North Carolina in the USA and at the Jalousie in St Lucia.

An article in the *Maldives Independent* in 2015 quoted an Environmental Impact Assessment (EIA) in relation to the development of an artificial beach in Male's western waterfront. The EIA suggested that if the project went ahead it could cause long-term water pollution, reef slope failure and damage the coastline. (Maldives Independent, 12 September 2015). Yet, a number of leading destinations around the world have previously invested in artificial beaches including Sentosa Island in Singapore which has three, using reclaimed sand from Malaysia and Indonesia.

In terms of the impact on the marine environment there may be a fundamental difference between seeking to create a brand-new beach where none has existed before and trying to artificially restore a beach which has been damaged by coastal erosion or storm damage. An article from the University of California, Santa Barbara 'Explore Beaches' projects highlighted a number of potential problems with beach nourishment including the following:

- *The sudden input of massive amounts of sand can kill the animals living on the beach*
- *During nourishment, the beach becomes a major construction zone. The heavy machinery used to truck in and distribute new sand also kills beach animals and disturbs wildlife*
- *The new sand may not be the same grain size or chemical makeup of the natural sand, changing the habitat that beach animals rely upon*
- *The resulting catastrophic loss of intertidal prey resources for wildlife such as shore-birds means these birds have to travel to another beach to find food*
- *Long-lived species that do not reproduce often, such as Pismo clams, may take decades to recover*
- *As the ocean starts eroding the introduced sand, the water offshore can become muddy, potentially smothering marine life and changing coastal water quality*

(www.explorebeaches.msi.ucsb.edu, 2019)

It is also important to recognise that tourism has an adverse effect on the meiofauna and nematofauna found on beaches whether they are artificial or natural (Gheskiere et al., 2005)

The planning and management of the ocean fringe

Having explored the four types of interaction between tourism and the ocean fringe, identified in Figure 6.1, it now seems appropriate to introduce some relevant concepts around the planning and management of the ocean fringe. These will be discussed in greater detail in Chapter 10.

First, in most countries, there are land use and new development **planning controls** in place that, at least, should extend to the shoreline of the ocean fringe.

There is also the opportunity to control marine leisure activities in the ocean fringe through regulations. The public sector should therefore be able to control some of the impacts in these areas, if not on the open sea.

Second, we need to develop some measures of the **carrying capacity** of the ocean fringe in terms of tourism activities, bearing in mind that no two locations are the same. And as we know from land-based tourism, carrying capacity is notoriously difficult to measure and implement in practical terms.

Given that many of the negative impacts of tourism in the ocean fringe are caused by conflicts between different leisure uses, and between leisure uses and the marine environment, there is a role for the **zoning** of ocean fringe space to reduce conflicts. The **designation of protected areas** could help to conserve some of the most vulnerable marine environments in the ocean fringe.

Finally, perhaps greater use needs to be made of **Environmental Impact Assessments (EIAs)** to inform public-sector decision-making when new developments are proposed. However, our lack of detailed knowledge of tourism impacts on marine environments is an obstacle and local politicians may well be tempted to ignore the findings of an EIA when they are faced with a proposed development that will create jobs and generate tax revenue for local government.

Putting things in perspective

This chapter has concentrated upon the impact of tourism on the ocean fringe marine environment, and the picture that emerges is a worrying one. However, it is important to set the impacts of tourism in context to get things in proportion.

First, coastal communities that serve as tourist destinations have resident populations as well as transient tourist populations. The lifestyles of these permanent residents also have an impact on the ocean fringe in terms of potential waste disposal and pollution, for example. Many destinations are attractive places with large numbers of people who choose to move there permanently when they retire. While visiting as a tourist years before may have put this idea in their heads; they are longer tourists once they settle in the destination. Second, other economic activities that take place around the ocean fringe, including commercial fishing and shipping, may also have a negative impact on the marine environment.

Third, coastal tourist destinations can be simply conduits for pollution that has its origins inland. The rivers and streams that flow through seaside destinations and into the sea frequently carry plastic pollution and agricultural chemicals that has a serious impact on the health of the marine environment in the ocean fringe.

Nevertheless, to repeat a point made earlier, this does not mean that we can afford to ignore the impacts of tourism which, after all, is not essential to human existence and can be seen as a 'luxury'.

Conclusions

We have seen that activities in that interface between ocean and land that we have termed the 'ocean fringe' are many and varied. And, while one or two of them may be beneficial for the ocean in the longer term, the over-whelming majority of them are negative and some can even be catastrophic. Perhaps we should be most concerned by the trend towards larger and larger projects which have correspondingly greater impacts, whether they be the construction of artificial islands or mega resort complexes. It may be that our ability to manage the impacts of ever larger projects is not keeping pace with the growing size of the projects. Of course, infrastructure projects should, ideally, be developed in the context of a functioning system of land-use planning and public sector regulation of development. However, corruption and the power of the tourism industry can undermine such systems even in those places where they exist.

Even if systems are in place to manage development properly, they may also be further undermined by our lack of detailed knowledge about the impacts of different types of development on different types of marine environment over different timescales. We also seem to lack data concerning how far the effects of impacts in the ocean fringe extend out into the open sea, if at all, or over the length of time such effects last in marine environments. This makes it difficult to undertake credible and convincing Environmental Impact Assessments. This is particularly worrying as climate change and global warming is going to put even greater pressure on coastal destinations in future in terms of sea level rise and flooding and coastal erosion. In 2012, Scott et al. suggested that around 29% of resort properties in the Caribbean would be fully or partially inundated by a one metre sea level rise while up to 60% of resort properties could be at risk of damage from beach erosion. (Scott et al., 2012)

Given this growing pressure on coastal zones it is worrying to note the lack of capacity for effective governance of coastal management seen in many parts of the world. We will return to this topic in Chapter 10.

Finally, it is important to place the impacts of tourism in context in relation to other human activities. A study in Mexico published in 2005, for example, suggested that '*the oil and related petrochemical industries as well as big tourist resorts were identified as the largest contributors to the degradation of the natural ecosystems in the Mexican coastal zone.*' (Ortiz-Lozano et al., 2005) And as far back as 2001 researchers were saying that '*the development of coastal cities in the Asia and Pacific region has strongly impacted marine and coastal environments, degrading their health, and modifying and destroying marine and coastal ecosystems.*' (Jiang et al., 2001)

References

www.artiproje.com.tr (2019) www.artiproje-com.tr/en/portfolio/detail/of-hotel [27 Nov 2019]

Bellan, GL, and Bellan-Santini, DR (2001) A review of littoral tourism, sport and leisure activities: consequences on marine flora and fauna. *Aquatic Conservation: Marine and Freshwater Ecosystems*. **11**(4), 325-334.

www.biologydictionary.net (2019) www.biologydictionary.net>Botany [20 Sept 2019]

Domroes, M (2008) Conceptualising state-controlled resort islands for an environment-friendly development of tourism: the Maldivian experience. *Singapore Journal of Tropical Geography* **22**(2)122-137.

Dyson, K (2020) Links between beach pollution and tourism. *KMI International Journal of Maritime Affairs and Fisheries* **2**(2) 57-78.

www.ecomena.org (2019) www.economena.org/environmental-impacts-of-seawater-desalination [22 Nov 2019]

www.explorebeaches.msi.ucsb.edu (2019) www. explorebeaches.msa.ucsb.edu/beach-health/beach-nourishment. [22 Nov 2019]

Gheskiere, T, Vincx, M, Weslawski, JM, Scapini, F and Degraer, S (2005) Meiofauna as descriptor of tourism-induced changes at sandy beaches. *Marine Environmental Research* **60**(2) 245-265.

Gormsen, E (1997) The impact of tourism on coastal areas. *Geojournal* **42**, 39-54.

Jiang, Y, Kirkman, H, and Hua, A (2001) Megacity development: managing impacts on marine environments. *Ocean and Coastal Management* **44**(5-6) 293-318.

Klein, M and Zviely, D (2001) The environmental impact of marine development on adjacent beaches: a case study of Herzliya Marina, Israel. *Applied Geography* **21**, 145-156.

www.marinesafe.org (2016) www.marinesafe.org/blog/2016/03/18/sunscreen-pollution/ [22 Nov 2019]

Ortiz-Lozano, L, Granados-Barba, A, Solís-Weiss, V and García- Salgado, MA (2005) Environmental evaluation and development problems of the Mexican coastal zone. *Ocean and Coastal Management* **48**(2) 161-176.

Phillips, MR, and Jones, AL (2006) Erosion and tourism infrastructure in the coastal zone: problems, consequences and management. *Tourism Management* **27**(3) 517-524.

www.phnompemhpost.com (2019) www.phnompenhpost.com/business/new-bokor-city-planned-high-end-tourist-destination. [28 Sept 2019]

Roberts, DA, Johnston, EL and Knott, NA (2010) Impacts of desalination plant discharges on the environment: a critical review of published studies. *Water Resources* **44**(18) 5117-5128.

Salahuddin, B (2006) *The Marine Environmental Impacts of Artificial Island Construction, Dubai, UAE*. Masters project s for the Master of Environmental Management degree in the Nicholas School of Environment and Earth Sciencesof Duke University

www.sciencing.com (2018) www.sciencing.com/effectssewage-aquatic-ecosystems-21773.html [20 Nov 2019]

Scott, D, Hall, CM, and Gössling, S (2012) *Tourism and Climate Change: Impacts, Adaption and Mitigation*. Routledge. London

Shaalan, IM (2005) Sustainable tourism development in the Red Sea of Egypt: threats and opportunities. *Journal of Cleaner Production* **13**(2) 83-87.

www.en.tempo.co (2019) www.en.tempo.co/764938/ten-new-tourist-destinations-planned-for-Indonesia [20 Oct 2019]

Thinh, NA, Thanh, NN, Tuyen, LT, and Hens, L (2019) Tourism and beach erosion: valuing the damage of beach erosion for tourism in the Hoi A World Heritage site, Vietnam. *Environment, Development and Sustainability*, **21**, 2113-2124.

Questions and exercises

1. Using a variety of sources, discuss the historical development of coastal tourism, highlighting how its impact on the marine environment has developed over time.
2. Critically evaluate the statement that 'the impacts of tourism on the marine environment are much worse in the ocean fringe than they are in the open sea'.
3. Working in a group, select a country or region with which you are familiar or for which you can find up-to-date information. Look at Figure 6.2 and draw up a list of examples of the six types of infrastructure project identified in the diagram. Look at the impacts of each of these projects and produce a report highlighting what you believe to be the strengths and weaknesses of each project.
4. Discuss the advantages and disadvantages of the construction of artificial islands from the points of view of effective marine environment conservation and the development of successful tourist destinations. Use a range of real-world examples to illustrate the points which you make.
5. Using published research papers, examine the impacts on the marine environment of leisure activities which take place largely on beaches or in the sea close to shore. Try to produce a diagram identifying those activities which have the greatest and the least negative impacts on the marine environment, using examples, wherever possible, to support your arguments.
6. Working in a group, devise a survey for people who take vacations in coastal destinations, designed to ascertain both their knowledge and their perceptions of the impacts on the marine environment in the ocean fringe, including issues covered in this chapter. Conduct your survey with a sample of tourists and analyse the results and outline your main findings in a report, noting anything that surprised you in relation to your findings.

7 The Consumption of Marine Resources by Tourists

Most of this book to up to now has focused upon tourists and tourism harming our oceans by putting undesirable things into the sea, whether that be plastics, fuel spilled from boats or even nasty chemicals from sunscreen lotions. However, in this chapter we will focus on the harm which tourists and the tourism industry do when they remove things from the sea and the ocean fringe. We will see that the consumption of marine resources by tourists and the tourism industry is adding to the problems faced by our oceans. I recognise that most of the damage being done to the marine environment from the consumption of resources in the sea and under the seabed is not due to tourism. The majority of it is due to over-fishing to meet the everyday needs and desires of the populations of towns and cities, and the exploitation of natural resources including, oil, natural gas and minerals.

However, tourist consumption contributes to the problems faced by the oceans and most of this consumption is for pleasure rather than necessity. Furthermore, most of the consumption of marine resources by tourists occurs in specific locations and this concentrates its impact on particular areas of the marine environment, around coastal destinations.

A typology of tourist consumption of marine resources

A simple attempt to identify the various ways in which tourists and tourism consume marine resources is presented in Figure 7.1. This is followed by a closer look at each point from that figure.

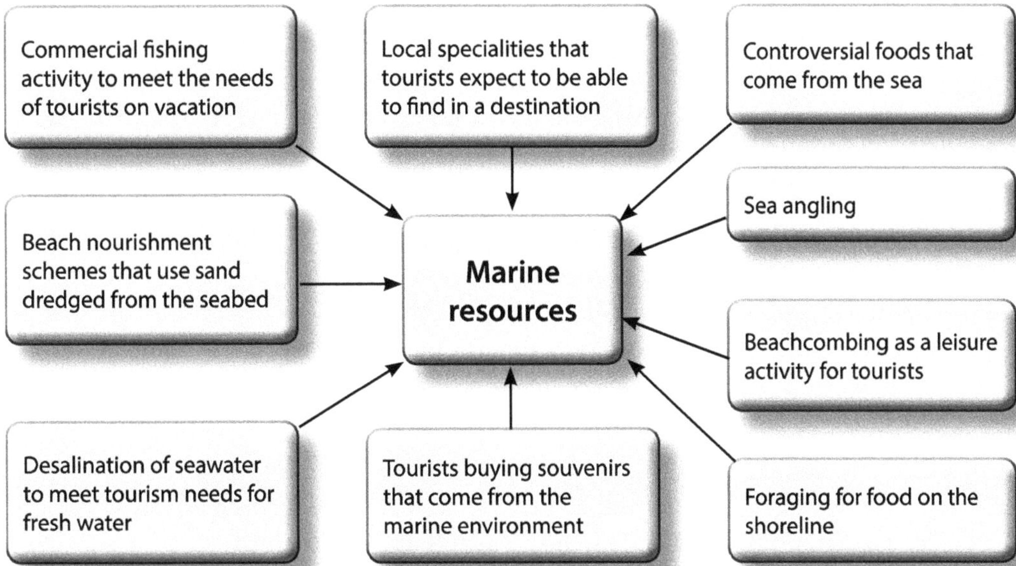

Figure 7.1: A simple typology of the ways in which tourists and tourism consume marine resources

■ Commercial fishing to meet the needs of tourists on vacation

In many coastal destinations, tourists expect to have fish on the menu in local restaurants and their hotels, or be able to buy locally sourced fish and seafood to cook in their self-catering accommodation. This can put great pressure on the local fishing industry to meet this demand, which is highly seasonal. However, these seasons may not coincide with the times of the year which are most productive from the point of view of commercial fishermen. Furthermore, many coastal tourist destinations are found in areas where over-fishing has depleted fish stocks so that the amount of fish available is declining year on year. This is certainly the situation around the Mediterranean Sea, which every summer probably attracts more tourists than any other area of sea in the world.

A 2017 report from the Union for the Mediterranean painted a clear picture of the challenges faced in relation to fish stocks. The report noted that 80% of Mediterranean fish species were being fished at or above their levels of sustainability. This was in addition to a European Commission report, also from 2017, that suggested that 93% of Mediterranean fish species were being over-fished. At the same time the report noted that international tourism alone in the region had grown from 58 million visits in 1970 to more than 350 million in 2016. It also noted that just five countries around the Mediterranean accounted for 80% of these tourists, suggesting that there is further scope for growth, particularly in the countries that fringe the southern half of the Mediterranean Sea. (Union for the Mediterranean, 2017).

Some tourists are attracted to Mediterranean destinations because of the health benefits of the so-called 'Mediterranean diet'. This diet, by promoting the eating of fish rather than red meat, may, inadvertently, be further contributing to the problem of over-fishing in the Mediterranean. This problem of overfishing, partly linked to tourist demand, is not confined to the Mediterranean region. It can also be observed in the islands of the Indian Ocean, the coastal zones in south-east Asia, and parts of Central America.

Much of the commercial fishing that takes place in the world is undertaken by some form of trawling which is indiscriminate in its impacts on the marine environment. As well as catching the fish being targeted, it can damage seabed habitats and catch and kill endangered species, something known as 'by-catches'. This may include species such as dolphins and sea turtles that are vital elements in marine ecosystems, but also key attractions for tourists in coastal destinations.

The pressure from tourist demand as well as the depletion of fish stocks can lead to three outcomes that can be a threat to the environment.

First, fishermen may be tempted to break the rules on quotas and minimum sizes of fish that can be caught, because of the pressures on them to earn a living. This can result in a reduction in future fish populations.

Second, fish and seafood may be imported, sometimes from considerable distances, if there is insufficient fish available in local waters. It may also reflect the desire of hotels and restaurants to offer exotic species not found locally or provide fish that guests are familiar with from home. Thus, the restaurant customer in London may be eating tuna steaks that have come from the Indian Ocean, while the hotel guest in Sri Lanka is eating smoked wild salmon from Alaska. Such meals have a significant carbon footprint and do little to improve the sustainability of global fisheries.

Third, destinations can be tempted to move into fish-farming when natural stocks become depleted. While the potential benefits of fish-farming seem clear when faced with depleted natural stocks, it brings its own problems, which have been well documented. These include the introduction and transmission of diseases to the marine environment, deterioration in water quality, water eutrophication and habitat modification. And they may not solve the problem of supply in tourist destinations as many tourists are likely to reject farmed fish, preferring something fresh and local.

Yet again, though, we have to note that demand in tourist destinations is not the main cause of global over-fishing although, again, we need to recognise that tourist consumption of seafood is for the pleasure of consumption rather than being everyday sustenance for people with few alternative sources of protein.

Ironically, in some places, such as Taiwan, commercial fishermen are diversifying into tourism to help mitigate the effects of declining fish stocks. (Chen, 2010)

Local specialities that tourists expect to be able to consume in a particular destination

Many destinations around the world use food and drink as an attractor for visitors and as a way of differentiating themselves from competitors. For coastal destinations this often means promoting local seafood specialities that are either unique or better quality than those elsewhere. Given that most people live some way away from the sea and will usually only get to eat processed or frozen fish, this can be a very powerful motivator for visiting particular destinations.

In Greece, for example, on the island of Rhodes, the website of the Rhodes Hotel Association promotes the idea of eating fish and seafood in Rhodes. It states that '*The rich Greek waters offer a very good variety of tasty fish that fed the Greek people throughout the centuries. Most of the villages in the Greek Islands and the coast of Greece are fishing villages.*' (www.rhodes.hotel.com, 2019) It then goes on recommend a number of local types of seafood such as tsipoura, a type of fish, and the Greek lobster, but it does offer a few words of warning that in the peak season demand may not be sufficient, and some fish may be imported to the island and some of this may be frozen. They also acknowledge that some of the species on offer are likely to have been farmed. This is not surprising as in recent years the news in Greece has included stories about depleted stocks, and Greenpeace has also been drawing attention to this issue in Greece and in the wider Mediterranean.

The desire to ensure that tourist demand for fish and seafood is met means, as the Rhodes example shows, that often fish has to be imported, sometimes from some distance away, so that each fish has its own carbon footprint! Or it has to be farmed which, as we have seen in the case of salmon farming, can lead to its own problems, including pollution and disease. In recent years, there has been a particularly worrying issue arising in Iceland, where an increasing number of tourists appear to want to eat whale as part of their vacation experience. It seems that these visitors believe this is what you should do in Iceland, despite the widespread international condemnation of whaling.

An article on the 'Culture Trip' website by Nikki Vargas made some interesting points about this phenomenon. She says '*Iceland's whaling industry has defied international outrage for decades, but what people don't know is that it's not Icelanders eating the whales they kill ....I am ...drawn to the misconception among Iceland tourists that whale meat...is a local speciality worth trying.....Within a few minutes (of walking down the main street) I spotted minke whale on a restaurant menu next to a selection of 'puffin tapas'.....With whaling and whale-watching taking place in similar areas, many tourists don't realise the same whale they ogled in the morning might very well end up on their dinner plate the following evening. ...whale meat is hardly a local Icelandic dish.*

According to a poll .... only 3.2% of Iceland's population eats whale meat ... six times a year or more. (www.theculturetrip.com, 2019).

So, tourists are apparently trying whale because they believe it is part of Icelandic food culture, but it is not. On a more positive note, around 60 restaurants in Iceland have pledged to be a 'whale-friendly' eatery, while by August 2016 more than 100,000 tourists and locals had signed a petition against eating whale meat (www.theculturetrip.com, 2019)

The future for those destinations that have traditionally used local seafood specialities as one of their main attractions seems rather bleak. I looked at several websites using the search term 'best places to eat seafood in the world' and the results included a number of well-known examples of places associated with particular types of seafood. These included:

- Maine and its lobsters
- Chesapeake Bay and its crabs
- Sicily and its sardines
- Chile and its locos or abalone and corvina, also known as Chilean sea bass

For each of these, I then looked at the status of the population of each of these species in these locations. In every case, the picture was one of declining catches and stocks, without exception. There can be little doubt that in these and other cases much of the reduction in stocks is the result of tourism in terms of either:

- Consumption in the destination by tourists encouraged to try 'the local speciality, or
- Consumption at home of specialities which were first experienced on a vacation to the place that is famous for that speciality.

Thus, the idea of promoting specialities, regardless of the sustainability of stock, can contribute significantly to stock depletion and harm to the marine ecosystem.

Controversial foods from the sea

This issue is, of course, closely related to the matter we have just discussed and there is certainly a grey area between the two topics. Perhaps the most controversial marine wildlife dish in the world is shark fin soup, a dish that is particularly popular within the Chinese gastronomic culture where it embodies the idea of luxury and may date back nearly a thousand years.

This dish is controversial for two main reasons as follows:

- It is based on a marine creature which is at the top of the marine food chain but has been decimated in recent decades by the actions of mankind. There are various estimates for how many sharks are killed each year, having been 'finned', but the consensus appears to be between 70 and 75 million per annum. This includes endangered species such as Hammerhead sharks

- The sheer cruelty and waste that is involved given the fin is stripped off the shark and it is then thrown back into the sea to die a slow and agonising death.

Ironically, the fin itself is tasteless and its role is primarily simply to provide texture and thickening to the soup.

Interestingly there are various reports that consumption of the dish has fallen in China itself by up to 80% in recent years. At the same time, though, consumption has apparently risen in Vietnam, Thailand and Indonesia. Given all three countries are popular destinations for the growing number of outbound Chinese tourists, I wonder if this growth in demand may have been driven by increasing numbers of visitors from China. As vacations are special occasions for everyone, it might not be surprising if tourists did not see it as the appropriate time to consume a dish that is often associated with luxury and hedonistic living.

While some countries have banned the import and/or export of shark's fins, the dish can still be found on the menus of restaurants in major cities and tourist destinations around the world. A quick Internet search in October 2019 produced lots of opportunities to try the dish in New York, for instance, and there were even lists of the 'ten best restaurants to try shark fin soup'. Interestingly, though, several hotel brands including Peninsula and Shangri-La have taken the dish off the menus in their hotel restaurants.

The turtle has been a food source for centuries and its meat and eggs are still popular with people in parts of central America and Asia, where they are believed to have a positive impact on human health. However, historically it has been best well known as the main ingredient in turtle soup, a dish that became fashionable amongst the affluent in Europe and its colonies from the eighteenth century. Their over-consumption of turtle meat started the process of depleting populations which continues today, although it is decades since it was a popular dish in Europe. In the USA turtle soup was popular in states like Louisiana, where it was made from the snapping turtle, but its popularity has waned in recent years. Unfortunately, where turtle meat is still consumed it is frequently endangered species which are targeted, particularly green turtles.

At the same time, there is cause for concern at the impact which the global boom in sushi has had on bluefin tuna stocks and prices. The response, primarily in Japan, to the increasing rarity of this fish has not been to seek to conserve it but to simply watch as prices rise, reflecting its rarity value, to the point where a single fish can be worth nearly $1 million!

Two points need to be made at this point, neither of which is being made for the first time in this book. First, the consumption of marine creatures we have discussed in this section occurs, probably, far more amongst people living their everyday lives at home than it does amongst people on vacation. Second, the enforcement of regulation remains problematic, not least because of the large amounts of money at stake, when a single fish can be worth hundreds of thousands of US dollars!

Sea angling

We looked at sea angling in Chapter 5, but here the focus is going to be specifically on its impact on fish stocks when the fish caught, from shore or boat, are taken from the sea by recreational anglers, either as food or as a 'trophy specimen'. For this discussion we will split recreational sea angling into shore and boat-based for 'normal' species, and sport and big-game fishing where the targets are specimen fish from the largest marine fish species.

In terms of the former, estimating its impact on fish stocks is hard to do but there was an interesting survey undertaken in England in 2012 which gives us some clues. Some of the findings were as follows:

- Shore fishermen caught an estimated 143 tons of bass and 138 tons of cod while the equivalent figures for privately owned and rented boats were between double and triple the figures for the shore anglers, although these figures were difficult to calculate with any accuracy. The equivalent figures for chartered boats, which usually have a skipper who has specialist knowledge, were 44 tons for bass and 175 tons for cod.
- The proportion of fish kept rather than being returned to the sea was highest for the private and rented boats and lowest for the shore anglers. However, this varied by species with shore anglers keeping almost all of the mackerel they caught but few of the bass and cod, whereas the opposite was the case for the anglers on the private and rented boats. This may reflect the fact that boats may be catching larger cod and bass and there are minimum size limits for taking fish.
- In the case of bass, the catches by sea anglers who kept their catch, rather than releasing it back into the wild, tended to be between 30 and 40% of those caught by commercial fishing boats in England, France and the Netherlands. In other words, sea anglers in the three countries were estimated to have caught and kept more than 1,000 tons of bass in 2012, a not inconsiderable amount.
- Catches of cod that were kept rather than being returned to the sea in other countries were just 30 tons for Germany, 226 tons for Sweden, 360 tons for the Netherlands, and 537 tons for Denmark. Again, these are significant quantities of fish being removed from the sea by recreational anglers.

(Armstrong et al., 2013)

The more controversial type of sea angling is sport or big game fishing which targets the largest specimens of the biggest species such as sharks, tuna, marlin, and swordfish. Some of these species are endangered or at least vulnerable and the motive is not primarily to eat the catch, so it is sometimes viewed in a similar way to big game hunting on land, as something that is not appropriate in a modern age. This form of sea angling has become a major income generator for a number of countries, particularly in the Indian Ocean including the Maldives and Mauritius, as well as in Central America and the Caribbean.

At one time, the catch would have been killed and photographed and hung up and displayed as a trophy before, perhaps, being sold for food if it was still in a fit state to eat. However, in recent years we have seen the development of two types of sport fishing, which has been mirrored in sea angling as a whole, namely:

- 'Catch and release', which is simply unhooking the fish and releasing it back into the ocean
- 'Tag and release', where anglers help researchers tag fish so their habits can be studied which can aid conservation initiatives.

While a positive development, this has not wholly silenced critics who talk about the trauma the fish suffers when caught and the potential injuries it can suffer from being hooked and landed on a boat.

Beachcombing

For many tourists, one of the joys of a seaside vacation is the opportunity to go 'beachcombing' after the tide has receded to see what exotic objects may have been left behind. Elsewhere in the chapter I look at collecting shells which is, perhaps, the most popular activity for most beachcombers, and note that the main issue is not to take shells that contain live creatures or could be used as a home by marine creatures, such as spiral shells which often provide shelter for hermit crabs. People also regularly take driftwood and even seaweed off the beach and there has been an increase in people doing this, worldwide, because of their potential use in crafts-making. However, it is important to recognise that the removal of such apparently rather insignificant items from the beach can have an impact on the marine environment because they can be the habitat for a number of small marine creatures.

In recent years, in some countries, we have also seen an increase in people taking pebbles off the beach. This can be really serious on heavily used beaches as pebbles are a natural part of the defences against coastal erosion and also represent the habitat of some very small marine life forms. There has also been an increase, in some countries, of people gathering stones and even boulders from the beach and arranging them in artistic piles, or 'cairns', above the high tide mark. Again, this can disrupt marine wildlife and undermine coastal defences.

In a number of countries there are regulations about beachcombing; in the UK, for example, the 1949 Coastal Protection Act made it an offence to remove pebbles and other material from beaches. In recent years there have been several cases where tourists have been threatened with a large fine if they did not replace pebbles they had removed from various beaches. However, the enforcement of such regulation around the world is generally lax so the activity continues pretty well undisturbed in most places.

Foraging for food on the shoreline

In recent years, it has become trendy to forage in the natural environment for cooking ingredients. Indeed, some top restaurants around the world have gained a worldwide reputation for using foraged ingredients, including Noma 2.0 in Denmark, O.M in Sao Paolo, Masque in Mumbai and Kungkas Can Cook in Australia. This has helped to make foraging fashionable and it has now extended to the foreshore. People have always collected edible shellfish from beaches in most countries, but foraging has now developed, in some places, to include seaweeds and marine plants such as samphire and sea buckthorn as these are seen as healthy and natural sources of vitamins. In Cornwall, in the UK, where I live, there are now 'foraging trips' organised, on a commercial basis, for tourists, that have proved very popular. This is, in addition, to the growing numbers of tourists who are foraging on their own.

Coastal foraging takes place in a wide range of coastal ecosystems including sand dunes, estuaries and mudflats, cliff faces and rock pools and beaches. Clearly, if people take things which are rare or endangered this can have a negative impact on the marine environment. However, foragers also need to be aware that some of the plants or creatures they find may not be edible or may even be poisonous.

Souvenirs from the sea

The desire to take souvenirs home from a vacation to keep alive memories of an enjoyable time is as old as tourism itself. Eighteenth century travellers taking the 'Grand Tour brought back works of art, early mass market beach tourists brought back unusual alcoholic drinks and stuffed toy donkeys and today's tourists bring back 'selfies'! However, from the early days of tourism, tourists have also plundered the natural resources of the ocean to remind them of their vacation by the sea. Museums are full of turtle shells and examples of fish which early adventurers and colonialists took back to their own countries. And local people in tourist destinations have, for decades, crafted souvenirs from sea creatures of all shapes and sizes.

One of the commonest forms of souvenir collecting when visiting the seaside is the gathering of **seashells.** In a number of coastal destinations in the UK, there are even buildings made of and/or lined with seashells that have become favourite attractions for visitors over decade. Collecting shells has always been seen as harmless fun but now that idea is under scrutiny and in some places, bans have been imposed on the collection of shells. It has been suggested that up to 5,000 species of molluscs are sold as curios and souvenirs around the world in a trade which is often undocumented. (Dias et al., 2011)

An interesting article by Rachel Nuwer, in 2014, reported research which claimed that the number of seashells on beaches in Spain had declined by 60% while tourism had grown by 300% over a 30-year period. (www.smithsonianmag.com, 2014). However, it seems that more empirical research is needed before a cause and effect relationship can be proven between a decline in seashells and an increase in tourism. Nevertheless, every shell taken by a tourist interferes, albeit in a minute way, with the functioning of the marine ecosystem. One type of shell that should never be taken from beaches are spiral shells as these are often taken over as homes by other marine creatures, most notably hermit crabs.

There is widespread recognition that **coral reefs** are struggling to survive in many parts of the world, yet they are still taken home as souvenirs by tourists, and the Internet is still full of advertisements for gifts made of coral. Corals may be broken off by divers as souvenirs or collected by souvenir sellers based in coastal destinations, who often turn them into ornaments or items of jewellery. This continues despite the fact that the taking or exporting of stony corals is illegal in a number of countries. One loophole with some laws is that they relate purely to exporting the item so may have no impact on domestic tourists. The other obvious problem with laws is whether or not they are effectively enforced.

There are a number of other species, some of them endangered, that are exploited as tourist souvenirs, including:

- **Sea turtles**, an endangered animal in many parts of the world, are regularly used for tourist souvenirs. Their shells are often used to make everything from jewellery, clips for the hair, the frames of sunglasses, combs and even musical instruments.
- Handicrafts can sometimes be made from marine animals, particularly their ivory, with the **walrus** being a particularly targeted species.
- **Seahorses**: it is believed that, despite being endangered in some places, between 100 and 200 million seahorses are captured annually and after being dried are used as souvenirs or in the making of traditional medicines.
- The teeth of predators such as sharks and particularly the infamous Great White Shark.

- Occasionally the pelts of marine creatures, notably **seals**, can find themselves on the wish lists of tourists.

One marine life form that it seems may be acceptable as a holiday souvenir is, perhaps, surprisingly, the natural **sea sponge**. In general, it is believed that the sponge is not endangered because of harvesting for souvenirs or the much larger trade in sponges for use in the health and beauty sector as an aid to skin care. Those who make a living from this trade are, understandably, keen to stress that its exploitation is sustainable as it can regenerate after 'cutting', depending upon the way in which the harvesting takes place and where the cut is made. There is even a suggestion that the sea sponge populations may actually grow after a period of harvesting. Not all harvesting seems to be conducted responsibly, though, and where this is the case the sponge colony can be harmed by the harvesting process. At the same time, the industry claims that sea sponges are more sustainable than synthetic ones. This is because the synthetic sponges are made with petroleum products and are often not biodegradable. On the other hand, perhaps questions remain over the removal of an element in the marine ecosystem so that people can wash their face with something that feels nice.

Pearls are a popular tourist souvenir as well as an important part of the fashion retail business. While the collection of natural pearls has been going on for millennia, pearl-farming is relatively modern and is now a major source of income for people in French Polynesia and the Cook Islands. The industry is keen to promote its efforts to make pearl farming more sustainable, not least because they say that sustainable production also leads to higher quality pearls. At the time of writing a news article from Indonesia announced the launch of a new sustainable pearl oyster fisheries operation whilst acknowledging that '*the industry's fast growth has taken a toll on wild oyster populations and there's also been a decline in the quality of pearls'.* (www.news.mongabay.com, 2019)

There are many local bans and regulations in place around the world relating to particular species. However, the European Commission considers the situation to be serious enough to place on its website information under the heading of 'The Wildlife Souvenirs Guide'. This seeks to make tourists aware of the laws in different countries and regions on what it is illegal to take out of the country as a souvenir. (www.ec.europa.eu, 2019)

It is important to recognise that the CITES (Convention on International Trade in Endangered Species of Wild Fauna and Flora) rules apply equally to marine creatures as they do to land-based species, such as elephants and rhinoceros, whose plights are better known.

However, it is also necessary to point out that that the damage caused to the marine ecosystem by the removal of creatures from it is only partly due to tourists and souvenirs. More, perhaps, is due to the beauty, health and fashion industries and to consumption by people in their everyday lives in their own homes.

It is now time to look at two issues which were already considered in Chapter 6, but this time we will consider them in the context of the consumption of marine resources.

Desalination

In many arid parts of the world, desalination is vital to the everyday life of local residents. However, it has also been used in certain coastal locations to facilitate the growth of tourism. There are many examples of this from around the world, including the islands of Aruba and Grand Cayman in the Caribbean, the Maldives, Lanzarote in Spain, Sharm-el-Sheikh and Hurghada in Egypt and Dubai.

The main challenge of desalination to the marine environment as a result of the extraction of the seawater is that a large number of tiny marine life forms are likely to be destroyed in the process. While these may not even be visible to the naked eye, they are an essential part of the marine ecosystem, not least due to their role in the food chain. The effects of water extraction on the marine environment can be mitigated through the use of technology but not eradicated altogether. And, as we noted in Chapter 6, desalination can also harm the marine ecosystem through the depositing of heavy concentrations of salt and an increase in the turbidity of the water.

Beach nourishment

It is clear that most coastal destinations rely on the quality of their beach, to a large extent, for their success. Increasingly, we are seeing established beaches being denuded of sand as a result of coastal erosion, floods and storms. Many affected destinations have felt the need to artificially replenish their beaches to safeguard their tourism industries. However, the process of 'beach nourishment' can have three main impacts on the marine environment, as follows:

- Marine plants and creatures will be disturbed and killed when the sand is pumped from below the surface of the ocean.
- The new material deposited on the beach can bury existing wildlife on the beach; there have apparently been cases where the eggs of sea turtles have been buried by beach nourishment operations.
- Nourishing one beach can have negative knock-on impacts on beaches further along the coast.

On the other hand, of course, a rejuvenated beach will provide a new habitat for some marine wildlife.

Coastal seabird egg collecting

Many places have a long tradition of people taking birds' eggs, usually from nests on sea cliffs, for food and in some cases that has led to some bird populations being decimated. There has also been a long history of eggs being taken by collectors or to sell to collectors and this practice continues today, despite it being illegal in most countries. Of course, most collectors want to get their hands on the rarest eggs and their rarity reflects the fact that the bird that laid it is rare. Any such egg that is taken, therefore is to be regretted, as it helps put the sustainability of a particular species in an area at risk.

Cases of theft for profit occur all over the world, with 2018-2019 seeing thefts of rare seabird eggs from Norfolk in the UK to the Lakshadweep Islands in the Indian Ocean.

Taking behaviour as a tourist home with you when the vacation ends

One of the most intriguing questions for tourist behaviour researchers is the matter of the extent to which people's vacations influence their everyday consumer behaviour. To what extent does a vacation in Thailand result in a desire to eat Thai food at home, or a trip to Mexico make a tourist want to learn Spanish. Similar questions can be asked in respect of how people are affected by their vacations to coastal destinations. Do they want to eat the same fish they ate in a resort, even if that means transporting it across the world, or do they order gifts made from shells that they saw on the beach in front of their hotel? Or do they return home with a greater awareness of the environmental issues facing the oceans and start to get involved with marine conservation work? The answers to these questions may start to help us answer a bigger question about the extent to which tourism has a negative impact on the marine environment.

Conclusions

In this chapter, we have looked at the impact on the marine environment that occurs when tourists, the tourism industry and the public sector consumes marine resources which usually involves removes things from the ocean, the shoreline and the beach. We have seen that these impacts tend to be concentrated in the relatively small areas where tourists are gathered together in and around coastal destinations.

As elsewhere in the book we have noted that in many places, regulations are in place that are designed to minimise the negative impacts of tourism on the marine environment. However, it is also clear that the implementation of these regulations is often ineffective, particularly when a lot of money is at stake as in the case of bluefish tuna and shark fins. In cases such as the pearl and sponge industries, the commercial interests are keen to highlight their commitment to sustainability.

Nevertheless, it appears clear that the consumption of marine resources discussed in this chapter has a generally negative impact on the marine environment.

However, yet again, the point needs to be made that tourism is only one cause of damage to the marine environment and that harm is also caused by the actions of those who live by the sea. It is also caused by those who live far from the sea but buy the products of over-fishing by the commercial fishermen and the gifts made from shells or coral and sold by irresponsible entrepreneurs.

References

Armstrong, M, Brown, A, Hargreaves, J, Hyder, K, Pilgrim-Morrison, S, Munday, M, Proctor, S, Roberts, A, and Williamson, K (2013) *Sea Angling 2012 - a Survey of Recreational Sea Angling Activity and Economic Value in England.* Department for Environment, Food and Rural Affairs. London

Chen, C-L (2010) Diversifying fisheries into tourism in Taiwan: experiences and prospects. *Ocean and Coastal Management* **53**(8) 487-492.

Dias, TLP, Leo Neto, NA and Alves, RRN (2011) Molluscs in the marine curio and souvenir trade in NE Brazil: species composition and implications for their conservation and management. *Biodiversity and Conservation* **20**, 2393-240.

www.ec.europa.eu (2019) www.ec.europa.eu/environment/cites/info_souvenirs_en.htm 2019 [25 Oct 2019]

www.news.mongabay.com (2019)

www.theculturetrip.com, 2019 www.theculture-trip-com/central-america-articles/watch-sea-turtles-hatch-central-america/ 2018 [18 Jan 2020]

www.rhodes.hotel.com, 2019 www.rhodes-hotel.com/rhodes_greece_fish_seafood.php [3 Jan 2020]

www.smithsonianmag.com, 2014 www.smithsonianmag.com/science-nature/beach-tourists-who-collect-shells-may-be-harming-environment-180949368/ [20 Oct 2019]

Union for the Mediterranean (2017) *Blue Economy in the Mediterranean*. Union for the Mediterrnean

www.smithsonianmag.com, 2014 www.smithsonianmag.com/science-nature/beach-tourists-who-collect-shells-may-be-harming-environment-180949368/ [20 Oct 2019]

Union for the Mediterranean (2017) *Blue Economy in the Mediterranean*. Union for the Mediterrnean

Questions and exercises

1. Look at Figure 7.1 and think about each of the nine ways in which tourists consume marine resources. Using evidence, suggest which of these does the most harm to the marine environment.
2. Using examples, discuss the statement that 'government regulations seem to provide little real protection against the harmful consumption of marine resources'.
3. In a group, design a survey to be conducted with people who have taken a vacation by the sea during the last twelve months. Ask them if they think anything they bought or ate on vacation may have had a negative impact on the marine environment. Write a report on your findings, noting anything about your results that you found surprising.
4. Select a country or region that promotes its seafood as one of its attractions for tourists. Then, using examples, discuss the extent to which its promotion of its seafood is compatible with the principles of sustainable tourism.
5. Choose a coastal destination where tourists or tourism consumes marine resources in at least two of the ways identified in Figure 7.1. Propose a set of regulations that should help to reduce any damage that might be caused to the marine environment and suggest how they should be enforced.

8 Oceans, Natural Disasters and Tourism

In recent years, the world seems to have seen increasing numbers of natural disasters, affecting coastal tourist destinations as well as places with no connection to tourism. Ritchie noted in 2008 that despite the apparent increases in natural disasters there still seemed to be a lack of research in the tourism field on the management of these disasters covering response, recovery, reduction and readiness (Ritchie, 2008). While things have improved since then there is still some truth in his contention. These natural disasters include extreme weather events such as hurricanes, storms, floods and landslides, and heatwaves as well as disasters in which the weather plays a part, such as wildfires, and disasters that are not linked to the weather at all, such as earthquakes, volcanoes, and disease outbreaks. These natural disasters not only appear to be becoming more frequent but also more severe and sometimes occurring in places which have not seen such things before, and sometimes in major coastal tourist destinations.

Over the past 15 to 20 years we have seen:

- Powerful hurricanes and typhoons in the Caribbean, the USA, and south-east Asia.
- Deadly wildfires in California and Oregon in the USA, southern Australia, New Zealand, Portugal, Greece and the Canary Islands in Spain.
- Outbreaks of potentially lethal diseases, sometimes spread by air travel, including SARS avian flu, Ebola, Dengue fever and Covid-19.
- Record-breaking heatwaves in Australia, parts of the USA and southern Europe.
- The catastrophic tsunamis in the Indian Ocean in 2004 and Japan in 2011.
- Major landslides in Afghanistan, the Philippines, the USA, China, India, Taiwan, Sri Lanka, Colombia and Madeira in Portugal.
- Lethal or highly disruptive volcanoes in Iceland, Italy, Japan, Guatemala and the Democratic Republic of Congo.

This list requires a little commentary to put it in perspective. First, many of these phenomena are regular occurrences in some parts of the world. However, we are seeing some of them become more severe and start to appear in places where they are not normally seen. Weather phenomena are also becoming more difficult to predict as the global climate changes; this is particularly true of the seasonal monsoon rains that affect parts of Asia. It is fair to say that most places affected by the natural disasters outlined above are not tourist destinations, and those affected are generally not tourists or those working in tourism. However, in the interests of accuracy, it is fair to point out that coastal tourist destinations do seem to have become more affected by natural disasters in recent years. In Europe, for example, we have seen record-breaking deadly heatwaves and widespread wildfires followed by floods and landslides. These have affected tourist destinations from the Canary Islands and Portugal to Greece. This has been happening at the same time as we have been becoming increasingly aware of a potential man-made, rather than natural, disaster that threatens the future of the planet, namely climate change and global warming.

While debate still rages, most scientific opinion suggests that some of the upsurge in natural disasters relates directly to global warming and that the oceans are at the heart of what is happening. Global warming is increasing sea temperatures and causing rises in sea levels around the world. As we saw in Chapter 2, what happens in the oceans largely determines the weather on land, so it should not be surprising that as global warming gathers pace, our weather becomes more extreme and flooding more widespread. It is also becoming increasingly clear that in coastal areas in particular, the impacts of some natural disasters such as wildfires, floods, landslides and even tsunamis are magnified due to the actions of man. In coastal tourist destinations these actions include building too close to the shoreline, clearing trees and mangrove swamps for new developments, and altering the composition of the beach.

The complex relationship between the oceans, natural disasters and tourism

In this section we are going to examine the complex relationship between the oceans, natural disasters, and tourism, especially the impact of such disasters on coastal tourist destinations. We will look at several aspects of this relationship including:

- The role which oceans play in the natural disasters that have an impact on coastal tourism destinations.
- Potential typologies of natural disasters such as those which are extreme weather events and those which are not and those which originate from the ocean and those which do not.

- 'Natural disasters' which are made worse by global warming, which is partly caused by tourism, against those natural disasters which are not the result of any human activities, including tourism.
- The impacts which natural disasters have on the oceans themselves through the effects of the impacts of the disasters on land, both on the coast and inland.

Figure 8.1 sets out a simple typology that endeavours to present a perspective on the relationship between oceans, natural disasters and tourism. It distinguishes between natural disasters that originate from the oceans and those which are land-based. Furthermore, it identifies those which are partly the result of tourism itself and those in which tourism plays no causal role whatsoever. It also suggests that the impacts of natural disasters can be business impacts as well as physical impacts on coastal tourist destinations.

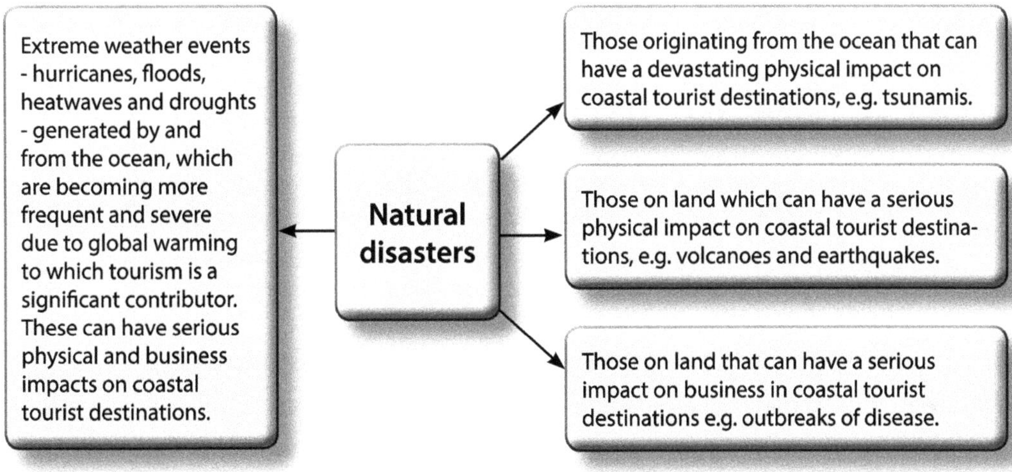

Figure 8.1: A simple typology of the relationship between oceans, natural disasters and tourism

In Figure 8.2 I attempt to model the relationship between the oceans, extreme weather events and tourism. Of course, I am fully aware that the model presented in Figure 8.2 is simplistic. This is not least because of its focus on coastal tourist destinations, and the exclusion of a consideration of other human activities and industries which have an impact on global warming and thus on the frequency and severity of extreme weather events.

However, it also excludes the impacts of natural disasters, of whatever kind, on inland tourist destinations, whether they be cities and historic towns, mountains or rural areas.

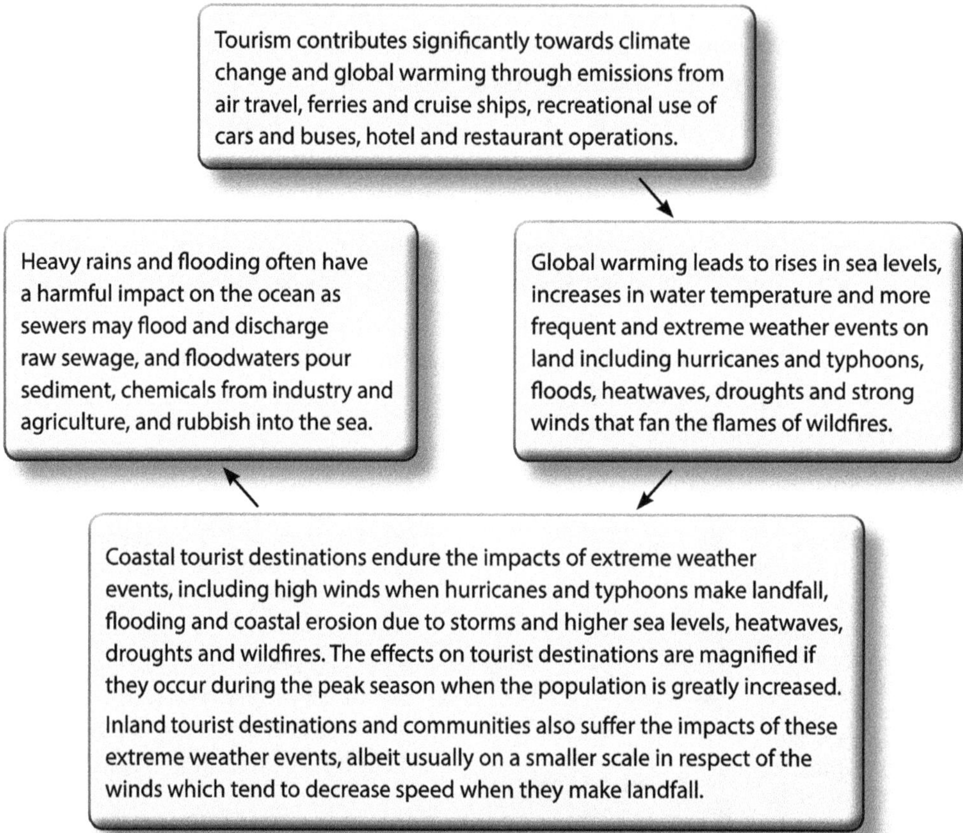

Figure 8.2: A model of the relationship between the oceans, extreme weather events and tourism

■ How tourism contributes towards natural disasters which originate in the oceans

In the first place, we need to recognise that tourism does contribute to some natural disasters that originate from the sea. These are generally the types of disaster that could be described as extreme weather events; our oceans are in crisis and most scientists are agreed that perhaps the most important challenge they face is global warming. As we saw in Chapter 2, the temperature of the oceans is increasing and the sea levels are rising as the warmer air temperatures melt the ice in the Arctic and Antarctic, as well as the ice in the glaciers of the world, which then flow down to the sea in the rivers. While scientists argue about exactly how quickly air and sea temperatures are rising and debate how rapidly the sea level is rising the fact that both are continuing to occur seems beyond dispute. There also seems a general consensus that the lifestyles of human beings and the activities of industries are the main cause of global warming. The finger of suspicion is pointed at everything from our use of private cars and hydrocarbon-based fuels to the emissions from a wide range of industries.

In recent years, the impact of tourism on global warming has attracted a lot of attention, with most of it focused on the airline industry. However, in the last few years, the cruise ship industry, marine leisure activities that involve engines, and hotels and restaurants have all begun to come under scrutiny in respect of their carbon footprints. This is, in addition, to the recognition that a significant minority of trips by private car are made for leisure purposes by tourists and day trippers. Concern over the impacts of travel and tourism on the climate are further highlighted when major new airports are built or huge cruise ships launched, and when roads to coastal destinations become ever more clogged by leisure traffic.

It is not surprising that tourism is coming under the spotlight as there is a growing recognition of the so-called 'climate emergency', and tourism is not a necessity of life but could be seen as something of a luxury.

In Chapter 2, we saw that most of the earth's weather is influenced by what happens in and above the oceans. Most commentators seem agreed that the increasing occurrence and severity of extreme weather events such as hurricanes and typhoons are due to the warming of the oceans. Furthermore, the ocean influences are also being seen as a significant factor, along with increases in air temperatures, in the incidence of climate-driven or facilitated natural disasters, such as heatwaves, droughts and wildfires. Meanwhile, rising sea levels, together with the growing number of severe storms, are increasing the frequency of coastal floods which also cause coastal erosion. Both of these are threats to the long-term sustainability of coastal tourist destinations in a number of parts of the world. It is ironic that tourism, through its contribution to global warming, may be putting its own future viability in jeopardy.

Finally, maybe, we need to rethink the term, 'natural disaster', for such extreme weather events appear to be becoming both more frequent and increasingly severe as a result of global warming, a process which seems to be driven by the activities of mankind. They are, perhaps, now as much human disasters as they are the work of nature.

The impact of natural disasters on tourist destinations and tourists

In this section, we will explore some of the different types of natural disasters and the impact they have on tourist destinations, focusing on coastal destinations. Figure 8.3 outlines some of the key characteristics of the natural disasters that affect coastal tourist destinations. While it is clearly neither comprehensive nor sophisticated, Figure 8.3 does attempt to identify some of the key considerations about the types of natural disaster that most affect coastal tourist destinations. An understanding of these is essential for any coastal tourist destination in terms of its emergency planning and its strategies for building resilience into local tourist destination management. Understanding the five types of natural disaster covered in Figure 8.3 will enable planners to ensure that destinations:

- Have early warning systems in place for natural disasters such as extreme weather events and tsunamis.
- Have realistic and effective crisis management planning systems in place.
- Have recovery plans in place covering the physical recovery and the rebuilding of the destination after any natural disaster.
- Develop marketing campaigns that are sensitive, appropriate and realistically timed to rebuild the image of the destination and local tourism business.

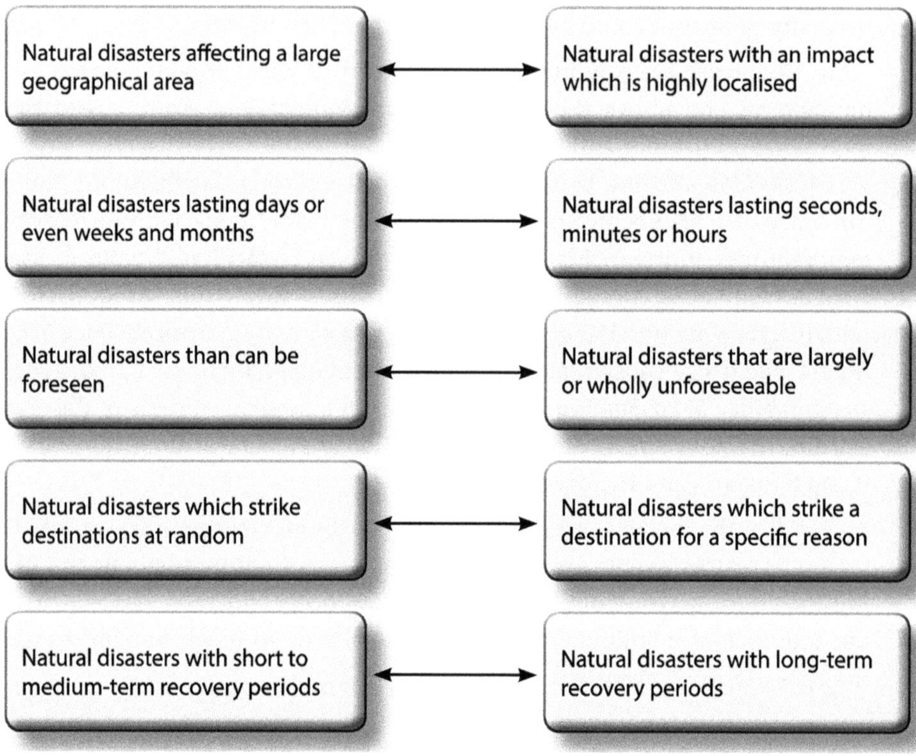

Figure 8.3: The key characteristics of natural disasters that affect coastal tourist destinations

The terms, hurricane, typhoon and cyclone, describe something that is basically the same phenomenon, but which is called one or the other depending on where you are in the world. As we noted earlier in the chapter these extreme weather events have their origins in the ocean, in areas of warm water, where its heat and evaporation create the conditions for their development. They are natural phenomena and are seasonal, being seen when temperatures are at their highest, so they occur in different months in the southern and northern hemispheres. In recent years we have seen more severe hurricanes, sometimes at times of the year which are a little outside the normal 'season' for them. They are highly destructive with winds ranging from 120 kilometres per hour to over 250 and they also tend to bring very heavy rainfall with them which can range from 15 to over 50 centimetres. They tend to lose strength when they travel inland, so their effects are worst when they first hit the coast, thus coastal tourist destinations normally suffer more than those inland. These extreme storms generally occur in particular parts of the world, notably the southern USA and the Caribbean, around the Indian Ocean, in south-east Asia and in northern Australia, although they can occur elsewhere at times if the conditions are right. In other words, they occur in areas which have high temperatures, for at least part of the year, and are, therefore, places that are highly attractive to tourists.

These phenomena have a major impact on the tourism industry and on tourism destinations, in four important ways.

- They have an impact on the tourism market as many tourists are nervous of visiting destinations when such extreme weather events are likely to occur. Indeed, many guidebooks advise against travel at this time of the year as these mega storms are also associated with particularly high temperatures and humidity which, in themselves, can be uncomfortable for tourists.
- However, in those places where the storm season coincides with major school holiday periods this can still be a peak time for tourism. And this poses another challenge to the tourism industry. Because of the risks posed by such extreme weather events, the industry has to respond when such severe storms are forecast, which means either cancelling flights to the destination or evacuating people from destinations where a storm is forecast. This is, understandably, unpopular with tourists, particularly if the track of the storm changes and it misses a particular destination altogether. It is also extremely expensive for airlines, tour operators and their insurers
- There is the physical damage which such a storm can do to a destination, particularly hotels and restaurants and airports and transport systems as well as to services such as water, gas and electricity. This can put a destination out of action for weeks or even months. The latest example, at the time of writing, of such a destructive storm affecting tourist destinations was Hurricane Dorian which wrought havoc across the Caribbean in 2019. Its impact was particularly devastating in the Bahamas, a popular tourist destination.
- Finally, cruise ships want to avoid sailing through such storms for reasons of both safety and passenger comfort. As a result, if one is forecast, they may well change their route to avoid it. This often means they are unable to follow the itinerary that passengers booked from the brochure, which can be very

> upsetting for them. When I was in Sydney in early 2019, I met a number of cruise passengers complaining that because of a cyclone their ships had been unable to dock in the ports of call they had expected to visit, including Cairns and Darwin. But, at least, they were safe!

And that last point is an important one. Because the tourism industry is well organised and well resourced, such storms kill and injure very few tourists. However, for those who live in coastal communities in the affected areas, the death tolls from such storms can be high, running into hundreds and even thousands of deaths. This fact needs to be borne in mind when listening to tourists complaining of how their trip was cut short due to the threat of a storm!

The storms we have just been looking at play a large role in our next category of natural disasters, namely **floods and landslides**, as the rainfall that comes from such storms is a major cause of floods and landslides. However, coastal flooding can also be the result of monsoon rain, a natural phenomenon, but one which has become much more difficult to predict in recent years as climate change modifies their long-established seasonal patterns. Coastal floods, like those seen in Venice annually, are also the result of high tides and particular wind conditions. Again, we see the impact of the oceans in tourist destinations with the sea playing a major part in the creation of monsoon rains and the floods in places like Venice.

Floods can have a devastating impact on coastal destinations, but their impact is also felt in the oceans as floodwaters and the resulting landslides sweep large amounts of silt, untreated sewage, plastics and other debris into the sea via the swollen rivers. This is harmful to marine habitats and to the quality of the water for bathers. We also need to note that the regular floods in the coastal areas of countries such as Bangladesh kill hundreds and even thousands of people every year.

However, we are now faced with something even more serious due to **rising sea levels**. Instead of occasional or seasonal flood events we may well be facing something even worse, namely the permanent destruction of low-lying coastal communities. Whole countries are at risk, most famously the Maldives, which could disappear within decades if sea levels continue to rise. This will, potentially, displace large numbers of people and remove popular destinations from the tourist map, of which the latter is the least important of the two, of course.

If we now look at the opposite extreme where the problem is one of **droughts and heatwaves** then, again, we can see the influence of the oceans, for one of the causes of the weather that creates both is high surface temperatures in the surface waters of particular areas of ocean. It is ironic that while people, at least those from cooler countries, hope for hot, dry weather on their vacations, the increase in temperatures from global warming has led to more and more heatwaves and droughts that have caused devastation to agriculture but also led to problems in coastal tourist destinations. Australia has been particularly badly affected by high temperatures and drought over the past few y ears. The heatwaves have also caused health problems and deaths in inland tourist destinations such as Paris in recent years.

The main problem in coastal destinations is the **lack of water** due to the rise in temperatures and seasonal droughts. This has caused some places, even in Europe, to turn to desalination as a way of trying to safeguard their water supplies and future-proof their tourism industries. However, as we have seen elsewhere in this book, desalination is a process which can have a negative impact on oceans both when the water is taken from the sea and from concentrations of salt put back in the sea after the process. Perhaps, the time has come to also look more closely at how tourism consumes fresh water and if it pays a fair price for the water that it uses. Some progress has been made on water management and conservation by the tourism industry but its use of it is still profligate in some places. Should tourism and leisure have priority over food production and water for local residents where this vital resource is in short supply?

Heatwaves and droughts are also linked to **wildfires** which have affected a number of coastal destinations in recent years, notwithstanding the fact that some fires are started deliberately by arsonists. And the sea creates many of the strong winds which fan the flames and worsen the impact of wildfires. In recent years we have seen distressing images from countries in the Mediterranean of people rushing into the sea to avoid the flames as wildfires reach the seashore, and many people have died in destination countries including Greece, Portugal and Spain.

As we have seen, the extreme weather events, and natural disasters caused by the weather, all appear to be caused by factors in which the oceans play a key role. And, they all seem to be being made worse by human actions which are causing global warming which is fuelling a growth in the frequency and severity of extreme weather events. **Tsunamis** are different. While they are ocean-based their origin would appear to have nothing to do with the actions of humans but is instead normally due to underground or major seabed disturbances in the oceans, usually due to under-sea volcanic eruptions or earthquakes. Tsunamis can generate waves which are tens of metres high. They are thankfully rare but when they do occur their effects on coastal communities are devastating, as we saw in Japan in 2011. However, it was the tsunami that affected the Indian Ocean basin of December 2004 that had the most impact of any tsunami in the past century. It is estimated that this single tsunami killed an 227,000 people of whom 168,000 were in Indonesia, mainly in the remote province of Aceh. This tsunami also had a massive impact on some countries that were popular tourist destinations with international travellers, including Sri Lanka, where some 35,000 people died, Thailand where around 8,000 lost their lives, the Maldives and the Seychelles.

Researchers around the Indian Ocean have studied the factors that meant that some communities were more badly affected than others, such as the work of Calgaro and Lloyd in Khao Lok in Thailand. These authors hoped that by identifying these factors, communities could be better protected in the event of future tsunamis or devastating floods. (Calgaro and Lloyd, 2008)

The western media focused much of its attention on the loss of life amongst foreign tourists and on stories involving tourists, even though the proportion of victims who were international tourists was low. Over 500 tourists each from Germany and Sweden died as a result of the tsunami, primarily in Thailand. The European media also contained many heart-warming stories of how local people had helped rescue foreign tourists, again largely these stories came from Thailand.

Many coastal communities in the Indian Ocean worked quickly to rebuild their tourism industries, not least because of their importance to the national and local economies. However, the process of rebuilding has not been without controversy with allegations. It has been alleged that:

- Having made statements that future building should take place further away from the sea for safety reasons, some governments have permitted building as close to the sea as it was before, under pressure from the tourism industry because it knows that tourists want to be close to the beach.
- There were 'land grabs', whereby entrepreneurs seized land previously owned by people killed in the tsunami so they could develop it for tourism purposes.
- Governments prioritised the interests of the tourism industry over local people in the allocation of resources for reconstruction work after the tsunami.

Those readers with a particular interest in tsunami impacts should also look at the case study about the 2004 tsunami which is to be found later in the book.

For some coastal destinations, potential natural disasters have been a constant threat in the form of **volcanoes**, of which the most famous example is probably Vesuvius in southern Italy. Indeed, for some destinations an active volcano is a tourist attraction in its own right for many visitors. This can be seen in places as far apart as Hawaii and La Reunion in the Indian Ocean, the Canary Island in Spain to Indonesia. However, as we saw in the case of the eruption of Eyjafjallajökull in Iceland in 2010, volcanoes can be highly disruptive to tourism with air travel across much of Europe being disrupted for many days. Ironically, though, this eruption increased awareness of Iceland around the world and helped stimulate a rapid growth in inbound tourism in the following years!

Earthquakes can also devastate places favoured by tourists, causing damage that can take a long time to repair. While they rarely seem to affect coastal tourist destinations specifically, the popular vacation islands of Bali and Lombok were struck by deadly earthquakes and aftershocks in 2018.

The final natural disaster we will look at is **epidemics of disease** which has little to do with the oceans directly but a lot to do with tourism. In the past 20 years the world, or at least parts of it, have suffered grievously from one epidemic to another including SARS, avian flu and Ebola. These directly involved tourism because their rapid spread from country to country is facilitated by air travel. But, ironically, this is being written at a time when the COVID-19 pandemic is ravaging the world, the most striking example the world has ever seen of how disease can be spread by air travel.

However, some diseases occur in places which are coastal destinations and affect tourists as well as local people. Many of these, such as Chikungunya and Dengue fever, are linked to insects and particularly the mosquito, and both have seen a growth in the number of outbreaks in recent years, often affecting tourist destinations. Global warming is leading to predictions that such tropical diseases will start to be seen more and more in more temperate lands. For the tourism industry, insect-borne diseases, in particular, are a problem because they make many tourists uneasy about choosing to visit destinations where they perceive the risk of catching such diseases to be high.

So, we have seen that the oceans play a key role in some of the natural disasters that affect the tourism sector and particularly coastal destinations. While we have focused on the impacts of natural disasters on coastal tourist destinations, we need to remember that such disasters can also affect places inland, and more importantly, they affect far more local people where they occur than they tourists. Furthermore, their impacts on local residents are usually much more profound and long term that those on tourists.

We will now look at some factors that can make the impact of natural disasters even worse, particularly in coastal destinations, before we go on to look at what can be done to make coastal communities more resilient in the face of disasters.

How tourism development can make the impacts of natural disasters even worse

At a time when we seem to be facing an increasing number of highly destructive extreme weather events, which usually have their roots in the oceans and/or have their worst effects when they reach coastlines, it is very concerning that humanity, and particularly, the tourism industry, continues to do things which make a bad situation even worse. These actions include:

- Continuing to build hotels, bars, restaurants and holiday homes – as well as the homes of local people – too close to the seashore or on low-lying ground in the coastal zone so that there is a high risk that they will be damaged or destroyed, and people killed, in the event of a tsunami, a tropical storm with a storm surge, or even flooding due to exceptional weather conditions. It also makes little sense when continued increases in sea level, as a result of climate change, will make living in such locations increasingly untenable.

- Destroying mangrove swamps and coastal vegetation including natural forests at the coast to make way for tourism developments as these protect the coast from damage from tsunamis, tropical storms and flooding. The Coastalcare website, in 2011, reported work by German agronomist, Georg Cadisch, which claimed that in the 2004 Indian Ocean tsunami, and in Aceh in Indonesia specifically, coastal vegetation could reduce casualties in coastal communities by around 5%, and up to 8% where natural forests existed on the coast (www.coastalcare.com, 2019).
- Undertaking activities such as infrastructure construction, beach nourishment, sea defence building, and so on, which can potentially cause coastal erosion as an accidental side effect. In times of extreme weather and flooding, coastal erosion makes matters worse because it can lead to enormous quantities of sediment being discharged into the sea and damaging wildlife while at the same time it reduces the ability of the coastal zone to absorb floodwater, thus making the impact of flooding worse than it needs to be.

Authors such as Bernard and Cook have clearly pointed out how unsustainable tourism development can exacerbate the flood risk for local communities, in a study of the Denarau island in Fiji. Talking about the resort they stated that '*the site selected was extensively transformed, re-claimed and re-engineered to accommodate international hotel chains. While this exclusive zone was exquisitely engineered and flood proofed, settlement proceeded largely unregulated in the adjacent town...and the already severe flood risk for local residents was exacerbated*'. (Bernard and Cook, 2015) This story could be repeated for many other locations around the world, sadly.

Making coastal destinations more resilient

First, coastal destinations can be made more resilient to a number of natural disasters by simply tacking the three issues outlined in the previous section, on the basis of the adage that 'prevention is better than cure'. This means having planning systems and building controls in place to prevent harmful developments that increase the impacts of extreme weather events. This will be discussed in much greater detail in Chapter 10. Leaving this aside though, and, recognising that most natural disasters cannot be stopped or even mitigated in their effects, the question is how to make coastal destinations more resilient to minimise the impacts they have on the environment, people and economies. This involves effective emergency planning and crisis management strategies which take account of the various characteristics of natural disasters such as those identified in Figure 8.3.

Such plans and strategies need several elements as follows:

- Having crisis management and emergency plans in place which list actions to be taken, and by whom, in the event of a range of natural disasters, based on detailed scenario modelling around what might happen if such and such a disaster occurs. One country where different states and destinations seem have developed impressive examples of such plans is Australia.
- Constantly updated early warning systems that give people and the authorities as much time as possible to take action to reduce the impacts, whether that means mass evacuations, for example. The need for such systems was highlighted by the 2004 tsunami in the Indian Ocean. Even a few minutes can reduce casualties significantly.
- Having emergency infrastructure in place to deal with the immediate impacts of disasters, such as places identified as emergency shelters.
- Putting in place clear guidelines on who deals with media enquiries and what the communications strategy should be in terms of the news media, the travel media, and tour operators.
- Anticipating the likely duration of the impact of different types of disasters and how long it might be before the time is right to seek to 're-launch' the destination in the tourism market in a way which is effective but does not seem insensitive or crass.

It is also increasingly being seen as vital that local communities are fully involved in resilience building and emergency planning in coastal zones, although this sometimes more of a theory than practice in reality at the moment.

Finally, in this section, one major lesson learned from the 2004 tsunami was the value of mangroves in protecting coastal communities from such events (Kathiresan and Rajeran, 2005). As a result, in some areas mangroves are now being protected or planted as part of the coastal defences.

The unfair geography of natural disasters

One thing that seems clear to anyone is that natural disasters, including those which are related to the ocean and affect coastal tourist destinations, are not equally shared around the world. The poorest parts of the world, in general, seem to suffer disproportionately heavily from so-called natural disasters. While terrible natural disasters have occurred in recent years in New Zealand, Japan, Australia and the USA, far more have occurred in the poorest countries in Asia, Africa, the Caribbean, and central and south America. These places are often ones which are heavily dependent, economically, on tourism and may have few other potential industries to turn to if tourism is harmed by natural disasters. Furthermore, local people, in such counties, cannot simply get in a car, or jump on a plane, to escape the consequences of natural disasters, like the tourists can, and usually do. Local people often have nowhere to escape to and have to cope, as best they can, when an extreme weather event or a volcano or earthquake destroys their home and their livelihood at the same time. At the same time, the impacts of natural disasters are often made worse in poorer countries because they lack the financial resources and high-quality emergency and health services which are required to effectively manage the impacts of natural disasters.

The feeling that the geography of disasters is intrinsically unfair is reinforced by the fact that global warming, which appears to be causing the increase in the number and the severity of extreme weather events, is largely a result of emissions from more affluent countries., or those undergoing rapid economic development. Yet the effects are being felt mostly in some of the poorest countries in the world and amongst the poorest sections of the population of those countries undergoing rapid economic development such as China and India.

Conclusions

This chapter has taken a broad view of the relationship between natural disasters and tourism but with an obvious focus on the marine environment and coastal tourist destinations due to the title of this book.

We have seen that some types of natural disaster originate from the sea, or are highly influenced by the sea, including tsunamis, and hurricanes, typhoons and cyclones. These are the disasters which cause the most destruction and loss of life in coastal tourist destinations. However, we have also noted that other natural disasters that affect coastal areas are linked to extreme weather events that are linked to oceanic influences on weather patterns including floods, landslides, heatwaves, droughts and wildfires.

At this point, it is important to recognise that the natural disasters discussed in this chapter tend to affect local people more than tourists and most of them will also affect inland tourist destinations and places with no real link to tourism, albeit, usually, less severely.

Other types of natural disaster can also be linked to the oceans including underwater earthquakes and volcanoes which can trigger catastrophic tsunamis that can devastate coastal areas.

As well as suffering from the effects of natural disasters, tourism in general, and coastal tourism destinations specifically can also contribute to creating them to the point where it is questionable whether some of them, such as extreme weather events in particular, can still be called 'natural' disasters any longer. Carbon emissions from the global tourism industry are playing a significant role in global warming which is leading to higher sea temperatures and rising sea levels. These in turn are leading to more and more severe storms being generated from the oceans that then come ashore as well as increased coastal flooding as sea levels rise. At the same time the tourism industry at the coast often makes things worse by building in unsafe locations, removing coastal vegetation that protects coastal communities, and undertaking projects that can cause coastal erosion.

And, while tourism does not create diseases, those which develop into natural disasters are often spread worldwide through air travel.

Returning specifically to coastal tourist destinations there is a need for some new attitudes and action on behalf of:

- The tourism industry which needs to recognise its responsibilities in terms of global warming and the role it plays in the impact of natural disasters on coastal destinations and realise that it is 'enlightened self-interest' for it to do more to protect both the oceans and the coastal destinations. This will be discussed in more detail in Chapter 9.
- Local and central government which needs to do more to protect local residents and tourists in coastal tourist destinations through more rigorous controls over coastal developments and better emergency and crisis management planning. This will be further discussed in Chapter 10.

However, unless we solve the global climate 'emergency' and the crisis facing our oceans it seems likely that coastal destinations, perhaps more than most places, will have to cope with an increasing number of natural disasters, most notably extreme weather events.

References

Bernard, K and Cook, S (2015) Luxury tourism investment in flood risk: case study on unsuitable development in Denaraau Island Resort in Fiji. *International Journal of Disaster risk Reduction* **14**(3) 302-311.

Calgaro, E and Lloyd, K (2008) Sen, sea, sand and tsunami: examining disaster vulnerability in the tourism community of Khao Lak, Thailand. *Singapore Journal of Tropical Geography*. **29**(3) 288-306.

Kathiresan, K and Rajendran, N (2005) Coastal mangrove forests mitigated tsunami. *Estuarine, Coastal and Shelf Science* **65**(3) 601-606.

Ritchie, B (2008) Tourism disaster planning and management: from response and recovery to reduction and readiness. *Current Issues in Tourism*, **11**(4) 315-348.

Questions and exercises

1. In a group, devise a survey designed to look at awareness amongst tourists of the relationship between tourism, the oceans and natural disasters. Make use of Figures 8.1 and 8.2 to help you identify the issues that your questions should address. When presenting your findings, note anything about your results that surprised or shocked you.
2. Select three types of extreme weather event that have affected coastal tourist destinations in three different countries in the past few years. Look at how each of them was reported in the news media in your own country. Do you notice any differences between the three events in terms of how they were covered by the media, and if so, what are the differences?
3. Using examples, critically evaluate the four types of natural disasters which were identified in Figure 8.1 and their impacts on coastal tourist destinations around the world
4. Select a coastal destination with which you are familiar or for which you can find information. Using examples of good practice from around the world, develop an outline of a set of regulations covering new developments and a natural disaster emergency plan for your chosen destination.
5. Discuss, using examples, the extent to which the different types of disaster discussed in this chapter can, accurately, continue to be described as 'natural'.

9 The Tourism Industry and the Marine Environment

Throughout this book we have been looking, primarily, at the negative impacts of tourism on the marine environment. It would appear reasonable to lay the blame for this at the door of the tourism industry, which facilitates the vacations that cause this harm and makes money from exploiting the marine environment and the other natural resources of tourist destinations around the world. However, things are not always that simple, as we will see in this chapter. For in tourism, and certainly land-based tourism, it has often been the tourism industry itself, and particularly the tour operation sector, which has often been leading the way on responsible tourism, often with little encouragement from the tourists themselves and usually no government regulation forcing them to act. Their motives may not have been altruistic but, nevertheless, they have taken action on a number of fronts, while events like the Responsible Tourism Days at World Travel Market in London have sought to share good practice around the tourism industry.

Despite the examples of good practice, though, it is far to say that the industry has paid far less attention to the oceans than it has to the impacts of tourism on land. In relation to the oceans, there are three main areas of concern, as follows:

- The tourism industry is a significant contributor to global warming due to the emissions from air travel and the leisure use of private cars. This is having a direct impact on the health of our oceans through increases in sea temperatures and rising sea levels.
- The cruise sector is having a negative impact on the oceans and the sector has grown steadily and has launched ever larger ships. Unlike the tour operation and hospitality sectors, the cruise operators, in general, have been slow to acknowledge their impact on the marine environment, and action to tackle the effects of cruising on the oceans has been disappointingly limited to date.
- In coastal tourist destinations, tourism continues to harm the marine environment, largely as a result of poorly managed marine leisure activities, beach

nourishment, de-salination and damaging construction projects, including the development of artificial islands as sites for hotels and airports.

It is vital that the whole tourism industry becomes more responsible in the way it manages its impacts on the marine environment, but it cannot do this alone. Tourists need to acknowledge the need to change their behaviour too, so they stop flying so often and so far, although this will adversely affect airlines and destinations that rely on tourists arriving by air. We also need tourists to start recognising, and abstaining from, those leisure activities which have a harmful impact on the marine environment, or at least practicing them more responsibly. And we need local and central government to be more willing to regulate tourism and more importantly perhaps, be more committed to enforcing the regulations.

I hope the reader will forgive me if I continues this chapter with two short sections that are based on an old British saying and an old movie title respectively! I believe that they are an idea way to raise two important issues.

Protecting the goose and the golden egg

There is an old British saying, which has an equivalent in many other cultures, that you should not 'kill the goose that lays the golden egg'. The meaning is clear and in the case of tourism, the goose is the environment and the places that motivate tourists to make trips, and the golden egg is the business and income that tourism brings. In the context of this book, the saying is most apt because much of the international tourism industry is dependent on the oceans of the world. Without the sea there is no cruise industry, and without healthy oceans there will be far fewer seaside vacations and less spending on marine leisure activities. The tourism industry has a vested interest in protecting the oceans and minimising their negative impacts on the marine environment. This should motivate it to be at the forefront of action to protect the well-being of the world's oceans.

The good, the bad and the ugly

I believe that this 1960s movie title sums up, nicely, the range of approaches and attitudes towards the health of our oceans taken by different tourism businesses around the world. Some are the 'good guys', aware of the issues and trying to take effective action in their desire 'to do the right thing'. Others make bad choices and do things which harm the marine environment, but this may often be due to a lack of awareness of the issues and alternative ways of working, rather than an uncaring attitude or ruthless pursuit of profit. But then there is the 'ugly', those companies that do not care what impacts their activities have on the marine environment, providing they are making money. And, there are also the corrupt politicians who turn a blind eye to bad practices and support irresponsible development projects in return for payments and favours.

The contribution of the global tourism industry towards global warming

While there are differences of opinion over the impact of travel and tourism on global warming, and thus on sea water temperatures and rising sea levels, no one suggests that it is not significant. The UN World Tourism Organisation, an organisation that exists to support tourism, claims that tourism as a whole contributes around 5% of global CO_2 emissions of which the majority come from transport (www.sdt.unwto.org, 2019). Meanwhile, an article in the *Independent* newspaper in the UK in 2018, quoting recent research, suggested that tourism was responsible for around 8% of global greenhouse gas emissions (www.independent.co.uk, 2018). Everyone seems agreed that air travel is the main culprit, but other forms of transport and the hospitality sector also make a significant contribution. Furthermore, currently, tourism is continuing to grow worldwide and there seems little prospect in the short term of major technological advances substantially reducing the carbon footprint of air travel.

Umair Irfan wrote an article on the Vox website in January 2019 that made some interesting observations about air travel, particularly in the USA. Using data reported by Atmosfair he suggested that air travel alone might be contributing up to 4.7% of global warming on its own. The article also reminded readers that air travel generates pollutants in addition to CO_2 that contribute to global warming including nitrous gases, soot, water vapour, sulphate and particles. It also provided a league table of the carbon emissions of different airlines, based on Atmosfair data. (www.vox.com, 2019). The Vox article drew attention to the fact that the emissions of newer aircraft are less than those of older aircraft, due to developments in engine and fuel technology, so they are preferable, but they still generate substantial quantities of harmful emissions.

In Europe in particular at the time of writing we are seeing 'flight-shaming', an anti-flying movement that originated in Sweden to encourage people to take fewer flights and take the train instead. While this is a perfectly viable approach in most of Europe, it is not possible for many of the world's most popular destinations, in Asia, Africa, and Caribbean and Indian Ocean islands, for example, which are reliant on tourists taking lengthy flights from their main source markets. Many of these places are in poorer countries and they are heavily dependent on tourism, so any reduction will have severe consequences for their local economies.

Some people try to find a compromise solution through carbon offsetting whereby the carbon emissions of an individual air passenger can be calculated and monetised as a donation the tourist makes on top of their ticket price that allows trees to be planted, or other relevant activities to be undertaken to 'offset' the carbon emissions from the flight. A number of airlines now offer this option, or a modest fee and British Airways announced it will introduce carbon offsetting on all its domestic flights from 2020. However, controversy rages over whether these schemes are a valuable tool in helping reduce global warming or just allow people to salve their consciences when they fly. And, if the carbon offset charges are to be set at a realistic level, they could add a lot to ticket prices. Researcher Professor Manfred Lenzen calculated that the cost of a scheme that would fully offset the carbon emissions for a return trip from the UK to Sydney would be £237 in 2019, a hefty supplement on top of the ticket price. (www.independent.co.uk, 2018) Frankly, it seems unlikely that many people, in a price sensitive market like air travel, will voluntarily pay such a levy.

Because it is almost inevitably harmful to the environment, airlines and the aviation industry have tended, understandably, to shy away from making any claims to being responsible or better than their competitors in respect of emissions. However, there was an interesting development in autumn 2019 when Irish budget airline, Ryanair started to advertise itself on its website and in its advertising as 'Europe's lowest emissions, lowest fares airline'. Using industry data, they talked about being 'number 1 for carbon efficiency' They claimed that their CO_2 emissions per passenger kilometre were 66g, 23% less than the average figure for the largest four competitor airlines in Europe (www.corporate.ryanair.com, 2020). This situation has come about because Ryanair has invested heavily in a fleet of very modern aircraft and has high load factors and no business class. Ryanair also publicises the fact that passengers can offset their carbon on Ryanair flights and that they support various environmental good causes. This is an interesting development particularly in relation to a short-haul airline in Europe, the birthplace of the 'flight-shaming' movement.

Unfortunately, the heavy focus of the media on air travel has kept attention away from the impacts of the rest of the transport sector, the cruise industry and the hospitality sector in terms of their impacts on global warming. Yet, we need to focus on every part of tourism that is contributing, in any way, to global warming which, as we have seen, is having a devastating impact on our oceans through increases in sea temperatures and rising sea levels.

■ The impact of the cruise sector on the oceans and coastal destinations

In Chapter 3 we saw that the cruise industry, currently, has a number of negative impacts on the oceans that have largely been overshadowed by the media focus on air travel when discussing tourism and global warming. These impacts include significant levels of carbon emissions; pollution from fuel spills; pollution from 'grey' and 'black' water; waste disposal at sea; pollution from sewage disposal; air pollution at sea; and wildlife habitat damage and injury to wildlife.

Yet until recently, the industry, has largely tended to either deny or ignore the existence of such problems. However, the pace at which the industry has begun to address sustainability issues has grown in recent years, not least because of some high-profile court cases and actions imposed on some cruise companies by judges, following infringements of various regulations regarding waste disposal at sea, for example. Nevertheless, in terms of responsible business, the cruise sector still seems to be lagging behind other sectors of tourism. It also remains very wary of raising awareness of the issues identified above with its customers. Perhaps this is because cruising sells itself very much around the ideas of pampering, glamour and luxury, and the companies fear that such consumers may not like their 'onboard fantasy bubble' to be pricked by such cold realities.

At the same time there has been growing attention paid to the impacts of cruise ships on their ports of call with accusations that they are a major cause of over-tourism in some places, with their impact magnified by the fact that they disgorge large numbers of people at a location for just a few hours. Furthermore, questions also continue to be raised about the ethics of some of the human resource management policies on cruise ships, most notably low pay, the issue of gratuities and the fact that certain roles on board seem to be reserved for particular nationalities.

Most cruise companies do now have sustainability policies in place although some are more convincing than others. Some companies have taken a similar approach to airlines, with annual reporting of performance in terms of measurable outcomes around environmental impacts and the use of resources. Like airlines, cruise operators have to abide by a set of international regulations that apply to the whole shipping industry, so this approach makes sense. The main cruise sector professional body, the Cruise Lines International Association promotes sustainability on its website and talks about things like reducing carbon emissions. However, these actions involve targets, which are meant to be industry-wide, but which seem rather too slow-paced in the light of the climate change challenge we face. Individual companies, notably Carnival, publish an annual sustainability report but again, the targets seem modest given the scale of the problems faced by our oceans and the planet as a whole.

While things have improved in recent years, corporate social responsibility reporting by cruise companies is still often less than satisfactory in comparison to some other sectors (Bonilla-Priego et al., 2014). In general, cruise companies are focusing their responsible business strategies on their environmental impacts rather than taking a holistic view that encompasses labour issues and their impacts on ports of call. This is, in contrast, to the approach taken by some hotel companies such as Accor which have a comprehensive policy on sustainability.

Finally, in contrast with many tour operators, cruise companies still seem to be reticent about talking about environmental issues and their sustainability policies in their brochures. Whereas, many tour operators now devote several pages in their brochures to sustainability and what they are doing about it, cruise companies tend to 'bury' mentions of such matters in the small print at the end of the brochures. Is this because they do not want to mention it for fear it may put consumers off taking a cruise, or is it because they think tourists want to ignore such things when they are on vacation? Either way, cruise companies, in general, are missing an opportunity to raise awareness amongst their customers of the crisis facing our oceans.

■ The impacts on the marine environment of the tourism industry in coastal destinations

It is now time to focus on the tourism industry in coastal destinations and the impact which it has on marine environment, through both the construction of new developments and its day to day operations. In many cases the harm done to the marine environment is not deliberate but is the result of bad practice. The damage is not inevitable as there are techniques that can be used, in most cases, to reduce or even eliminate the negative impacts on the marine environment.

Figure 9.1 seeks to identify the main sources of such impacts and each of these will then be discussed in a little more detail, although they have previously been covered in great depth in other chapters.

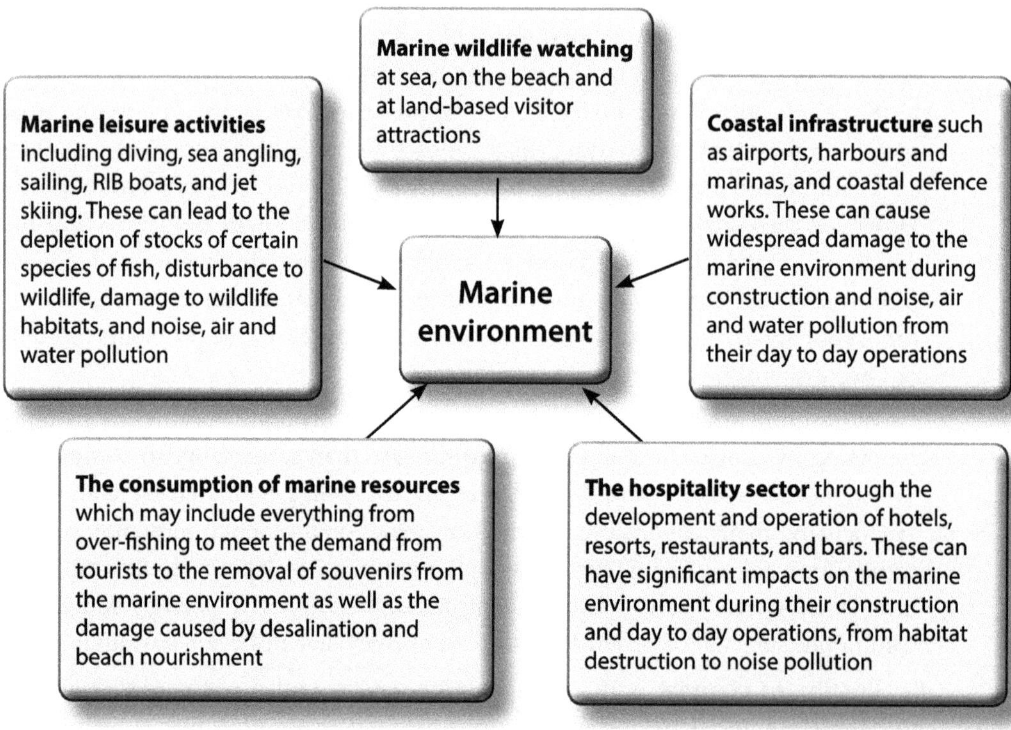

Figure 9.1: Sources of impacts on the marine environment from coastal destinations

Wildlife-watching at sea and on land

In Chapter 4 we focused on wildlife-watching in the sea but also to some extent on land, whether that involves birdwatching on the cliffs, seal-watching from the beach or the controversial practice of keeping marine mammals in zoos and theme parks. Marine wildlife-watching has seen a great increase in recent years involving both keen enthusiasts with an interest in particular species and ecosystems but, perhaps, more from ordinary tourists tempted to visit a place because they have been made aware of the opportunities for marine wildlife-watching it offers. There can be little doubt that this interest in marine wildlife has been strongly influenced by the travel media and television programmes such as the *Blue Planet II* series made by Sir David Attenborough, which has been seen all over the world.

In some ways, marine wildlife-watching has had a positive impact, and few would argue with the idea that watching whales in the wild is better than hunting them to extinction. Nevertheless, the growth of wildlife watching has had some negative impacts on marine wildlife including:

- Disturbance to mammals such as whales and dolphins when tourist boats get too close, as well as problems caused by the noise and vibration of boats.
- Harm caused to sea turtles when they come ashore to lay their eggs and when their hatchlings first make their way to the ocean, if tourists get too close when observing events.
- The interest in marine bird life can and does spill over into the theft of eggs, which damages the reproduction of marine bird species.

At the same time, it needs to be recognised that tourism can also help marine wildlife, such as when revenue from tourists supports wildlife conservation projects, such as sea turtle sanctuaries and hatcheries. However, even this is not straightforward for some sanctuaries can be of very limited quality and may even be established to make money, while often the sanctuaries might not be required if it were not for the impacts of tourists on the sea turtles!

In many places, there are voluntary codes of conduct while in others there are regulations in place, which may or may not be effectively enforced. In most destinations it would appear that the lead on more responsible marine wildlife watching is largely coming from individual operators, who seem to be doing it because of their personal ethical beliefs. While they are to be applauded, they may be at risk of losing out, in business terms, to less scrupulous operators who cut corners to impress consumers, by 'chasing' and getting too close to the whales for example. Perhaps regulation is needed, if only to create a 'level playing field' for all marine wildlife watching operators.

Marine leisure activities

Participation in marine leisure activities has grown enormously in recent years, and new ways of enjoying the sea are being invented, most of which are variations of traditional marine activities. At the same time, their impacts are also being seen in new parts of the world as tourism appears to reach virtually every place on earth. There is a widely held view that some marine leisure activities are low impact while others are high impact and there is some truth in that, of course. Wild swimming in the sea is likely to be less harmful to the marine environment than jet skiing or snorkelling on coral reefs, for example. However, two points need to be made at this point, namely:

- The idea that some leisure activities undertaken in the sea have no impacts at all are undoubtedly false; whenever humans enter the water or use the sea for a leisure activity, they have an impact on the marine environment and these impacts are nearly always negative.
- The impacts of activities such as diving, boating or sea angling, for example, are not all the same or equal in scale. They depend on a number of factors, but perhaps most importantly, they are influenced by the behaviour of the participant, which is a result of both their knowledge about, and their attitude towards, the marine environment.

There are differences between the various leisure activities in terms of who, if anyone, is leading the move to reduce their impacts on the marine environment. In sea angling and sailing it is the well-established professional bodies, while in surfing it is the participants themselves through the efforts of groups of activists. In most cases, everything depends on goodwill and voluntary codes of conduct, although there are some regulations controlling some activities in some locations.

More detail on this issue may be found in Chapter 5, while the next two sections cover topics featured in Chapter 6.

Coastal infrastructure

Coastal infrastructure is a broad term which covers marinas and harbours, coastal defences designed to prevent flooding and coastal erosion, airports and roads in the coastal zone. Here the damage to the marine environment can be caused during the construction phase as well as during day to day operations. In both cases these impacts can be substantial, including habitat destruction, the death of marine creatures and long-term changes in water quality. Such coastal infrastructure can be developed in ways which minimise its harmful impacts during the construction phase as techniques exist to do this. However, whether they are used depends on the relevant government agencies and the level of rigour in the planning and development control systems they have in place, including the role and quality of Environmental Impact Assessments that should be required for all major infrastructure projects. This is particularly the case with the increasing number of artificial islands that we see being created which can have a major negative impact on the marine environment.

There are some excellent examples of coastal infrastructure although these are perhaps more common in richer countries or those countries where the regulation of new development is less likely to be influenced by corruption. In countries where tourism is the main industry and corruption is rife, the loud cry of money is likely to be heard more than the quiet whisper of the marine environment!

The hospitality sectors

Like coastal infrastructure, the impacts of the hospitality sector on the marine environment can occur during the construction of new units and from the day to day operation of enterprises. Construction can involve the destruction of habitats such as mangrove swamps and pollution of the sea, while operations can cause everything from disturbance of wildlife through water noise pollution, to the consumption of marine resources such as fish, to the discharge of waste into the sea. This is in addition to the visual pollution, when new units are sited in prominent positions and built in styles and from materials that are not appropriate to the location. Fortunately, specialist architectural texts on how to design resorts that are sustainable are available including one relating to tropical regions specifically, by Bromberek, published in 2009.

One issue with new resorts and hotels is often their scale which can lead to large-scale impacts being focused on a tiny geographical area; some resorts now cover areas as large as a village and some are even being built on artificial islands that are created just for them. Again, the key to reducing the impacts of the hospitality sector are local planning controls on new developments and regulations covering hospitality operations.

However, some new hotels and resorts are being designed in a more sustainable way in terms of location and materials and with an eye on their impact on wildlife habitats. Some are even using this to attract visitors by describing themselves as 'eco' or 'green' hotels or resorts. I would like to make two observations about this before moving on, as follows:

- While some accommodation units are now being designed to minimise their impact on the environment on the seashore, they are not always being designed with their potential impacts on the marine environment in mind.
- In most cases, there is no external verification of units that call themselves 'eco-resorts' or 'green hotels', so there is no guarantee they are any more environmentally friendly than any other unit.

The hospitality sector in coastal destinations is also becoming increasingly aware of how it could be impacted by extreme weather events and sea level rise and, in some cases, is trying to increase its resilience in the face of these challenges. In 1996, Burby and Wagner noted that the level of preparedness in terms of potential storm damage varied significantly from one hotel to another, in a study of New Orleans. (Burby and Wagner, 1996) Nine years later Hurricane Katrina devastated the city! Some good work has been done following the 2004 Indian Ocean tsunami to ensure that hotel guests and staff are better protected from extreme weather and natural disasters, but much remains to be done.

Consumption of marine resources

We saw in Chapter 7 the ways in which consumption of marine resources by tourists and for the benefit of the tourism industry can have a serious impact on the marine environment. This includes everything from tourist souvenirs such as shells, bits of coral, stones off the beach and dried seahorses, through eating fish which are endangered, to the harm done to the ocean by beach nourishment and desalination. Regulation of this consumption of marine resources is not common, with the exception of fishing, and the taking of certain marine creatures and even pebbles from the beach, and there are often not even codes of conduct in place in relation to things that 'beachcombers' remove from beaches.

Summary

Before going on, I would like to just summarise some of the main points from this section which has looked at the five main sources of tourism impacts on the marine environment. These are as follows:

- Almost all of the impacts on the marine environment from tourism are negative.
- Taken together, the different types of impacts can have a significant detrimental effect on the marine environment.
- As the impacts tend to be concentrated in and around coastal destinations they are often highly concentrated in small geographical areas.
- Attempts to reduce the negative impacts are taking place at different levels, from local to international, and they are being led by a range of different actors depending on the type of impact and its source. Sometimes it is an enlightened local company or operator, while at other times it might be a sector-wide professional body, or a voluntary organisation set up by participants in a particular leisure activity.
- Overall, there is more reliance on voluntary codes of conduct rather than government regulation in relation to leisure activities and the consumption of marine resources.

- In the cruise sector and in relation to infrastructure and the hospitality sector in coastal destinations, public sector planning and regulation does not appear to be reducing the negative impacts as fast as one would hope.

Indeed, the slightly depressing conclusions one might reach would be that things are improving, but not fast enough, and that too often the economic benefits of tourism are being allowed to outweigh the needs of the marine environment when decisions are taken. Let us now look at how the tourism industry tends to respond when it is criticised or challenged about its practices and their impacts on the environment, before we go on to look at the vexed question of whether self-regulation by industry or government regulation is the best way to make tourism more sustainable.

Industry responses to criticism

When the tourism industry, or other industries for that matter, are challenged about the impacts of their activities they tend to adopt particular responses, which can be placed on some form of continuum. In Figure 9.2 I present a model of this continuum which is adapted from one I developed some years ago.

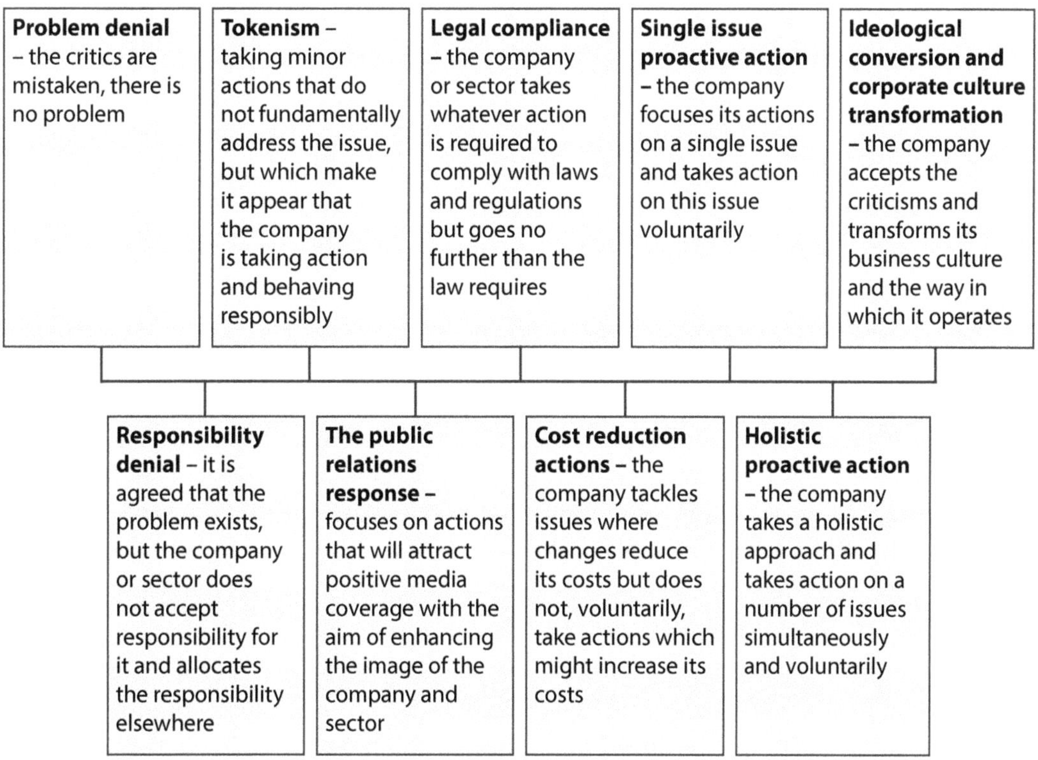

Figure 9.2: The industry responses to ethical challenges continuum

Within the tourism industry as a whole we see examples of all of the nine types of response shown in Figure 9.2. In relation to the impact of tourism on the marine environment we see many examples of some of these responses. Tokenism is very common and is often linked to the public relations approach such as when a hotel is built on a beach where sea turtles lay their eggs and the hotel encourages its guests to observe the turtles laying their eggs or the baby turtles hatching and making their way to the sea, even if this means floodlighting the event! This can all have a serious impact on the welfare of the sea turtles, when the best thing is simply to leave them in peace to lay their eggs and allow the hatchlings to make their way to the sea. Instead, many hotels continue to encourage irresponsible behaviour by their guests but simply make a nominal donation to support sea turtle conservation charities or sanctuaries, ignoring the fact that if the hotels were not there or their did not encourage their guests to indulge in behaviour that harms the sea turtles, there may be less need for such charities and sanctuaries.

A particularly controversial issue is the question of carbon offsetting and air travel, with a growing number of airlines offering passengers the chance to pay to 'offset' the carbon footprint of their flights. There is a suggestion that this is just a way of reducing the feelings of guilt felt by some who fly and that it is no solution to the environmental impacts of flying. In 2017, Becken and Mackey suggested that around a third of airlines at that time were offering voluntary carbon offsetting schemes for their customers. They commented that airline information on offsets was inconsistent and sometimes inaccurate. (Becken and Mackey, 2017) Zhang et al., in 2018, stated that '*airlines have introduced voluntary carbon offsetting... but literature suggests a low adoption rate partly because of low perceived credibility*'. (Zhang et al., 2018) This issue is discussed in more detail in Case Study 24.

At the same time the commonest other responses we see are cost reduction measures by hotels, restaurants and resorts that are certainly beneficial to the environment, such as reduced use of energy and water and a reduction in waste. The legal compliance approach is very often used, particularly by cruise companies defending themselves against criticism. They regularly confirm that their ships meet all legal requirements, although some recent court cases have brought such claims under increased scrutiny. Legal compliance remains an interesting response, as in a democratic society where laws are made by elected representatives it is morally defensible to argue that legal compliance is an ethically sound approach. However, it is not the fault of the companies that in relation to some crucial issues, legislation and regulation is often lacking.

In the field of tourism, and in the business world as a whole, we have seen the rise of CSR (corporate social responsibility), a business philosophy that acknowledges that businesses have responsibilities towards more than just their shareholders and customers. These responsibilities extend to the environment and society as a whole, locally, nationally and internationally. The tourism industry, in general, has been at the forefront of this development, although not all sectors have engaged equally with CSR, with tour operators and hotels being most engaged and the cruise lines the least. Often those companies which have engaged have acted on issues where there was no pressure to do so from either government or consumers. Some have used CSR, rather cynically, as a way of trying to achieve short-term competitive advantage in a highly competitive market. Some have regretted this when their activities have been seen not to live up to the marketing hype. Others have implemented CSR quietly, aware that they cannot claim that nothing they do has a negative impact on the environment or society. Many companies have adopted CSR in the belief that it will prevent competitive disadvantage in the medium to long term, as consumer interest in environmental issues grows, and governments consider or implement regulation, an issue which we will consider shortly.

Industry self-regulation versus government regulation

In terms of the sustainable or responsible tourism on land, in recent years there is no doubt that whatever progress has been made has been the result of voluntary action undertaken by either individual companies, groups of companies working together or sector-wide industry bodies.

Governments and supra-governmental bodies such as the United Nations World Tourism Organisation and the European Union have written endless documents exhorting the tourism industry to behave more responsibly. However, it is hard to detect much evidence that such documents have had any real impact on either the tourism industry or tourists. Perhaps, this is not surprising given that such bodies either do not have the power or lack the will to legislate or regulate the tourism industry to reduce its negative environmental impacts. At the same national governments have been generally unwilling to legislate to control the impacts of an industry which is seen as a success story, generating millions of jobs and billions of dollars of tax revenue to governments.

From the 1980s to the early 2000s, the tourism industry, and most notably the tour operation and hospitality sectors, implemented interesting initiatives such as the International Hotels Environmental Initiative and the Tour Operators Initiative. These were voluntary sector-wide projects that were based on the sharing of good practice across a sector to improve its overall performance in terms of reducing the negative impacts and maximising the benefits of tourism. I believe that the reason the tourism industry did this was to show governments that the industry recognised its negative impacts and could be trusted to reduce them and put things right through its own efforts, it did not need or require government regulation. Industry believed such regulation might be onerous, bureaucratic and might be accompanied by new taxes. It therefore used these initiatives to 'sell' the idea of industry self-regulation as an alternative to government regulation. And as this was the era of de-regulation and freeing up markets in many parts of the world, the idea of industry self-regulation matched the mood of the time, and so was generally accepted. Since then, we have seen the rise of CSR which again has been partly about reassuring government that regulation is not necessary because industry can be trusted to 'do the right thing'.

However, the reality is that even in relation to land-based tourism, let alone marine tourism and the marine environment, industry self-regulation is not working. Improvements are being made but not quickly enough to keep pace with the emerging crises we face thanks to global warming, in particular. Furthermore, industry self-regulation only works if every company participates constructively, behaves responsibly and improves its performance in respect of its impact on the marine environment. Sadly, the simple fact is that many businesses do not do this either because they lack the resources or the knowledge or the will to change, not least because being irresponsible can be more profitable in the short term! If industry cannot up its game across the board, and increase the pace of change, voluntarily, then governments will be forced to use regulation and tax to tackle the negative impacts of tourism in the air, on land, and at sea.

One interesting example of regulation and self-regulation working alongside each other is Antarctica where, since 1991, responsibility for the management of tourism in the marine environment is shared between the governments who are the Antarctic Treaty Parties and the International Association of Antarctica Tour Operators. (Landau and Splettstoesser, 2007)

The relationship between the tourism industry and the marine environment in the future

In Figure 9.3, I endeavour to present a vision of how the relationship between the tourism industry and the marine environment might look like in the future. It can be seen as an idealised picture of something that will never happen. However, it is my contention that something along these lines needs to happen in the near future to avert a disastrous outcome that will further harm our oceans and ultimately undermine the tourism industry and particularly coastal tourist destinations. A number of pre-requisites will be needed if the ideal situation shown on the right-hand side of the diagram is to become reality, namely:

- Continued and increased coverage of the challenges facing the marine environment and the contribution of tourism towards the global crisis of our oceans, in both the travel media and the news media.

- More pressure being put on industry and governments by pressure groups and activists.
- Greater knowledge of the impacts of tourism on the marine environment that can inform policymaking and decisions about regulations and proposed developments.
- Sector-wide culture change in which professional bodies such as the Cruise Lines International Association have a vital role to play.
- Political will and willingness to use 'carrots' such as fiscal incentives and 'sticks' like the strict enforcement of regulations to improve industry performance.

The current situation		The potential future situation
The global tourism industry seems not to appreciate its impact on our oceans through its contribution to global warming and thus rising water temperatures and sea levels, as well as the impact of coastal tourism on the marine environment.	→	The global tourism industry fully acknowledges its impact on the marine environment through the effects of global warming. It recognises the need for further regulation and that growth in demand for air travel is not sustainable.
The cruise industry is improving its environmental performance at a slow pace, is not making its consumers aware of the plight of the oceans, and is causing environmental problems in its ports of call.	→	The cruise industry sets itself, or is set through regulation, tougher targets to reduce its harmful impacts on the marine environment and the ports of call and substantially. It also needs to raise the awareness of its customers about what they can do to make cruising more sustainable.
There is generally little regulation of marine leisure activities and the consumption of marine resources and a reliance on voluntary codes of conduct instead. This is not providing sufficient protection for the marine environment	→	More comprehensive regulations are developed by governments, in consultation with key stakeholders, and effective enforcement is put in place. Voluntary codes of conduct encourage good practice that go beyond what is required by the regulations
Too many developments are taking place in the coastal zone which are harmful to the marine environment in the construction phase and once they are completed and in operation	→	Developments in the coastal zone are only approved following a full Environmental Impact Assessment and with conditions to minimise or eliminate negative impacts on the marine environment
Tourist awareness of the impact of cruising and coastal tourism on the marine environment appears to be very limited and this is leading to too much behaviour which is not responsible	→	All tourists are aware of the potential impacts of tourism, and their behaviour, on the marine environment. They use this knowledge to modify their behaviour for the benefit of the marine environment

Figure 9.3: How the relationship between the tourism industry and the marine environment might change in the future

It will also mean the tourism industry having to abandon the comforting idea that technology will find solutions so that its impact on the marine environment will cease to be an issue and it can carry on with 'business as usual' and focus on growth. At the same time, it will not be enough to defend itself by saying that the worst impacts on the ocean are caused by other industries and economic activities, even though there may be some truth in that view. As something that is not essential to human life and could therefore be seen as a luxury, in a time when the planet faces its worst ever crisis, tourism will need to play its part in solving the crisis, if that is still possible.

It will also need to recognise that in some cases it will have to accept some limitations on where it can operate and where tourists can be taken. Some marine environments places are just so fragile or endangered, or tourism so damaging to them, that the only solution is to ban trips to them altogether. Some other areas may need a moratorium on tourism to allow them to recover from the effects of tourism before visits can resume. Just being able to afford the price may not be enough in future. This is, of course, in addition to the need to reduce emissions from air travel to curb global warming, which is having such a terrible impact on marine environments around the world.

Conclusions

In this chapter, we have seen that the tourism industry is making progress on tacking its environmental impacts but that their efforts are too small-scale and too slow given the crisis we are facing, particularly in respect of the oceans. Furthermore, I would suggest that the industry is doing far less about its impact on the marine world than it is in relation to its impact on the land and the people who live in tourist destinations. Given, as we have seen, that what happens in the oceans also has a huge impact on the land and the people, notably through extreme weather events, we cannot allow this situation to continue.

It is clear that voluntary industry self-regulation on its own does not work because many companies do not participate and may indeed benefit from not being responsible. At the same time, those who do the right thing and take action to reduce their environmental impacts may actually suffer and lose out to their less responsible competitors. Greater regulation might, at least, protect the 'good guys' by creating a more 'level playing field'.

While we have talked about the industry, we cannot ignore the fact that the industry does what it does to meet the desires and demands of tourists, their consumers, without whom they have no business. Therefore we need to create a virtuous circle where industry raises consumer awareness of issues affecting the marine environment, tourists modify their behaviour based on this knowledge and this encourages industry to make its activities less harmful to the marine environment and communicate what it is doing to its customers. However, there seems relatively little evidence of this virtuous circle being created very often at the moment so it perhaps needs a helping hand, whether that comes in the form of government incentives for industry, a requirement for more coverage of environmental issues in industry brochures and websites, greater education about tourism impacts in schools and colleges, or more campaigning by activists and pressure groups.

We also need to encourage tourists to recognise the need to get away from the idea that the 'customer is always right' and accept the need for them to self-regulate their own behaviour by choosing to avoid certain places voluntarily, as well as avoiding some leisure activities or gastronomic experiences, or not collecting sea shells or buying bits of coral as gifts. This means a complete change of tourist consumption culture in effect. However, if this does not start to happen very soon, the only option will be greater regulation if we wish to avoid the disastrous consequences of irresponsible tourism on our marine environments.

To date, what little has been achieved in reducing the negative impacts of tourism as a whole has largely been the result of initiatives taken by the industry itself and individual companies. However, this will not be enough if we are to effectively counter the crisis facing our marine environments. Tourists themselves will have to change their behaviour and modify their expectation that they can travel anywhere and do anything if they can pay the ticket price. And, governments will need to have the courage to regulate tourism and tourist behaviour and to enforce regulations strictly and fairly.

If this does not happen the crisis of our oceans will get worse and worse and many marine environments will face total destruction. If that happens it will threaten the sustainability of the tourism industry itself and of the communities that live in coastal zones across the world.

If we accept the need for radical government action to safeguard our marine environments then it is clear that the public sector, from local to supra-governmental government, needs to become much more active in both tourism and the planning and management of marine environments. We will look at this issue in more detail in the next chapter.

References

Becken, S and Mackey, B (2017) What role for offsetting aviation greenhouse gas emissions in a deep-cut carbon world? *Journal of Air Transport Management* **63**, 71-83.

Bonilla-Priego, MJ, Font,X and Pacheco-Olivares, M del R (2014) Corporate sustainability reporting index and baseline data for the cruise industry. *Tourism Management.* **44**, 149-160.

Bromberek, Z (2009) *Eco-Resorts: Planning and Design for the Tropics*. Routledge. London

Burby, RJ and Wagner, F (1996) Protecting tourists from death and injury in coastal storms. *Disasters* **20**(1) 49-60.

www.corporate.ryanair.com (2019) www.corporate.ryanair.com/environment/ [2 Feb 2020]

www.independent.co.uk (2018) www.independent.co.uk/travel 19 Dec 2018. [4 Jan 2020]

www.sdt.unwto.org (2019) www.sdt.unwto.org/content/faq-climate-change-and-tourism [20 Nov 2019]

Landau, D and Splettstoesser, J (2007) Management of tourism in the marine environment of Antarctica: the IAATO perspective. *Tourism in Marine Environments* **4**(2-3) 185-193.

www.vox.com (2019) www.vox.com/energy-and-environment/2019.1/11/18177118/airlines-climate-change-emissions-travel [12 Dec 2019]

Zhang, B, Ritchie, B, Mair, J, and Driml, S (2018) Is the airline trustworthy: the impact of source credibility on voluntary carbon offsetting. *Journal of Travel Research* **58**(5) 715-731.

Questions and exercises

1. Critically evaluate the statement that 'the reliance on industry self-regulation in tourism represents an abrogation of responsibility on the part of governments'.
2. Working in a group, select a number of tourism companies or organisations that have faced criticism over the impact of their activities. Research how they have responded to the criticism. Write a report, using the categories in Figure 9.2 as a framework, identifying how they responded and why you believe they responded in a particular way.
3. Using examples and evidence that you are able to find, write an essay about the history to date of the International Hotels Environmental Initiative/International Tourism Partnership and critically evaluate its achievements.
4. If the negative impacts of tourism on marine environments are to be reduced, is it most likely to happen as a result of government regulation, voluntary action by individual tourism organisations or consumer pressure from tourists?
5. Discuss the factors that will determine if the potential future situation outlined in Figure 9.3 becomes reality or not.

10 Key Challenges in the Planning and Management of Tourism in Marine Environments

Throughout much of this text to date I have suggested that the public sector needs to be doing more to plan and manage tourism to reduce its negative impacts on the marine environment. However, I have never said this would be easy and in the Preface, I outlined some of the complexities involved in the planning and management of tourism of the oceans, from the beaches and shoreline to the most remote areas of open sea. In this chapter we will explore some of these complexities and endeavour to look beyond them to see how we might try to develop a model of best practice for the future. Planning and management is obviously a wide field so I will be focusing upon three main areas of activity as follows:

- The planning process for tourism including master planning of new resorts, zoning of land and sea for various uses and the system for evaluating and making decisions on proposed new developments.
- The ways in which the operations of the tourism industry are managed by the public sector, including legislation and regulations and their enforcement as well as policies on infrastructure development and management.
- Emergency planning and crisis management in the event of various kinds of natural disasters which is an important issue in relation to tourism.

We will focus upon what happens in the coastal zone, the inshore waters and the narrow belt of land that fronts on to the sea, and its impact on the marine environment. We will also consider what happens in the open seas where the scope for planning and management is more limited and what does take place is usually based on international agreements which are rarely focused on tourism. In terms of the open seas, the focus will, naturally, be on the cruise industry.

We will concentrate on those tourism activities which have a direct impact on the marine environment rather than those with indirect impact, such as air travel's emissions which increase global warming that ultimately increase sea levels and water temperatures. A number of the factors we will be discussing in relation to tourism also apply, in various ways, to other economic activities that based in marine environments such as shipping, mineral, oil and gas extraction, and commercial fishing. However, it is fair to say that these industries do seem to attract more interest from regulators and policymakers than tourism for some reason.

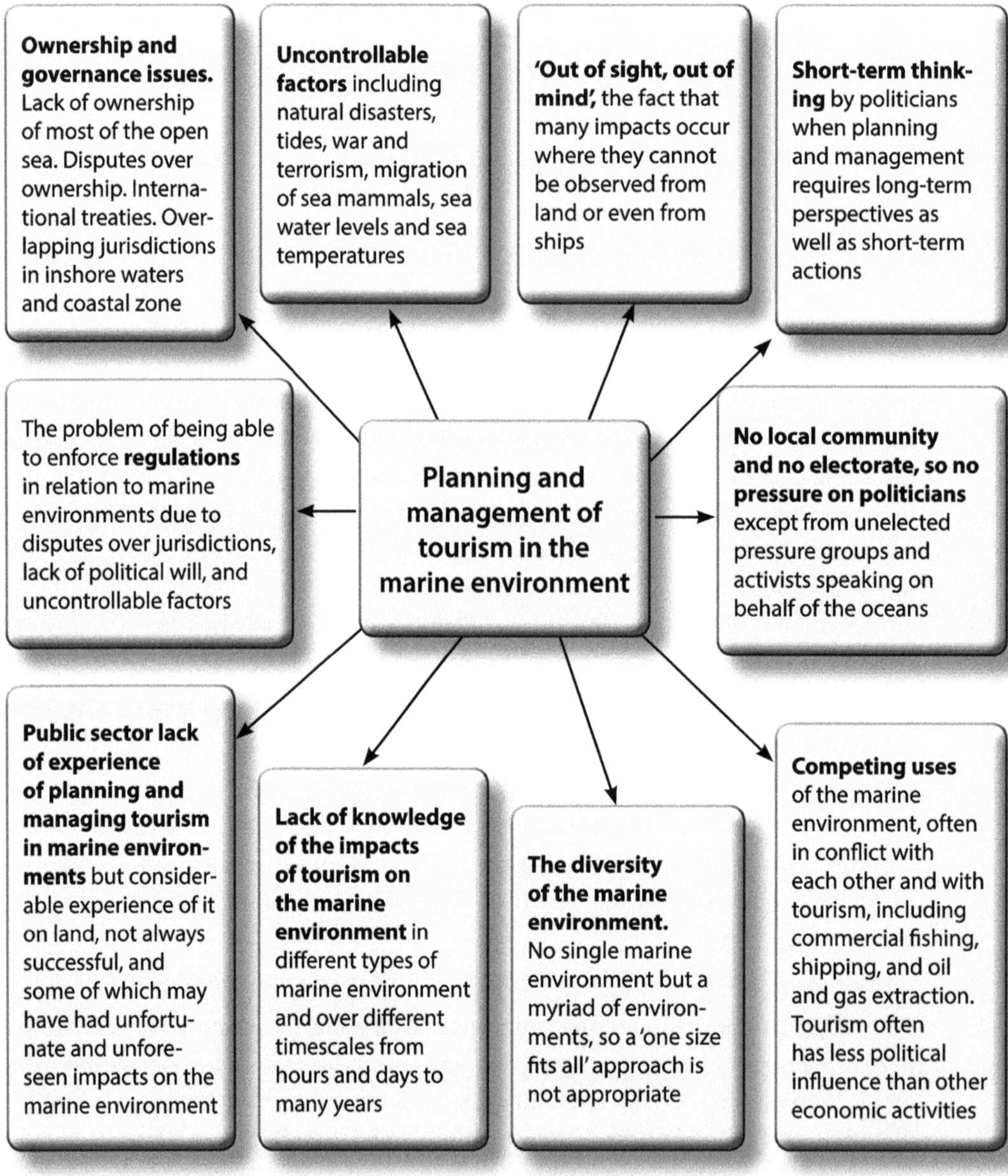

Figure 10.1: Key challenges in the planning and management of tourism in marine environments

In Figure 10.1 I identify what I consider to be some of the main complexities involved in the planning and management of tourism and its impact on the marine environment, and these will then be discussed in more detail later in the chapter. Before that readers will notice that there are also connections between a number of the factors highlighted in the diagram, which are important to an understanding of the challenges involved in trying to plan and manage tourism impacts on the marine environment.

Lack of clear ownership and governance responsibilities

I was a student of urban and regional planning some years ago, and clearly recall being told very clearly that to plan and manage somewhere effectively you had to either own it or have a form of clear legal jurisdiction over it in respect of new development. The latter means having the statutory right and authority to regulate development within a defined geographical area. On land, this authority is usually vested in central or local governments, which are, in most countries, usually elected bodies responsible and answerable in some way to their electorate. Theoretically, at least, every hectare of land in every country is the responsibility of at least one government body. The same public sector bodies are usually also empowered to regulate a wide range of other things which impact on the environment, including industrial pollution, waste disposal, and traffic management, for example. However, the situation in the oceans is very different in a number of important ways.

First, in complete contrast to the situation on land, no country or body owns the sea as such, even the sea which just a few hundred metres offshore.

Second, despite no one owning the sea, the ships of all nations have the right to 'innocent passage' through any waters under the United Nations Convention on the Law of the Sea, which was ratified by most of the countries of the world in 1982. This is important because 'innocent passage' would include cruise ships going about their business, and this might make it very difficult to prevent them from going to places with fragile marine environments, unless they agreed.

Third, not owning the sea does not mean that national governments have no rights in relation to the oceans and the marine environment. The same UN Convention allows countries to introduce a 200 nautical mile Exclusive Economic Zone (EEZ) an area in which theoretically *'the coastal nation has sole exploitation rights over all natural resources'* (www.globalresearch.ca, 2019). This is interesting because it permits exploitation but puts no responsibility at all on the country to conserve or protect the marine environment. There is no such equivalent on land to my knowledge. This implicitly suggests that the marine environment exists to be economically exploited, which is not conducive to the idea of sustainable development. Interestingly, tourism is not explicitly acknowledged to be an exploiter of the natural resources of the ocean in this context, although it clearly is. Readers will probably already have recognised another potential issue with the EEZ concept, namely that that many countries with a coastline are less than 200 miles from another country with a coastline, which provides great scope for conflict. Such a situation underpins the fisheries disputes that regularly arise within the European Union.

Fourth, because of the riches of the oceans in terms of fish and natural resources, disputes can arise anyway between nations over who has the right to claim ownership, or at least control over, thousands of tiny islands and swathes of remote ocean, so they can exploit those resources. These disputes can threaten international stability, as we see at the moment in the South China Sea. The disputes are also extending into areas with no permanent settled population but lots of natural resources, such as in the Arctic and the Antarctic, the two parts of the ocean where global warming is causing massive changes that will ultimately impact the whole planet.

Fifth, in the coastal zone there can be overlapping jurisdictions between different tiers of authority, including national, region and local government bodies, as well as public sector utility organisations, which can lead to confusion and duplication.

Sixth, in some countries, notably the UK, ownership of the coastal zone can be largely in the hands of a particular entity; in the UK the Crown, owns the seabed up to 12 miles out to sea. Therefore, the area of seabed that may be most affected by tourism activities and by what happens on land in terms of waste disposal, sewage, and pollution, is owned by an agency which is not democratically controlled and has no formal link with local government, which is responsible for the day-to-day management of tourism in coastal destinations. This is a further potential source of confusion, and fragmented planning and management.

Finally, due to the nature of oceans, the effects of the planning and management of its area of jurisdiction by one authority usually have an impact far beyond its own geographical area of responsibility. In other words, areas of the ocean, perhaps hundreds or even thousands of kilometres away, can be affected by decisions over which they have no influence.

In conclusion it has to be said that the current situation is not conducive to the effective planning and management of the impacts of tourism, or any other economic activity for that matter, on the marine environment. The jurisdictions and responsibilities of governments in respect of the oceans and the coastal zone are either confused or seem to be designed to stimulate conflicts. Furthermore, even the United Nations in its Convention on the Law of the Sea seems to emphasise the exploitation rather than the conservation of the oceans. However, in fairness, the United Nations is concerned about marine ecosystems and sees opportunities for action to conserve them, but from its website it is mainly talking about the 'Areas Beyond National Jurisdictions' (ABNJs), in other words the seas beyond 200 nautical miles of any country. This makes it appear that national governments are largely in a position to use their EEZs pretty well as they wish even though what is done in an EEZ can have impacts in ABNJs or in the EEZs of other countries.

Uncontrollable factors

Those authorities endeavouring to plan and manage tourism and its impacts on the marine environment have to contend with a number of uncontrollable factors that can have a major impact on their efforts. These include:

- Extreme weather events, including hurricanes, typhoons and cyclones, which can destroy, or cause severe flooding to, coastal accommodation, attractions and infrastructure.
- Tsunamis caused by underwater earthquakes and volcanic eruptions that can devastate whole swathes of the coastline.
- Tides which can cause significant cause coastal erosion when they are coming to shore and can sweep plastic pollution and sewage out to sea when they go out.
- Changes in the migration and breeding patterns of marine wildlife, which can create or end wildlife-watching opportunities for any coastal destination.
- Wars, terrorism or international conflicts which can focus on the oceans as well as having an impact on tourist flows to coastal destinations.
- Global warming and its impact on the water temperature and sea levels which poses a threat to marine ecosystems and coastal zones around the world.

While these factors are largely uncontrollable by the individual national and local governments that are responsible for planning and managing tourism and the marine environment, most of them can be foreseen and anticipated. Therefore, it is clear that emergency planning and crisis management has a vital role to play when dealing with things as volatile as the natural world and the forces of nature.

In recent years there have been some interesting studies on emergency planning and tourism. These include Becken and Hughey who in 2013 published a paper based on their work in the Northland region of New Zealand (Becken and Hughey, 2013).

Finally, on the subject of natural disasters and emergency planning, readers might be interested in Case Study 21, which looks at hurricanes and the tourism industry in the Caribbean, and Case Study 34, which discusses approaches to emergency planning in Australia.

'Out of sight, out of mind'

This challenge relates to the fact that many of the negative impacts of tourism on the marine environment are largely invisible to those living in the coastal zone or inland. Usually, no one sees the plastics being ingested by the fish and mammals, or the dolphin that dies after being in collision with a cruise ship, or the shark dying slowly after its fins have been ripped off to make soup! At the same time many of the impacts happen gradually over time and are not dramatic or easy to see until it is too late, whether it be the decline in wildlife populations or erosion of coasts and damage to corals caused by diving or the taking of bits as souvenirs. All of this means that it is easy for people to be totally unaware of what is happening unless they spend time immersed in the ocean. Things are largely different on land where changes and impacts are generally more visible to more people

Short-term thinking

This is a common problem with any planning and management when it is the responsibility of the public sector, which is the norm. Given that decisions are taken by politicians, they are likely to be based on short-term thinking which may often be limited to the date of the next elections. This is unfortunate as sustainable development requires a long-term perspective. Ironically, it also means the best chance of government decision-making having such a perspective occurs in countries where hereditary rulers are in power or where there is little real democracy so that decision-makers do not need to be concerned with election cycles.

No local community, no electorate, no pressure on politicians

The simple fact that the oceans have no resident human population, and therefore no electorate, also means that that traditional ideas of democratic local planning and management are largely irrelevant. The ocean has no voters to put pressure on politicians to act to conserve the marine environment. However, in coastal destinations, there is likely to be a strong tourism lobby with a loud voice that might well sway politicians to favour development over conservation. On the other hand, the ocean does have a number of non-governmental organisations and groups of activists advocating on behalf of its well-being although they may lack the influence and resources of the tourism lobby.

Diversity of marine environments

Of course, planning and managing tourism and its impacts on the marine environment would be relatively straightforward if marine environments were all the same, but as we saw in Chapter 2, they are not. Indeed, they are all unique and their differences are based on variables that include:

- Whether the water is cold or warm or something in between.
- The range of marine wildlife species that inhabit a particular marine environment.
- Whether or not the marine environment is on the migration route of specific species or plays a key role in the breeding activities of one or more species.
- The depth of water, from a few centimetres to many thousands of metres.
- Whether the area of sea is subject to strong tides or relatively slight tides.
- The coastal vegetation, including mangroves and coastal stands of trees.
- The material of which the seabed is made up, whether it be rocks, sand or gravel, for example.
- The physical environment of the coast and shoreline, which may consist of broad estuaries or narrow coves, mudflats or cliffs, sand dunes and beaches or pebbles and rockpools.
- The extent to which sunlight permeates the water, based on its depth and the amount of sunshine in the local area.
- How much, and what type of, human settlement there is on the shore, and what impact this has on the marine environment

Even this selective list easily demonstrates the fact that every marine environment is different, which means that if you are a planner or manager of tourism, you need to be aware that standardised 'one size fits all' type solutions are likely to be ineffective

Competing uses

As if the diversity of types of marine environment did not make things complicated enough, we also have to bear in mind the many competing uses that need to be reconciled in the marine environment. First, we have to reconcile tourism uses and the use of the sea by cruise ships with non-tourism uses such as freight shipping, commercial fishing and aquaculture, oil and gas extraction, and military operations and exercises, for example.

However, we also have to manage computing tourism uses of the oceans and the marine environment such as jet skiing and sea angling, powerboats and wildlife-watching, and noisy beach parties and the breeding activities of marine wildlife that come ashore to lay eggs or give birth to their young.

This competition for a finite resource further complicates the planning and management of marine and coastal tourism.

Lack of knowledge of tourism impacts on marine environments

Effective planning and management of tourism in relation to the marine environment is certainly made more difficult because of our general lack of empirical evidence about the impacts of tourism on the oceans. We have a general understanding of what can happen but what policy-makers need is information about how tourism does or will affect a specific marine environment and over what kind of time scale. It would also be helpful if we knew how far out to sea the various impacts of tourism extend and again, over what kind of timescale.

Lack of experience of planning and managing tourism in marine environments

We have a significant amount of experience of planning and managing tourism on land, although not always successfully. However, we have very little experience of doing the same in respect of the marine environment. We have a number of tried and trusted techniques that we use on land and we will look at some of them shortly to see how they work in relation to marine ecosystems. Indeed, one or two of them are already being utilised, as we shall see. However, we need to learn the lessons from land-based planning if we are to successfully plan and manage marine tourism and tourism impacts on the marine environment in a sustainable way. Three of these lessons are as follows:

- Master plans where large scale developments and land-use zones are plotted on maps at a particular point in time, based on knowledge and conditions at that time, tend to be unsuccessful in that they either do not get implemented, or they do and the outcomes are not as anticipated.
- Organic, incremental change that responds to changes in the physical and business environment tends to be more effective.
- Poorly researched plans tend to have unforeseen consequences which usually seem to be negative.

The planning and management of marine environments requires great sensitivity because most ecosystems struggle to cope with sudden radical change, and apparently small developments can lead to major changes in the ocean over the longer term, again most of them negative.

In recent years we have seen a growing literature around a concept called Marine Spatial Planning which *'can mitigate conflicts between tourism on the coast and in the marine environment and other human uses'*. (Papageorgiou, 2016)

Regulation and enforcement

Throughout most of this book to date I have suggested we need more regulation to protect the oceans from the impacts of tourism rather than simply relying on voluntary codes of conduct and industry self-regulation. However, as we have seen above, regulation of tourism and its impacts on the marine environment will face a number of obstacles including the lack of clear ownership and management responsibilities, a lack of political will, the power of the tourism industry lobby, and a lack of knowledge to inform good policy-making and decision-making.

Having looked at all the challenges1, as summarised in Figure 10.1 we will now go on to look at how these challenges might be overcome. We will first explore how techniques used on land are being, and might be, adapted and applied to the oceans and then I will endeavour to suggest what a model of best practice for the planning and management of tourism and marine environments might look like.

Marine Protected Areas

On land there is a long history of rulers and elected governments designating certain areas for protection on the basis of their beauty or their resident wildlife, particularly where it was felt that this could be threatened by developments whether they be agricultural, industrial or residential. The modern era of such designations probably began in the USA with the designation of the Yosemite National Park in the late nineteenth century. Since then national parks have been designated around the world, although they often function in very different ways. In some cases, the state or a governmental body owns all the land while in others the land largely remains in private hands. Some national parks offer visitors free access to all areas, while in others access to particular parts of the park may be restricted. Limited development is permitted in most parks, but some have zones where no development is permitted.

However, the idea is usually quite similar, namely, to manage development and access with the aim of protecting the natural environment. The success of national parks on land has been mixed, with some actually attracting more visitors because the designation raises their profile. Perhaps the biggest challenge for these parks has been trying to get the balance right between tourism and recreation,

In recent decades many governments around the world have designated selected marine environments as protected areas, although the titles given to these areas vary, and include national park, marine reserve, marine park, nature reserve and state park. There are dozens of such parks around the world and at the time of writing when I searched online for information the countries with the largest number of such designations included the following:

Australia	50
Canada	4
Hong Kong	6
Indonesia	8
Malaysia	10
Mexico	4
New Zealand	3
Norway	4
Philippines	3
USA	4

The dominance of Australia is probably ot be expected, but it is a little surprising that India only has two marine parks, China apparently none, and there are also none in a number of European countries. At the same time, it has to be said that the way in which these parks operates varies dramatically as does the level of financial resources they have and the amount of regulatory power they wield.

In the UK the government has taken a different approach to marine conservation and has developed a framework within which, at the time of writing, 91 Marine Conservation Zones (MCZ) have been designated. together with the other types of marine protection designation, around a fifth of all UK waters are now covered by some form of protection. In September 2019, the UK government announced a new type of marine designation with the creation of the Plymouth Sound Marine Protection Area. This recognised the biological diversity of what is one of the largest natural harbours in the world but may also be an acknowledgement of the marine science expertise of the University of Plymouth, whose scientists have played a leading role in research on oceanography and plastic pollution in the ocean. Its designation has been welcomed by local government in the area which sees it as a vital element in the regeneration strategy, and the vision of the future, for the city.

The implementation of MCZs is based on voluntary partnerships with stakeholders, such as user groups, and negotiation rather than on regulatory powers and ownership. In November 2013 the UK government published a document outlining how MCZs were to be designated and managed; this included the following:

- *Generally, each MCZ has one conservation objective. The objective applies to all of the features being protected. The objective is that each of the features being protected be in favourable condition*
- *A range of public authorities have responsibility for regulation of activities occurring in the sea and on the coast* (it then names a number of central government departments, quasi-governmental organisation and local governments). *Where the functions of a public authority have the potential to impact on an MCZ, the Marine and Coastal Access Act of 2009 created an obligation on the authority to carry out its functions in a manner that best furthers the conservation objectives of the MCZ.*

Where this is not possible, the public authority is required to proceed in the manner that least hinders the achievement of the MCZ's conservation objectives.

- *Successful management of marine protected areas requires stakeholders working in partnership and cooperating with each other, conservation advisers and public bodies.*
- *Stakeholders may be engaged through informal and formal consultation, and in the development of voluntary measures, which may be a suitable approach where these can be shown to be an effective* [means of achieving the conservation aims].
- *The Marine and Coastal Access Act places an obligation on DEFRA* (a central government department of state) *to publish a report every six years with an assessment of how well MCZ's are achieving their objectives.'*
(DEFRA, 2013)

The MCZs are not backed by either substantial funding or strong legal powers, relying on the powers which public authorities already have in terms of developments in the coastal zone and activities carried out at sea.

Having looked in detail at coastal protected areas around the world while researching this book, I would like to make a number of observations about them:

- First, while the staff of these parks are clearly committed to marine conservation they are often hampered by their lack of financial resources.
- Second, many of them have limited legal powers and need to rely on negotiation and voluntary action by various marine environment users.
- Third, their terms of reference often seem almost as concerned about tourism as they are about conservation, so often there is a tension between the two and when that happens the voice of tourism is often the louder.
- Fourth, the designation of a marine protected area may actually attract tourists and thus add to existing pressures on the marine environment.
- Fifth, it is hard to see how effective they can be when the ocean is an open, rather than a closed, system and when there are so many uncontrollable factors as we saw earlier.
- Sixth, to be effective, the marine protected area, in some ways, need to be linked to planning on land given the impact of coastal zone developments on the marine environment. However, this is not always the case with existing marine protected areas.
- Finally, as we saw earlier, whereas in many national parks on land the public sector owns the land this is not the case at sea where no one owns the oceans and ownership of the seabed is limited. This makes conserving the marine environment a much more difficult task for the park authority.

The challenges facing protected areas at sea are clear when we look at the Great Barrier Reef Marine Park, which was designated in 2004. Despite having regulatory powers and dedicated staff, the Great Barrier Reef continues to deteriorate at a rate which is alarming scientists. The forces that are causing the problem to seem to be too great for the park authority, not least because some of them have their source well beyond the marine park itself.

Criticisms have been voiced about the management of Marine Protected Areas, and the broader activity of marine conservation, and the extent to which they involve and work with local communities. Critical research in this area includes:

- Christie describing them as '*biological successes but social failures*' in southeast Asia (Christie, 2004)
- An exploration of the negative as well as the positive impacts of Marine Protected Areas based on a literature review of 32 studies from the 1980s to 2017 (Kamil et al., 2017)
- The issue of the rights and role of indigenous peoples in the management of marine conservation has been raised in relation to the Bajau people of southeast Asia (Clifton and Majors, 2012)
- Oracion et al. asked the question '*Marine Protected Areas for whom*' in a paper based on a study of Luzon in the Phillipines (Oracion et al., 2005)
- The challenges facing the growing number of Marine Protected Areas in China (Li and Fluharty, 2017)
- The conflicts between economic activities, such as fishing and tourism, and conservation in Marine Protected Areas in Brazil (Lopes et al., 2015)

Nevertheless, the designation of protected areas at sea at least it is a valuable first step, albeit a small one, towards conserving the marine environment.

Readers who are particularly interested in Marine Protected areas might also like to look at Case Study 25, which looks at the management of the Great Barrier Reef Marine Park in Australia, and Case Study 26, which offers an overview of Marine Protected Areas in Asia.

Zoning and competing uses

There is a history of using zoning in planning to manage potential problems with conflicting uses or users who may come into conflict. Conventional urban planning would usually seek to separate industry from residential uses, while in countryside management zoning might be used to separate user groups such as walkers and off-road vehicle drivers, anglers and boat owners on rivers and canals.

Zoning has also been used in coastal zones in an attempt to reconcile conflicting uses. In 2017, a paper by the UNEP-DHI suggested that:

'Coastal zoning... allows multiple uses of a coastal area. In a specific zone, certain activities can be allowed, allowed with permission, or forbidden, and the zoning can be geared to a range of different conditions... The zoning is generally managed, and enforced,

by public authorities and can be developed on consultation with different coastal stakeholders. (It) *requires a relatively high degree of public management, monitoring and enforcement to ensure it remains functional. Different users with conflicting interests can violate. regulations.'* (www.unepdhi.org, 2017)

Agostini et al. published an interesting study about the development of the marine zoning strategy for St Kitts and Nevis which illustrates some of the relevant issues (Agostini et al., 2015). Things are particularly complicated when there are Marine Protected Areas which cross national areas of marine jurisdiction as has been explored by Portman in relation to the case of the Red Sea Marine Peace Park (Portman, 2007). In many countries, such zoning, if it exists, is not always enforced while in other places it is purely voluntary and, unfortunately, not every marine environment user is a responsible citizen who follows exhortations to confine themselves to a particular part of the sea or shoreline for the benefit of the marine environment.

Case Study 30 looks in rather more detail at the subject of the zoning of different marine leisure activity uses of the coastal zone.

Environmental Impact Assessments

Environmental Impact Assessments (EIAs), as their name suggests, are evaluations of potential new developments, designed to identify their likely environmental impacts to aid decision-making about the proposal. Today, they are widely used and are part of the legal planning framework for new developments in many countries including: China, Egypt, India, Iraq, Malaysia and Sri Lanka, as well as the countries of the EU which has a range of directives around EIAs. There are several points worth making about EIAs including:

- The rules vary significantly between countries about when an EIA is required based on the size of a development and whether or not it is in a protected area.
- Due to the influence of industry lobbies, and even corruption, an unfavourable EIA may not be enough to stop a development going ahead.
- EIAs can only be based on the latest knowledge we have of impacts and this is often imperfect.

EIAs have been used for projects which may impact on the marine environment, but again their use varies between countries and in many countries, they are not yet part of the land-use planning system.

Readers may be interested to know that Case Study 29 also focuses on EIAs.

Carrying capacity

The concept of carrying capacity has been around in tourism for some years and it seems very sensible, for intuitively we imagine that everywhere must have a capacity. However, implementing this idea in practice is fraught with difficulties. First, there are potentially a number of measures of carrying capacity for the marine environment and coastal zones including:

- **Physical** carrying capacity, the point at which an area cannot physically accommodate any more tourists.
- **Ecological** carrying capacity, the point at which the volume of tourists begins to damage the ecosystem.
- **Economic** carrying capacity, the point at which any further increase in tourist numbers will harm the other economic activities which rely on the marine environment.
- **Perceptual** carrying capacity, the point at which the number of tourists begins to affect the quality of the tourist experience and the sense of place. Of course, this is highly subjective, and opinions will differ from one tourist to another

Second, we often lack the empirical data that would allow us to accurately measure carrying capacity for a location. This is not an easy task and depends very much on the type of marine environment and how resilient it is to tourism pressures, as well as the types of tourist activity that are taking place in the location.

Finally, even if we can measure the carrying capacity accurately, the challenge is how to keep numbers below the capacity. Politicians are wary of restricting the freedom of movement of tourists, and it would also require sophisticated visitor management techniques that are rare in coastal tourist destinations.

More information about carrying capacities and coastal and marine tourism can be found in Case Study 31. The reader will also find some interesting papers about the carrying capacity for marine leisure activities in the References and Further Reading section. These are works by Chen and Teng, 2016; Zhang et al., 2016; Ríos-Jara et al., 2013; Quicoy and Briones, 2009; Roman et al., 2007; Davis and Tisdell, 1995; and Pearce, 1986.

Towards a model of best practice

Having considered all the complications of planning and managing the impacts of tourism on the marine environment, it is now time to make suggestions as to what best practice might look like. That should help us judge what action needs to be taken in order to put in place a system of planning and management that reduces these negative impacts and ensures that tourism can play a useful role in the conservation of the marine environment.

Figure 10.2 endeavours to identify the pre-requisites would need to be in place for the development of an effective system of planning and management of tourism and the marine environment.

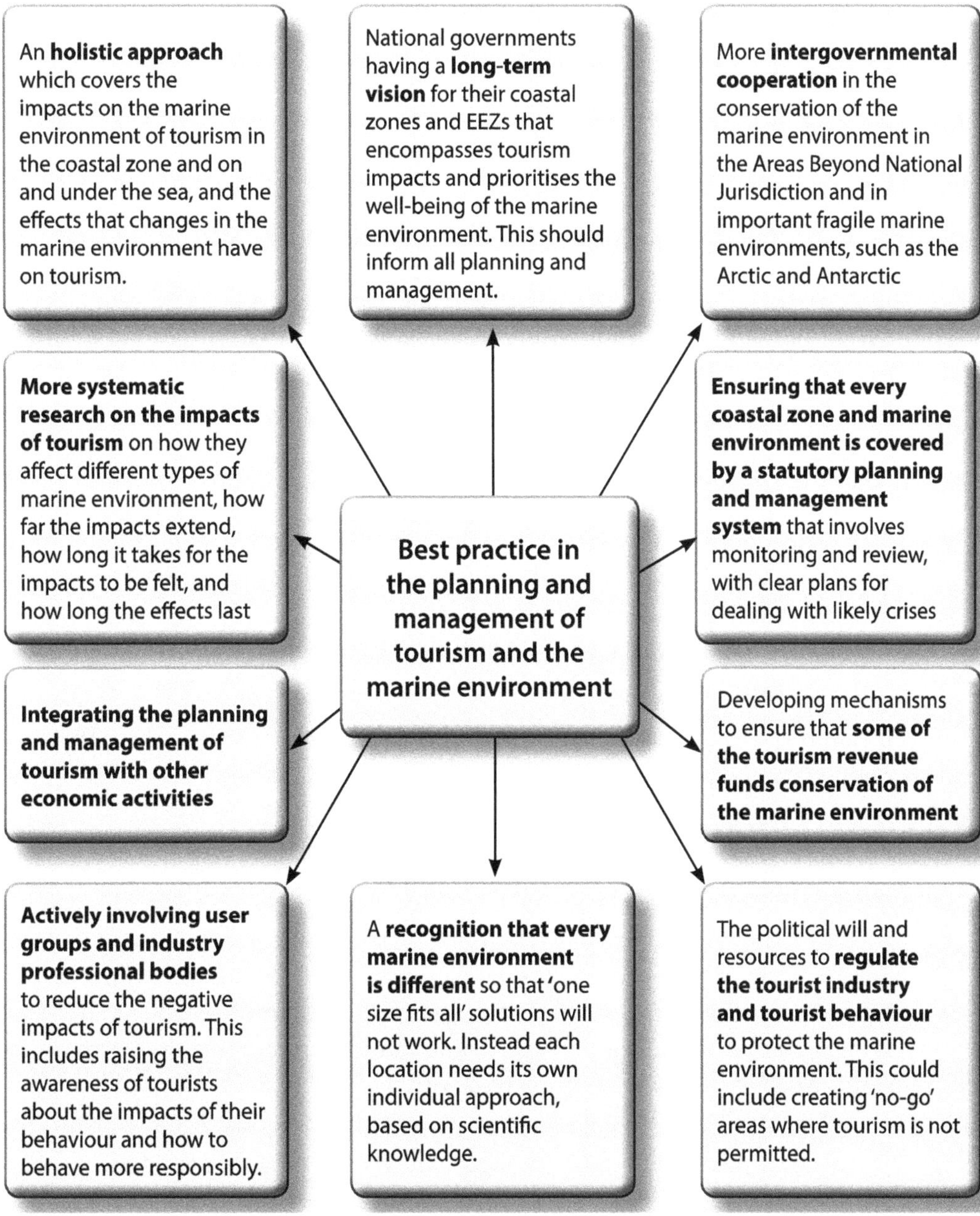

Figure 10.2: Pre-requisites for a model of best practice for the planning and managing of tourism impacts on the marine environment

Clearly, this is an idealised view that may seem unlikely to be realised any time soon. However, we need to understand the urgency for action given the crisis facing our oceans, which is partly the result of the impacts of tourism. And as we have seen any crisis in the oceans means a crisis for the world's population given how much our lives are dependent on what happens to the oceans.

I am aware that Figure 10.2 excludes the impacts of air travel and the carbon footprint of the global tourism industry as a whole on global warming and thus on the marine environment all over the world. It is crucial that this issue is tackled but I feel that that is beyond the scope of this text which focuses mainly on the impacts of tourism which take place on the shore of the sea or on and under the surface of the oceans.

The planning of tourism and its impacts on the marine environment also needs to take into account the broader issue of climate change and its future impact on coastal communities, both in terms of tourism and broader economic and social issue. It needs to focus on reducing vulnerability and increasing the resilience of these communities. There are a number of interesting studies addressing this issue in the literature including, for example:

- Moreno and Becken, 2008, which explores methodologies for assessing the vulnerability of coastal communities and tourism in relation to climate change.
- Hoogendoorn and Fitchett, 2018, which looks at adaption strategies to climate change in Africa, with a focus on the poorest residents who are likely to suffer most from the effects of climate change.
- Scott et al., 2012, which focuses on the vulnerability to climate change and sea level rise of coastal communities in the Caribbean.
- Larsen et al., 2011 about resilience building in coastal communities in Thailand.

The special case of the cruise industry

Figure 10.2 also does not explicitly talk about the cruise sector and that is because I believe that this sector needs to be dealt with separately as something of a special case. It is the only part of the tourism industry that operates across the surface of the open sea as well as in coastal waters and it has probably been the fastest growing sector of tourism in recent years. Sadly, it appears that it is perhaps the sector which is currently least engaged with the principles of responsible tourism and its environmental and social impacts, although that is slowly changing.

Given its potential to cause damage to the marine environment, I believe there is a strong case for further regulation of cruising that incorporates the following elements:

- The imposition of more ambitious targets for the cruise industry in terms of reducing its emissions and the other forms of pollution that come from the operation of cruise ships currently. It is improving its environmental performance but not quickly enough given the scale of the crisis facing our oceans.
- Permanent or temporary bans on all cruise ships within a given distance of marine environments that are particularly fragile and endangered, rather than

voluntary arrangements with the cruise operators. There are some areas of the marine world that should probably become 'no-go' areas for all ships over a certain size. This idea, while sensible, is in some ways in direct conflict with the idea of freedom of navigation, which is enshrined in the 'law of the sea'.

- Restrictions to reduce damage at ports of call caused by cruise ships berthing in places with vulnerable marine environments, of which Venice is a good example. This may involve limiting the size and/or number of vessels entering.
- Ensuring that cruise passengers are made fully aware of the potential negative impacts of cruising on marine environments and how they can reduce these impacts by modifying their behaviour on board and during shore excursions.

I realise that the cruise companies will not welcome further regulation, but I am confident that such successful and innovative companies would be able to manage the challenge of further regulation and remain profitable.

Conclusions

As we saw in Figure 10.1, the planning and management of the impacts of tourism on the marine environment is very complex due to a number of factors. It is not an activity that the public sector has a lot of experience of in comparison to its experience of planning and managing on land. We need to get better at planning new developments, managing the impacts of the day to day operations of the tourism industry, and making plans about how to manage crises and emergencies that either have their origins in the oceans but affect the land too or those which have their origins on land but cause damage to the marine environment.

Within the Economic Exclusion Zones we must ensure that conservation has greater priority rather than the focus simply being on the exploitation of the zones for economic gain. This needs to happen in parallel with increased international cooperation to help conserve the open seas. And, when we designate protected areas of marine environment, we have to make sure that there is sufficient regulation and resources in place to mean that they fulfil their objectives.

In Chapter 9 we looked at the role of the private sector and in this chapter, we have focused on the public sector. However, if we are to plan and manage tourism more effectively to protect the marine environment we will need to see public-private sector cooperation as well as the active engagement of tourists to encourage them to modify their behaviour for the benefit of the oceans which in turn will benefit the tourism industry. I have to admit that in the current climate this all seems a long way away from where we are at the moment, but as we become increasing aware of the scale of the climate emergency and the crisis facing our oceans, it is to be hoped that everyone will get the message and take action.

References

Agostini, VN, Margles, SW, Knowles, JK, Schill, SR, Bovino, RJ, and Blyther, RJ (2015) Marine zoning in St Kitts and Nevis: a design for sustainable management in the Caribbean. *Ocean and Coastal Management* **104**, 1-10.

Becken, S, and Hughey, KFD (2013) Linking tourism into emergency management structures to enhance disaster risk reduction. *Tourism Management* **36**, 77-85.

Chen, CL and Teng, N (2016) Management priorities and carrying capacity at a high-use beach from tourists' perspectives; a way towards sustainable beach tourism. *Marine Policy* **74**, 213-219.

Christie, P (2004) Marine protected areas as biological successes and social failures in southeast Asia. *American Fisheries Society Symposium* **44**, 155-164

Clifton, J and Majors, C (2012) Culture, conservation and conflict: perspectives on marine protection among the Bajau of southeast Asia. *Society and Natural Resources*. **25**(7) 716-725.

Davis, D and Tisdell, C (1995) Recreational scuba-diving and carrying capacity in marine protected areas. *Ocean and Coastal Management* **26**(1) 19-40.

DEFRA (2013) *Marine Conservation Zones Designation Explanatory Note*. Department for Environment, Food and Rural Affairs. London

www.globalresearch.ca (2019) www.globalresearch.ca/who-owns-sea/5689740 [17 Oct 2019]

Hoogendoorn, G and Fitchett, JM (2016) Tourism and climate change: a review of threats and adaption strategies for Africa. *Current Issues in Tourism* **21**(7) 742-759.

Kamil, KA, Hailu, A, Rogers, A and Pandit, R, (2017) An assessment of marine protected areas as a marine management strategy in southeast-Asia: a literature review. *Ocean and Coastal Management* **145**, 72-81.

Larsen, RK, Calgaro, E and Thomalia, F (2011) Governing resilience building in Thailand's tourism-dependent coastal communities: conceptualising stakeholder agency in social-ecological systems. *Global Environmental Change* **21**(2) 481-491.

Li, Y and Fluharty, DL (2017) Marine protected area networks in China: challenges and prospects *Marine Policy* **87**, 8-16.

Lopes, PFM, Pacheco, S, Clauzet, M, Silvano, RAM and Begossi, A (2015) Fisheries, tourism, and marine protected areas: conflicting or synergistic interactions? *Ecosystem Services* **15**, 333-340.

Moreno, A and Becken, S, (2008) A climate change vulnerability assessment methodology for coastal tourism. *Journal of Sustainable Tourism* **17**(4) 473-488.

Oracion, E, Miller, ML and Christie, P (2005) Marine protected areas for whom: fisheries, tourism and solidarity in a Philippine community. *Ocean and Coastal Management* **48**(3-6) 393-410.

Papageorgiou, M (2016) Coastal and marine tourism: a challenging factor in Marine Spatial Planning. *Ocean and Coastal Management* 129, 44-48.

Pearce, D (1986) Carrying capacities for coastal tourism. *Industry and Environment* **9**(1) 3-7.

Portman, ME (2007) Zoning design for cross-border marine protected areas: The Red Sea Marine Peace Park Study. *Ocean and Coastal Management* 50(7) 499-522.

Quicoy, AR, and Briones, ND (2009) Beach carrying capacity assessment of coastal ecotourism in Clatagan, Batangas, Philippines. *Journal of Environmental Science and Management* **12**(2) 11-27.

Ríos-Jara, E, Galván-Villa, CM, Rodriguez-Zaragoza, FA, López-Uriarte, E, and Mũnoz-Fernández, VT (2013) The tourism carrying capacity of underwater trails in Isabel Island National Park, Mexico. *Environmental Management* **52**, 335-347.

Roman, GSJ, Dearden, P and Rollins, R (2007) Application of zoning and 'limits to acceptable change' to manage snorkelling tourism. *Environmental Management* **39**, 819-830.

Scott, D, Hall, CM, and Gössling, S (2012) *Tourism and Climate Change: Impacts, Adaption and Mitigation*. Routledge. London

www.unepdhi.org, 2017 (8 Jan 2020)

Zhang, L-Y, Chung, S-S, and Qiu, J-W (2016) Ecological carrying capacity assessment of diving site: a case study of Mabul Island, Malaysia. Journal of *Environmental Management* **183**(1) 253-259.

Questions and exercises

1. Using examples from a number of countries, discuss how effective you believe the various kinds of marine protected areas have been in reducing the negative impacts of tourism on the marine environment

2. Using examples, critically evaluate the statement that, 'Strict regulation is the only effective way to protect marine environments from the negative impacts of tourism'.

3. Working in a group, select a place with which you are familiar which could be a particular beach or an area of sea. Seek to apply the idea of carrying capacity to your selected place, noting how easy or difficult it would be to measure its carrying capacity, recognising that there are different types of carrying capacity we can use.

4. Using Figure 10.1 if it helps, discuss what you consider to be the three most important obstacles to the effective planning and management of tourism and the marine environment.

5. Prepare what you believe should be the outline of the contents of an EIA for either a new marina or a large-scale hotel which is to be sited on a beach.

11 Case Studies

In this section I offer a range of short case studies to illustrate issues covered in the text and provide detailed examples of what is happening in the relationship between tourism and the marine environment around the world at the time of writing, between autumn 2019 and spring 2020. As far as possible the sequencing of the case studies below seeks to mirror the structure used in the book starting at Chapter 1.

1: The impacts of tourism on the marine environment in 1820, 1870, 1920 and 1970

This book has focused, of course, on the impacts of tourism on the marine environment at the time of writing in 2019 and 2020. However, it is interesting to look back and compare the situation today with the past to see how much things have changed over the past two hundred years. In Table 11.1, I endeavour to show how these impacts developed over time

Table 11.1: The evolution of the impacts of tourism on the marine environment over time

1820	1870	1920	1970
The consumption of marine resources by tourism and tourists			
Very low levels of tourism so little consumption of marine resources by tourists	Domestic tourism to the coast is growing so consumption is growing but at a relatively slow rate	International travel is increasing, and domestic tourism is high in developed countries. Consumption by tourists is growing but is still at a modest level	Rapid increase in international travel increases the consumption of marine resources by tourists in the main tourist destination countries
The cruise industry			
No cruise industry but sailing ships were transporting business travellers and emigrants	Cruise industry has developed but is on very small scale and just for the economic elite	Cruising is fashionable with the economic elite but is still on a modest scale	Cruising has been in decline for several decades but is about to be reinvented and grow at a dramatic rate

1820	1870	1920	1970
Coastal tourist resorts			
Few established resorts and most have little purpose-built infrastructure	Growing number but with limited purpose-built infrastructure	More resorts exist and some have sophisticated purpose-built infrastructure. Visitor numbers rising	A large number of destinations are emerging, some of which did not exist before as settlements. Visitor numbers growing rapidly
Marine leisure activities			
None apart from bathing and swimming	A few rich people go boating	The economic elite go sport fishing and snorkelling	Surfing and all water-based leisure activities grow as the number of people vacationing at the coast rises rapidly
Marine wildlife-watching			
No organised marine wildlife-watching; marine wildlife is studied by travellers and specimens are collected	No organised marine wildlife-watching but individual travellers study marine wildlife and collect specimens	Land-based wildlife watching is popular with the economic elite although they often prefer shooting it! Sport or big game fishing likewise popular.	Tourists are beginning to show an interest in marine wildlife. The first whale-watching trips are launched in the 1950's
Geographical spread of the impacts			
Very limited to a few localities	Still very limited in geographical extent	The impacts spread to the Mediterranean and the coasts of Florida, as well as to the destinations favoured by the economic elite in Asia, Africa, and South America	The effects of tourism reach most inhabited parts of the world. The scale of the impacts is becoming substantial in more developed countries which are popular with both domestic and international tourists
Carbon emissions from tourism and their impact on global warming			
None – there are no aircraft, cars or trains and ships are powered by sail	Very low, but there are now cruise ships and trains powered by fossil fuels	Starting to grow significantly after the invention of cars and aircraft, and through the growth in tourism	Greatly increased due to the rapid rise in air travel and leisure use of private cars

Obviously, this is a very simplistic picture, but hopefully it makes the point that between 1820 and 1920 the impacts of tourism environment were low and increasing relatively slowly. The impacts then grew at a significantly faster pace between 1920 and 1970. However, I want to make the point that from 1970 to 2020 the impacts have grown at an incredibly rapid rate, not least because the total number of international tourists grew from around 166 million in 1970 to some 1.4 billion by 2018!

We also need to recognise that the increases in impacts have also been a direct result of the rapid growth of the cruise sector, the popularity of marine wildlife-watching, and the invention of new marine leisure activities. And, compared to 1970, the impacts of tourism on the marine environment are now being felt literally everywhere in the world, virtually from the North Pole to the South Pole and all points in between!

It is worth saying that whereas in the past the impacts were largely based on ignorance, we no longer have that excuse; we know much more about how tourism can affect the marine environment so now we have to act on that knowledge.

2: What future for the Maldives?

The Maldives is the lowest lying country in the world with an average height above sea level between 1.5 and 2.4 metres, depending on which sources are used. It is an archipelago of some 1,190 islands, of which only a minority are inhabited. Some 87 islands have been set aside as resorts for tourists. Indeed, tourism represents 66.4% of the total GDP of the Maldives (www.knoema.com, 2019), meaning the country depends heavily on tourism for its economic survival. Economically, at least, tourism in the Maldives has been a success story. It began in 1972 with just 60 beds available for tourists, but by the first quarter of 2019 there were no fewer than 47,000 beds. Over the first three months of 2019, tourism generated tax revenues for the Maldives government of no less than £102 million (www.tourism.gov.mv, 2019). The Maldives was attracting 1.4 million tourists per annum by 2018 and has set an ambitious target of 2.3 million arrivals by 2023 (www.maldives.net.mv, 2019).

However, the very future of the Maldives and its people are under threat from rising sea levels due to global warming. In 2008-2009 articles were written saying the Maldives could disappear within ten years, and its President was talking about buying land in another country and moving the whole population there and abandoning the islands to their fate. Many readers will remember the government of the Maldives holding a cabinet meeting under water in 2009 to draw attention to the threat the country faced due to rising sea levels. This led to a spate of 'see it while you can' comments in guidebooks and travel features based on the idea that the Maldives would not be around for much longer.

In recent years, attitudes on the Maldives have changed and the government now seems committed to trying to save the islands from rising sea levels, using geoengineering projects. *New Scientist* reported in 2017 that some islands were being 'rented out' and the proceeds used to reclaim, fortify and even build new islands. A key part of the strategy is the construction of new islands which are higher above sea level than the natural islands (www.newscientist.com, 2017).

Of course, as we saw in the main text, the construction of artificial islands can cause significant damage to the marine environment and this could in turn harm the tourism industry in the Maldives. The government's intention to continue to grow tourism could also pose a threat to the natural environment. However, faced with the possibility of the Maldives disappearing under the waves, the government's desire to grow tourism is perhaps understandable, not least because its revenues can help fund some of the engineering projects mentioned above.

So, it remains to be seen if the Maldives can survive the impacts of global warming, but there are still plenty of commentators talking about the islands disappearing within 30 to 100 years. Ironically, by encouraging tourists to take long haul flights to the Maldives the government is actually contributing to global warming but in the circumstances, it is hard to blame them for they seem to have few alternatives if they are to save their country from destruction.

3: The plight of the coral reefs

Coral reefs are living organisms which provide a vitally important habitat within the marine ecosystem as well as protecting coastlines from strong currents and powerful waves. At the same time, they are a major attractor for tourists because of the opportunity to dive on them to see the marine wildlife that lives on and around them. However, our coral reefs are under threat globally. A random online search in December 2019 produced a number of articles with frightening headlines that suggested that:

- 50% of the coral reefs that existed in the mid-1980s had been lost by 2015.
- More than 90% of all remaining coral reefs could be lost by 2050.
- 80% cent of coral reefs in the Caribbean had been lost 'in recent years'.
- Coral reef loss is occurring at about twice the rate of the destruction of rainforest

While some of these figures may be exaggerations no one disputes that coral reefs are being destroyed and damaged at an alarming rate. The US National Ocean Service has produced an infographic that links global warming to the threat to coral reefs. It argues that *'climate change dramatically affects coral reef ecosystems'* citing the following:

- *a warming ocean*, [which is a cause of coral bleaching]
- *sea level rise*
- *changes in storm patterns*
- *altered ocean currents*
- *ocean acidification*

(www.oceanservice.noaa.org, 2019)

11

All of these effects of global warming are harmful to coral reefs but in 2019 the International Coral Reefs Initiative (ICRI) identified a number of other ways in which humans were damaging the coral reefs including:

- *overfishing*
- *destructive fishing methods*
- *coastal development*
- *pollution*
- *the global aquarium trade.*

(www.icriforum.org, 2019)

They also identified unsustainable tourism as a major threat to the coral reefs. We have already noted in the book the various ways in which tourism contributes towards the destruction of coral reefs including: irresponsible diving and snorkelling; boating and anchor damage; seafood consumption and recreational fishing; sightseeing boats and cruise ships, souvenirs for tourists; the introduction of invasive species, and: human interactions with wildlife including fish feeding. Yet the tourism industry has a vested interest in conserving coral reefs because by attracting tourists they have a monetary value for tourist destinations. While it is difficult to quantify this value, a research team led by Mark Spalding in 2017 suggested a figure of US $36 billion per annum globally. The research also concluded that 70 countries had reefs that generated at least US $ 1 million per square kilometre in tourist revenue each year although they noted that only 30% of coral reefs across the world are currently being exploited because the rest are too difficult to reach (www.zmescience.com, 2017).

Other later case studies will look at voluntary codes of conduct for diving on coral reefs and the threats facing perhaps the world's most famous coral reef, the Great Barrier Reef in Australia.

4: The value of mangrove forests

Mangrove forests are a crucially important part of the marine environment, but they are also highly valuable to humanity in a number of ways. They protect shorelines from the effects of storms, provide timber and fish for many coastal communities and filter water running off the land to reduce the pollution of inshore waters. Yet, like coral reefs, they are under threat across the planet. It has been estimated that in the past two decades the world has lost some 35% of its mangroves (www.tourismimpacts.tumblr.com, 2019). According to the American Museum of Natural History: '*by some estimates less than 50% of the world's mangrove forests were intact at the end of the 20th century, and half of those that remain are in poor condition … Thailand has lost 84% of its mangroves … while the Ivory Coast, Guinea*

Bissau, Tanzania, Mexico, Panama, Malaysia, Myanmar, Pakistan and the Philippines have each lost more than 60% of their mangrove forests (www.amnh.org, 2019).

The destruction of mangrove forests has a number of causes. Aquaculture, particularly the raising of shrimps, involves clearing mangroves and replacing them with ponds, and also brings pollution of the marine environment due to the waste from aquaculture units. Mangroves have been cleared by agriculture to create rice paddies and palm oil plantations, while agricultural run-off polluted by fertiliser chemicals has done further damage to mangrove forests.

Tourism has played a major role in the destruction of mangrove forests in two main ways. First, the development of coastal resorts and infrastructure has led to the clearing of many mangrove forests. Even where mangroves are protected, some developers simply destroy them and pay a fine if they have to, because the fine is likely to be much less than the profits they will make. This demonstrates the weakness of many regulatory systems, where the penalties do not deter irresponsible development. Second, tourists visiting mangroves on boats cause damage from noise, engine fumes, fuel spillages, waste and light pollution at night.

Yet, as with the coral reefs, mangroves have an economic value due to tourism such that the industry has a vested interest in their protection. It is estimated that 'mangrove tourism' involves some 4,000 'attractions' in 93 countries and that in Florida alone recreational fishing in mangroves is worth some $1 billion per annum to the local economy (www.oceanwealth.org, 2019).

Research conducted by the World Conservation Union or IUCN suggests that in the tsunami of 2004, mangrove forests saved lives in some places by absorbing the energy of the storm (www.bbc.co.uk, 2005). In recent years, the message that mangroves can help mitigate the effects of tsunamis and extreme weather events in coastal areas has encouraged governments to take measures to protect mangroves in countries such as Sri Lanka, Indonesia, Madagascar and the Philippines.

However, every day more and more mangroves are being lost with the construction of resort complexes and tourism infrastructure.

5: The operation of cruise ships in Alaska

The wilderness of Alaska in the USA has become a major destination for the cruise industry with an estimated 1.30 million tourists expected to visit on some 567 separate cruises in 2019, compared to 1.09 million two years earlier (www.seacc.org, 2019). The Cruise Lines Industry Association estimates that cruise passengers spend brought an estimated $2.2 billion dollars to the economy of Alaska in 2019 (www.cliaalaska.org, 2019). However, the South East Alaska Conservation Council (SEACC) published data in 2019 that highlighted the risk which cruise ships may pose to Alaska through the dumping of wastewater. They claimed that up to 84,000 gallons of 'black water – treated sewage and medical effluent – could be potentially dumped per ship per week. The figure for 'grey water'– from showers, kitchens, laundries – was said to be up to 1,260,000 gallons per ship per week. Data was not available for 'Scrubber water' which includes sulphur dioxide emissions and exhaust gas cleaners. SEEAC claimed this water disposal could contaminate fisheries, cause the growth of algae and harm aquatic life (www.seeac.org, 2019). The State of Alaska Department of Environmental Conservation announced a new research project on the impacts of cruise ships on air quality in Juneau because of their concerns, in 2019 (www.dec.alaska.gov, 2019).

There are regulations in Alaska regarding the dumping of treated sewage and 'grey water' but they fall short of an outright ban. Ships can legally dump when a certain distance from the coastline or if they are travelling at more than six knots. Cruise ships are also required to comply with air quality regulations when in Alaska. However, it is clear that not all cruise lines are obeying the regulations, for in September 2019 it was reported that the Holland America Line had been fined $17,000 after one of its ships discharged 22,500 gallons of 'grey water' illegally in the Glacier Bay National Park (www.cruiselawnews, 2019). The ship concerned has capacity for nearly 2,000 passengers so the fine represented under $9 per passenger on a single voyage.

The cruise sector contributes greatly to the economy of Alaska, but at a time when the phrase 'over-tourism' is being widely used, some are asking if some ports of call in the state of Alaska are beginning to reach their capacities. Local media in the port of Juneau was certainly discussing such matters in 2018-19. In July 2019 alone, no fewer than 121 cruise ships were scheduled to visit Juneau for stays of between six and a quarter and fourteen hours. On some days no fewer than six cruise ships were due to be visiting the port (www.crew-center.com, 2019). There are also concerns about the impact of cruise ship tourism on local communities and their cultures in Alaska, as well as ethical questions about whether it is right that ships with capacity for more than 5,000 passengers should be sailing through this wild area of ocean.

6: Shark cage tourism in South Africa

Over the past 15 years or so South Africa has become the market leader in a new form of wildlife-watching, where tourists in diving gear observe sharks up close from within a cage. The coastal town of Gansbaai has become the 'capital of shark cage tourism' in South Africa although shark cage experiences are also available to tourists in Australia, New Zealand, Mexico and the USA. Shark cage tourism is an interesting mix of wildlife watching and adventure activity as the sharks get very close to the tourists and have on occasions 'attacked' the cages. It is far more interactive than passive whale-watching from a boat but less interactive than swimming with dolphins, for obvious reasons!

A leading operator, White Shark Diving Company, was offering a range of one day shark cage experiences in 2019-2020 at prices ranging from 2,050 Rand to 3,300 Rand or between around £105 and £175 (www.sharkcagediving.co.za, 2019). A year earlier, an article in *The Independent* newspaper in the UK had published an article which looked at the cases for and against shark cage diving, based on an experience of the product offered by Marine Dynamics (www.independent.co.uk, 2018). The argument in favour of it was that some of the revenue generated by shark cage operators was being used to support conservation projects, such as the Dyer Island Conservation Project. The operators also work with scientists to help with research about sharks. Most of the main operators certainly use their websites to communicate messages about the number of sharks being killed every year and the need to conserve them.

The case against shark cage tourism appears to consist of two inter-related arguments. The first is that shark cage operators are changing the behaviour of sharks because of the fact that they use decoys, hooks baited with dead fish and a fishy concoction called 'chum' to attract the sharks to the cages. Several researchers have suggested that this practice is affecting shark behaviour.

Some critics of shark cage tourism claim it makes shark attacks on humans more likely, although at the time of writing there does not appear to be any firm evidence to support this claim. Operators seem conscious of this fear that shark cages might increase shark attacks. The White Shark Diving Company had an infographic on its website in December 2019 outlining what was being done to protect humans from shark attacks under a banner headline that read '273 million sharks are killed every year - help save our sharks' (www.sharkcagediving.co.za, 2019).

At a time when using animals to entertain tourists on land is becoming less and less acceptable, the growth of shark cage tourism is an interesting phenomenon. However, with shark attacks on humans apparently increasing it seems that this type of tourist experience is likely to remain controversial for some time to come.

7: Whale-watching around the world – the good, the bad and the ugly

Few would argue with the idea that it is better to watch whales than hunt them and the evidence seems to indicate that a live whale that can be watched has a higher financial value than a whale that is caught for food. From the early days of whale-watching it has been recognised that irresponsible whale-watching could also be harmful to whales, though, in terms of noise and disturbance that can affect feeding and breeding behaviour as well as injury from collisions.

A survey conducted in 2012 by Dr Carole Carson, sponsored by the International Whaling Commission, noted that some 50 countries had already developed regulations and/or guidelines for whale-watching. The majority had voluntary guidelines but there were many examples of statutory regulation, including the licensing of whale watching operators, although often the licensing process seems to be largely an administrative procedure. The International Whaling Commission has its own *General Principles for Whale Watching,* which date back to 1996 (www.iwc.int, 2020). The World Cetacean Alliance has also produced guidance on best practice in whale watching (www.worldcetaceanalliance.org, 2018).

Parsons noted in 2012 that *'the introduction of guidelines and regulations for whale watching has been the most common method of trying to mitigate the impacts of boat-based whale watching. However, there is great variety in the comprehensiveness of guidelines, and even if operators have guidelines, compliance with them can be poor'* (Parsons, 2012).

In the USA there is a mixture of regulation and guidelines in relation to whale-watching which cover watching from boats, from shore and from the air. In terms of the distance a boat should be from the whales, the emphasis appears to be on voluntary guidelines rather than legally binding rules. There are guidelines for viewing different species of cetaceans such as Humpback Whales and Killer Whales (www.noaa.gov, 2020).

In 2018 Rebecca Walker studied marine mammal watching legislation and codes of conduct in South Africa and Canada. Her conclusion was that *'regulations are essential. Regulations are clear and unambiguous … and make codes of conduct enforceable and standardised. I also came to the conclusion that …the issuing of licenses works. It can provide a control on the number of vessels allowed in an area and/or the number of hours spent with animals'* (Walker, 2019).

However, it is clear that despite all the regulations and codes of conduct irresponsible whale watching is still common. A random search of TripAdvisor found the following comments about Mirissa in Sri Lanka – a country that does have regulations for whale watching – from 2016:

'Sad to see that most of the whale watching companies completely ignored the simple rules in place to protect the beautiful endangered Blue Whales that we saw today. A group of 6 large boats …chased a mother and calf at full speed to try to get as close as possible. Our experience was marred by some…irresponsible …tourists who had chartered a local fishing boat to come whale watching. The fishermen had absolutely no respect for the animal, charging up and down and scaring the whale into taking short surface breaks.'

(www.tripadvisor.co.uk, 2020)

Whale watching has now been taken a step further to encompass swimming with whales. For example, a San Francisco based operator, *Humpback Swims,* offers the opportunity to *swim up close and personal with humpback whales,* in Tonga. It acknowledges that *swimming with whales is illegal in the United States* (www.humpbackswims.com, 2020). The tourism authorities in Tonga promote this form of tourism and has a licensing system for operators. Given that swimming with whales brings no benefit to the whale and may well cause them harm in ways we do not yet even understand, this feels like exploitation of marine creatures for the enjoyment of privileged tourists. Interestingly, while the questions on the websites of operators frequently ask about the danger which whales pose to swimmers, there are rarely any questions about the impact swimmers might have on the whales! Yet again, I am left feeling that the future of marine wildlife will not be a bright one if it relies on industry self-regulation, voluntary codes of conduct and weak or poorly enforced government regulation.

8: Marine wildlife watching – personal experiences of the author

While writing this book I have been reflecting on some of my own wildlife watching experiences over the years. The opportunity to watch marine wildlife has never determined my choice of vacation destination and, with a few exceptions, all of my marine wildlife watching experiences have been largely coincidental and unplanned. The exceptions were largely when my son was young, and I thought it would be interesting for him to see marine wildlife in its natural habitat. I took him on a whale-watching trip in 2004 when he was 13 years old from the former whaling port of Husavik in northern Iceland. We travelled in an old wooden fishing trawler and the sea was rough; it was summer, but we still saw hail and snow as well as sun and rain on the trip. We saw a single humpback whale which was amazing, but the boat crew were very responsible and did not try to get too close. Ironically, most of our fellow passengers never saw the whale as they were too busy being seasick! On another occasion, we visited a sea turtle beach conservation project in Sri Lanka which was widely recognised for the good work it was doing. Visitors were kept a safe distance from the mother as she made her nest and laid her eggs, and no torches or camera lights were permitted, and the tourists had to be silent. I have to say I found it a very emotional experience but an uncomfortable one because I felt none of us should really have been there; it felt very voyeuristic.

In La Jolla Cove, near San Diego in California, sea lions and pelicans were part of daily life and we came across them by accident while out for an evening stroll. They seemed undisturbed by the presence of crowds watching them although they did look less happy when a few visitors got too close. In 2019, I had the chance to see a colony of sea lions on the beautiful coast of Kangaroo Island in South Australia. People were kept at a safe distance from them by the guides and there was an excellent visitor centre with information about them. In the same year, we saw an incredible scene when we stopped for lunch in Hamelin Bay in Western Australia. A small group of stingrays were swimming amongst the legs of swimmers, including children and they were being given food purchased from a nearby seller. This sems to have been a well- established phenomenon and no one seemed worried that their children were sometimes touching these potentially dangerous creatures.

Two of my most pleasurable marine wildlife encounters I can recall both happened purely by accident. In the Isles of Scilly off the coast of Cornwall, while taking a ferry between the islands we came across a large colony of seals basking on the rocks in the sun. And while taking a ferry from the Barbican to Royal William Yard in Plymouth our boat was accompanied by a dolphin swimming alongside.

I am afraid I am a bit of a dogmatist when it comes to marine wildlife-watching. I do not agree with people swimming with marine wildlife; as a sign I saw in Australia said, 'keep the wildlife wild'. Nor do I approve of people 'chasing' whales and dolphins in RIB boats or the use of artificial feeding to attract marine creatures for the enjoyment of tourists. I do not like to see any creature in captivity and would never go to see a whale or dolphin show. However, I need to recognise that I have the good fortune to be able to go and see wildlife in their natural habitat if I so choose. Many people do not have that opportunity, so maybe the aquaria found around the world do a valuable job raising awareness of conservation amongst those who may never otherwise see marine wildlife that is under threat such as sharks, rays, sea turtles, and some species of penguins, for example.

9: Sea turtle breeding beaches in Central America

Given our fascination with marine wildlife there is no doubt that the presence of sea turtles is a major attraction for many coastal destinations around the world. However, tourists are not just interested in seeing sea turtles in the water, they want to see the females come ashore to lay their eggs and be there when the tiny hatchlings begin their perilous journey to the ocean for the first time. Tour operators and destination marketers ensure that tourists are made aware of when this happens so they can visit at the right time of the year to see these spectacles.

For the countries of Central America, the sea turtle has become a vitally important part of their tourism offer, and the travel media has done much to promote 'sea turtle tourism' in the region. In 2018, for example, the online travel site, Culture Trip listed the best places in Central America to watch sea turtles hatch.

It said *'head down* [to Ambergris Caye in Belize] *between June and October to see the babies hatching and scuba dive with the adults'* (www.theculture.trip.com, 2018). The travel media has also promoted sites for sea turtle nesting and hatching in Panama, Guatemala, Costa Rica, and Nicaragua.

The tourism industry has sought to suggest that 'ecotourism' and responsible sea turtle watching can help protect sea turtles, not least by providing funding to support sanctuaries and conservation projects. However, it is not always clear how effective many sanctuaries and projects are, or even if their main motivation is conservation rather than income generation. This point is made by the tour operator Kuoni on its website, where it says about some so-called sanctuaries in Sri Lanka, that *'these places may be operating with good intentions, but their practice of retaining thousands of cute baby hatchling turtles in tanks to show paying tourists seriously compromises the life and future of the hatchlings. Turtle hatchlings must make their ways to the sea immediately after hatching....and the longer they are kept away from the sea in their first few days, the lower their chances of survival'* (www.kuoni.co.uk, 2019). However, there are a number of successful sea turtle conservation projects in Central America particularly in Costa Rica.

There is no doubt that responsible tourism can help with the plight of the sea turtles in ways which go beyond simply providing money to fund projects. In the Azuero region of Panama, tourists can volunteer to help with the nightly patrols to protect the sea turtles on the beaches and with scientific research to help conservationists improve their knowledge of sea turtles (www.realdealtours.com, 2015). On the other hand, some local entrepreneurs are still using turtle shell to make souvenirs which some tourists are still buying, sadly. However, it is unfair to blame tourism for all of the problems facing turtle populations; in some coastal communities in the region, the eggs and meat of the turtle are seen as a source of food, and irresponsible commercial fishing still injures and kills many sea turtles.

For tourists who want to watch sea turtles responsibly there is plenty of advice online such as at www.seeturtles.org. Despite this there is still too much irresponsible watching in Central America and the world as a whole ,with the egg laying and hatching being disturbed by tourists getting too close, or strong lights to help the tourists see what is going on and get good photographs.

10: Stingray interaction experiences – Hamelin Bay, Australia and Stingray City, Cayman Islands

The Margaret River area of Western Australia has become a fashionable place to visit for domestic and international tourists over the past decade or so because of its vineyards and coastal scenery. However, another motivation for visitors is the opportunity to see and interact with stingrays from the shore at Hamelin Bay. This is heavily promoted by the local tourism industry and has been communicated to a wider audience by a number of travel writers and television programmes. The stingrays come close to the beach during the Australian summer and tourists can stand in the water as the stingrays pass over their feet and swim amongst them. They gather around the jetty where fishermen dump offal when bringing their boats ashore after recreational fishing trips. Some tourists feed them bits of fish and when I was there in February 2019, I was surprised to see children feeding them. In some ways it is nice to see human beings enjoying being in the presence of these potentially lethal creatures after the negative publicity surrounding the death of the Australian naturalist and TV presenter, Steve Irwin, who was killed by a stingray in Queensland in 2006.

The local media in 2011 reported that fishermen with spear guns had killed the oldest resident stingray, 'Stumpy' in full view of tourists and their children. While the fishermen were not doing anything illegal as this was not against the law, there was an outcry, and in 2012 regulations were introduced to protect the stingrays at Hamelin Bay. One cannot avoid thinking that these regulations were perhaps introduced as much to protect the local tourism industry as the stingrays. One often reads promotional material for Hamelin Bay describing the stingrays as 'gentle' and 'friendly' and suggesting they 'enjoy' interacting with humans. The latter may be true but may reflect the fact that they associate people with being fed. Researchers have suggested that currently the behaviour of the stingrays is being affected by their interactions with humans and the practice of people feeding them, but that the stingrays do not yet seem to be dependent on humans for their food.

While Hamelin Bay is an informal opportunity for tourists to interact with stingrays, Stingray City, in the Cayman Islands, is in effect a visitor attraction which focuses on organised interactions with stingrays, which are fed to ensure they are present, and where tourists are able to swim and dive amongst stingrays and other marine creatures. In 2003 it was estimated that Stingray City was attracting 100,000 visitors (www.onlinelibrary.wiley.com, 2019). A number of criticisms have been made about Stingray City in terms of them feeding stingray all day when they would normally feed only at night, sometimes feeding them food that would not normally be part of their diet, and making it too easy for them to find food, thus weakening their hunting instinct.

This case raises again the ethical issue of human interaction with wild creatures and the impact this may have on them and their behaviour in the longer term.

11: Marine wildlife in captivity in theme parks, aquaria, and zoos

In Chapter 4 we explored the issue of captive marine wildlife, and particularly killer whales, performing for visitors, focusing on two theme parks in the USA and Spain. We saw how this has become less popular and more controversial as many people have become uncomfortable at the sight of whales, in particular, performing tricks to amuse tourists. As a result, Sea World in Florida has said it will modify its whale 'shows' from 2020 to have a greater focus on conservation. Lora Parque in Spain was still advertising on its website in January 2020 a total of twelve 'shows' featuring whales (www.loraparquetenerife.com, 2020). Some tour operators are now bowing to pressure from consumers and conservation groups and ceasing to offer trips to attractions with performing whales.

However, there is also the question about whether or not marine wildlife should be kept captive in theme parks, aquaria or zoos at all, even if they are not being made to put on 'shows' for visitors. This is clearly part of the wider debate about keeping wildlife of any kind in captivity. Those who support keeping marine wildlife in captivity stress the role of zoos, aquaria and some theme parks in conservation work while opponents claim that the main motive of such operations is to generate income through attracting and entertaining visitors. They also claim that the wildlife suffers from being kept in captivity. According to a UK pressure group there were some 60 captive killer whales or orcas in zoos and theme parks around the world in August 2019, of which half were captured from the wild and half were born in captivity (www.uk.whales.org, 2019) These are to be found in the USA, Spain, Argentina, China, France, Canada, Japan and Russia.

Interestingly, in many aquaria around the world visitors can swim with the captive wildlife. For example, the Sea Life Centre in Skegness, in the UK, in January 2020, was promoting 'diving with sharks' experiences on its website for between £85 and £130, which gave visitors thirty minutes on the reef amongst the sharks. At the same time the Blackpool Sea Life Centre was offering the chance to snorkel with sharks and sea turtles for £50. Many zoos and aquaria also offer marine wildlife feeding experiences that include interactions with sea turtles, sharks, and penguins. It seems that even when marine wildlife is not putting on actual shows for visitors they still have to be used to entertain and satisfy the desires of visitors.

But it is dolphins that seem to be the marine creature that is most exploited for the entertainment of humans. In specially developed 'dolphinariums', dolphins are still putting on regular shows for visitors, being swum with and interacted with while in captivity. While some claim dolphins like this human interaction, the fact remains that unlike in the wild, the dolphins do not have the option of swimming away and avoiding interactions if they so wish. These 'dolphinariums' exist in dozens of countries including Australia, Bahrain, Belgium, Canada, China, Germany, Indonesia, Japan, Mexico, Russia, Singapore and the USA.

While many zoos and aquaria are doing valuable conservation work, it is hard to see many of them as anything other than places of entertainment for visitors where marine creatures are kept in artificial surroundings and are fed rather than hunting for their own food.

12: Seal-watching in Cornwall and the Isles of Scilly

Cornwall and the Isles of Scilly is home to 40% of the world population of grey seals and the largest breeding colony of them is found in Cornwall and the Isles of Scilly, with up to 500 grey seals living there all year round. In summer they tend to be found in the open sea but in autumn they come ashore to give birth and spend much of the winter looking after their young on beaches and in sheltered coves. They have become a popular attraction for tourists all year round and tourists can see them in a number of ways. In the tourist season in the summer there are marine wildlife 'safari cruises' from ports such as Penzance, St Ives, Padstow and Falmouth which take people on trips of between one and three hours duration to see marine wildlife that may include seals as well as sea birds, basking sharks, dolphins, sunfish and even whales. A two-hour trip cost around £40 in 2019. On the Isles of Scilly, the boatmen organise trips to the seal colony on the Eastern Isles. In recent years, commercial operators on St Martins, one of the Scilly Isles, have advertised the opportunity to 'swim with seals' for around £50; this has been featured in national newspapers, which has increased interest from tourists

Tourists can also see seals when they are on land particularly during autumn and winter. There is a large colony at Mutton Cove near Godrevy Point. In December 2019, I counted some 180 seals on the beach there at one time. Fortunately, this site is owned by a non-governmental conservation body, the National Trust, and tourists can only watch the seals from the clifftops where they are asked to be quiet to avoid disturbing the seals. Elsewhere people can get close to the seals and that is when most problems occur. Every year seal conservation groups highlight examples of inappropriate behaviour including:

- Tourists getting too close to seals so they can photograph them and in doing so disturbing and frightening them.
- People feeding seals which can harm them as well as making them dependent on humans rather than their own hunting skills.
- Tourists touching baby seals which can lead to them being rejected by their mothers.

- Visitors thinking the seal is ill and would be better off in the water encouraging or pressurising it to go into the sea which stresses the seal.
- People doing activities such as jet skiing too close to seals causing them stress.
- There are also cases where dogs attack and bite seals when owners let them off their leads.

Some say the tourists do this out of ignorance, yet many beaches have large and clear signs telling people how to behave around seals. The sad fact is that some tourists just seem interested in doing what they want regardless of its impact on the seals. Seal conservation groups try to 'educate' visitors about how to watch seals responsibly. They also tell them that if they are concerned that a seal may be injured or ill to report it to the various organisations that exist that have the expertise to rescue and rehabilitate injured or sick seals. They tell them not to approach the seal as that can make matters worse.

13: Coral reef diving voluntary codes of conduct

As we saw in Chapter 5, irresponsible diving can cause great damage to coral reefs which are already facing other threats. The diving community and the tourism industry is well aware of the impact which tourism can have on this vitally important element of the marine environment. As a result, there are a number of schemes underway based on voluntary self-regulation by dive tour operators using codes of conduct.

In Indonesia, an upmarket refurbished Buginese schooner offers trips involving diving. The company has a three-page long code of conduct, which it describes as rules for diving, of which two pages relate to safety, but the first page is entitled 'Conservation awareness'. It contains points about not removing any living or non-living natural resource from the sea, keeping away from any contact with the coral, not dumping rubbish and forbidding fishing during diving trips (www.thesevenseas.net, 2017).

In North Queensland, Australia, CHARRAO, the Cod Hole and Ribbon Reef Operators Association, has a Code of Practice *'with the aim of ensuring long-term sustainable usage of the Great Barrier Reef and Coral Sea reefs, preservation of the reef and safe vessel and diving procedures'* (www.charroa.org.au, 2019). The code contains 22 points, of which seven relate to the conservation of the marine environment. One of these points limits fish feeding to two kilograms on any dive trip, which shows that even responsible operators artificially 'bait the water' to attract marine wildlife for the enjoyment of divers.

In Bali, Indonesia, there is a code of conduct for diving amongst two specific species, the oceanic sunfish, or mola mola, and the manta ray. This code was developed jointly by the government, the industry and donor countries. The code contains over 30 points for divers and a further 13 for dive operators (ww.bali.com, 2019).

There are also global initiatives to make diving around coral reefs more responsible. In November 2019 *The Guardian* newspaper in the UK reported that a new partnership had been created involving PADI, the world's largest dive training organisation, and the UK-based but UN supported Reef-World Foundation. The partnership aims to '*encourage the industry to sign up to the Green Fins scheme, which helps dive centres to reduce their environmental impact and mitigate the damage the burgeoning sport causes to coral reefs worldwide*' (www.theguardian.com, 2018). Green Fins is a well-established organisation with its own internationally recognised standards based on a 15-point code of conduct. Its network includes more than 550 dive centres in 56 of what it describes as 'diving hotspots' across the world (www.greenfins.net, 2019).

The tourism industry trade press has had stories in recent months about the need to make diving more sustainable and help tourists to find the most responsible dive operators. An article in *Travel Trade Gazette* in October 2019 encouraged travel agents to support dive centres within the Green Fins network (www.ttgmedia.com, 2019). In the following month *Focus on Travel News* had an article entitled, '*How tourists and operators can protect Egypt's coral reefs*' which again promoted the Green Fins Network (www.ftnnews.com, 2019). The Green Fins scheme seems to be the nearest we have to a global code of conduct for diving, but it has only 550 dive centres in its network while PADI alone has 6,600 members, and many dive centres are not even members of PADI. There is clearly a long way to go and given the continuing damage to reefs, one is bound to ask whether regulation might be needed rather than relying on voluntary codes of conduct alone.

14: Jet skis and the marine environment

Jet skis are perhaps the most popular type of personalised water craft. Millions of people worldwide use them every year to give them an adrenalin rush, as they can travel at up to 100 kilometres an hour or more.

Invented in the 1970s they have been controversial from the beginning and have been banned in some locations, although often only temporarily. Bans have tended to be imposed because of the impact of irresponsible jet-skiing on those using the ocean for other leisure activities rather than because of their potential impact on the marine environment.

However, their impacts on marine wildlife have been recognised for some time. In 2014, an online article by a group representing jet ski users in the UK quoted several pieces of research which showed that jet skis:

- Used two stroke engines that poured out gallons of unburned fuel annually.
- Helped create photo toxicity in the sea which affected mussels, sea urchins and certain species of fish.
- Caused noise pollution below the surface of the water that disturbed marine wildlife and disrupted communication for whales and dolphins that use sound waves to communicate.
- Injured and killed marine creatures when they struck them.

(www.solentskiers.org.uk, 2014)

Bodies that represent jet skiers and some operators have their own codes of conduct or rules for the use of jet skis. However, most of them say little or nothing about how to minimise the impact of jet skis on the marine environment. Instead they focus on the safety of the jet skiers themselves and not disturbing other water users.

A random online search about jet skiing in Thailand in January 2020 produced some alarming results. There were reports of a 'shoot out' between rival operators on a beach in 2018 and a dangerous incident in 2019 involving irresponsible Chinese tourists on jet skis. On TripAdvisor one reviewer said it was great to be able to stop for a beer midway through a jet ski trip in Koh Samui!

The regulation of jet skiing varies enormously around the world. Many countries do not allow jet skis to be used within a certain distance of the shore, usually between 200 and 500 metres, although it is not clear how this is enforced. I surveyed the regulations in some 19 countries around the world, none of which had regulations specifically addressing the impact of jet skiing on the marine environment, although two countries specified that jet skis could not be used in marine protected areas. However:

- In Croatia there is no speed limit for jet skis more than 300 metres from shore.
- In Egypt there is no legislation in place for the use of jet skis and there is no license required to drive them unlike most other countries.
- In the US Virgin Islands anyone over 14 years of age can operate a jet ski if they have taken a safe boating course.
- On the other hand, jet skis are not allowed at all in Anguilla and the British Virgin Islands.

(www.jetskiandpowerboattraining.co.uk, 2019)

One is left with the impression that regulators have failed to catch up with the growth of jet skiing and its impact on the marine environment in many parts of the world.

15: 'Surfers against sewage' in the UK

'Surfers Against Sewage' (SAS) is a voluntary organisation in the UK which was set up by a group of surfers concerned about the pollution of the sea by sewage, something they personally experienced when they went surfing. Their early work focused on reducing sewage pollution in the marine environment, something that was assisted by the EU Directive on Bathing Water Quality that was introduced in 1991. Over time, as it grew, SAS broadened its area of concern to include all types of marine pollution and the conservation of marine wildlife, internationally as well as in the UK. Its activities now include:

- Organised beach cleans by volunteers, both locals and tourists.
- Education work particularly with schools.
- An All-Party Parliamentary Group on the theme of 'Protect our Waves' which consists of 23 Members of Parliament from different political parties.
- International campaigning including the first ever 'Global Wave' conference held in Cornwall in 2015.

In recent years they have become heavily involved in campaigning for plastic-free oceans. Indeed, they are the force behind the 'Plastic Free Communities' movement in the UK, where communities commit to moving towards being plastic-free through various measures such as no longer offering plastic bags to shoppers. At the time of writing there are 648 communities involved in the scheme, which includes verification and certification. They also campaign on keeping plastic out of the oceans, suggesting that in 2019 there were thought to be on average some 3,000 plastic items found in the sea in each kilometre of the UK coast every year. (www.sas.org.uk)

They receive no government funding and have to raise money from memberships, donations, sale of merchandise and applying for various grants.

Surfers Against Sewage have been active in Cornwall from the beginning, as it is one of the top surfing destinations in Europe. In Cornwall there are a number of other voluntary and not-for-profit sector organisations that work to protect the marine environment and beaches. These include local groups such as the Friends of Portheras Cove, just a short walk from my home, wildlife sanctuaries such as the Gweek Seal Sanctuary and the National Lobster Hatchery at Padstow which breeds lobsters to release into the wild. There are marine conservation groups in coastal settlements in places such as Looe and Falmouth while the Cornwall Wildlife Trust runs 'Seaquest Southwest', a citizen science marine environment recording project.

I believe that such voluntary citizen activism is crucial if we are to develop more responsible tourism and protect our marine environment.

16: Sport fishing in Africa

In recent years the continent of Africa has become a popular destination for so called 'sport fishing' or 'game fishing'. Here the emphasis is on catching record sized specimens of the largest species rather than catching fish to eat. While some fish may be retained by the boat operators and sold ashore for food, the main aim of the angler is to have a photograph taken with their specimen. I find it interesting that 'big game fishing' has taken off on the continent where big game hunters destroyed so much land-based wildlife in the past.

Game or sport fishing itself has a long history, the main governing body, the International Game Fishing Association was founded back in 1939. To many people this 'sport' seems cruel, using big hooks and vicious looking 'gaffs' to catch and land fish weighing hundreds of kilos for sport rather than food. However, in Africa it has become a major component of the tourism offer, in the oceans and the major rivers. It has developed its own infrastructure of tour operators, boat operators, fishing lodges, and fishing camps and as a sector now employs thousands of people and generates millions of dollars of revenue each year.

In late 2019, the website of one tour operator, Sportquest Holidays, talked about the opportunities for game fishing off the coasts of Guinea, Bissau, South Africa, Gabon, Kenya, Namibia and Madagascar. While most of the species mentioned are not currently on the endangered list, the bronze whaler shark is on the list, but anglers were being encouraged to fish for them from the shore in Namibia (www.sportquestholidays.com, 2019). Given the decimation of shark populations, it is particularly sad to see sharks still being targeted by sport anglers.

A fishing blog in 2019 identified 'Anglers' Ten Most Wanted Fish in South Africa' (www.fishfishme.com, 2019), of which experts say that one is 'vulnerable' in terms of its population while another is said to be near to being under threat. It was depressing when I searched a random selection of the websites of operators who offer game fishing experiences that hardly any mentioned conservation or the protection of the marine environment. On the other hand, the governing body, the International Game Fishing Association does have rules about fishing tackle and techniques. Furthermore, in terms of records, the association will not consider any fish for inclusion in the record books unless it is released back into the wild in good condition. However, in recent years, criticism has grown of such 'catch and release' schemes, claiming that it still inflicts cruelty on the fish and leads to them being harmed even though most will survive the experience.

Criticism is also growing about the accidental harm caused by game fishing to marine creatures which are accidentally 'hooked', such as sea turtles. I wonder if one day catching big fish for 'sport' will be as socially unacceptable as big game hunting on land seems to have become.

17: Cruise company private islands – a step too far?

A major theme of this text has been the growth of cruising globally, its success as a sector and the suggestion that the sector has not yet fully engaged with the idea of sustainable or responsible tourism. We have noted that cruise ships can have negative impacts on the ocean and on the ports of call where they bring large numbers of people into a small geographical area for intense experiences over just a few hours. However, in the past 40 years we have seen a new phenomenon, whereby cruise lines have developed private islands for the exclusive use of their passengers. To date this has development been focused in the Caribbean region as can be seen from Table 11.2.

Table 11.2: Cruise line private islands in the Caribbean, 2019

Name of island	Location	Cruise Line
Amber Cove	Dominican Republic	Carnival
Castaway Cay	Bahamas	Disney Cruise
CocoCay	Bahamas	Royal Caribbean
Great Stirrup Cay	Bahamas	Norwegian Cruise Line
Half Moon Cay	Bahamas	Holland America
Harvest Caye	Haiti	Norwegian Cruise Line
Labadee Hispaniola	Haiti	Royal Caribbean
Ocean Cay MSC Marine Reserve	Bahamas	MSC
Princess Cays	Bahamas	Princess Cruises

(www.cruisecheap.com 2019 and www.cruisecritic.co.uk 2019)

'These islands are owned by the cruise lines and are for the exclusive use of their guests for the day ... A cruise line private island is its own little city with full-time inhabitants, a power plant, water treatment facilities and even a post office. The islands have bars, restaurants, restrooms, shopping, first aid stations, water-sports and tour huts. Each day, before a ship arrives, the island crew prepares the island by making sure the beach is raked, beach chairs and umbrellas are in position and the floating mats are ready for their guests for the day... All of the best cruise ship private islands include open air dining areas.... These all-you-can-eat buffets include everything you'd want in a private buffet.' (www.cruisecheap.com, 2019) They also feature mini theme parks and waterslides and are like all-inclusive resorts that welcome visitors for a day and then return them to their ship to continue their vacation.

While most private islands to date have been in the Caribbean, MSC operates one in Abu Dhabi and two off the coast of Mozambique while Paul Gauguin Cruises has one in French Polynesia (www.cnntraveller.com, 2019). A report in 2018 suggested that Royal Caribbean were scouting for potential private islands in Asia (www.asean.cruising.com, 2018).

The Ocean Bay MSC Marine Reserve is an interesting project, where MSC claims to have created a private island which has a focus on conservation, with the island being surrounded by 64 square miles of protected waters. They have earmarked funds for a coral nursery to help protect the existing reef (www.cnn-traveller.com, 2019).

Since the 1990s academics have been writing about this phenomenon and suggesting it poses a threat to the marine environment around these islands although there seems to have been little empirical research conducted to date. However, it is hard to believe that so many people arriving on large cruise ships does not have a negative impact on the marine environment, no matter how well managed the island may be. Some of the islands were formerly industrial sites rather than pristine islands, so negative impacts on the surrounding ocean will not be a new experience for them. Finally, these islands also raise an ethical issue around the extent to which a commercial organisation should be able to have exclusive use of an island for the hedonistic pleasure of its customers.

18: Artificial reefs in the USA

According to the National Ocean Service in the USA, artificial reefs are '*manmade structures that may mimic some of the characteristics of a natural reef*'. The same source acknowledges that '*submerged shipwrecks are the most common form of artificial reef* [and]... *marine resource managers also create artificial reefs in underwater areas that require a structure to enhance the habitat for reef organisms*' (www.oceanservice.noaa.gov, 2019). In other words, artificial reefs can be created accidentally through shipwrecks or as a result of a deliberate action to sink a vessel or create a purpose-built structure underwater.

Artificial reefs are not a new phenomenon, having existed in Japan for 400 years with the aim of protecting fish stocks. However, in recent decades their purpose has broadened to include:

- Providing new wildlife habitats that will attract fish thus providing new opportunities for recreational angling.
- Attracting visitors to a destination by offering a range of opportunities for diving and wildlife-watching.
- Trying to compensate for the destruction of natural reefs.

The first two examples demonstrate that a major motivation for the creation of artificial reefs today is the desire to attract tourists because of the economic benefits they bring. Artificial reefs have been developed in many popular tourist destinations from New Zealand to the Cayman Islands, Malta to Thailand. However, it is perhaps in the USA that their development has reached an unprecedented level.

In 2016, the Florida Fish and Wildlife Conservation Commission estimated that there were some 3,000 artificial reefs along its coastline of around 1,350 miles, varying in depth from around one metre to over one hundred metres. The development of artificial reefs has been encouraged by local government in Florida and other states. North Carolina for its part had some 42 artificial reef systems in 2019. The USA has a *National Artificial Reef Plan,* developed in 1985 and reviewed in 2007. This is a document that sets out '*guidance on various aspects of artificial reef use, including types of construction materials, and planning, siting, designing and managing artificial reefs for the benefit of aquatic life*' (www.bsee.gov, 2019). Interestingly, the Plan says little about tourism, despite the fact that the economic benefits for tourism have been a major motivator for new artificial reefs to be developed along the US coast.

On its website, the Florida Keys National Marine Sanctuary noted that '*the negative ecological impacts of artificial reefs may outweigh potential economic gains ... if artificial reefs are not carefully planned or constructed, they can actually damage natural habitats...monitoring observations indicate that many artificial structures are quickly becoming habitat ...for invasive species*' (www.floridakeys.noaa.gov, 2019).So, while artificial reefs could help to reduce visitor pressure on natural reefs, they can also harm the marine environment, particularly if those who develop them lack the resources and knowledge to maintain them and monitor their impacts effectively.

19: How sustainable are coastal 'eco-resorts'?

It is clear that the term 'eco-resort' is fashionable and has market appeal; a search for the term on Google in January 2020 produced 174 million results! The first page of results included headlines such as '*World's best eco-luxury resorts*', and '*20 world's most beautiful eco-resorts*'. At this point it is worth mentioning that' eco-resort' is a term for which there is no officially recognised definition and no agreed set of criteria that needs to be met. In other words, anywhere can be marketed as an 'eco-resort'. The term sits alongside eco lodge and eco hotel in the vocabulary of the tourism industry. One website which promotes eco-resorts says '*the biggest difference between eco resorts and eco lodges and an eco-hotel is the setting. Eco-resorts and eco lodges tend to be more remote, located in relatively pristine natural environments such as exotic islands, forests and mountains. Eco hotels, on the other hand, are more often associated with cities and towns*' (www.greenglobetravel.com, 2020). This perfectly captures the point that many eco-resorts on islands or on the coast are developed in places that were wholly natural and undeveloped before. So, no matter how sensitively designed they are, their creation will have had negative impacts on the natural environment. At the same time, eco-resorts or lodges are most commonly found in destinations that involve lengthy flights from the main source markets.

This must call into question whether they can ever be seen as sustainable regardless of how responsibly the unit itself is managed. An eco-resort should surely be judged on the carbon footprint of its guests, including their journey to and from the resort rather than just being based on their time in the resort?

It is also clear that most eco-resorts are sold as luxury experiences, where the use of the term 'luxury' is taken to include qualities such as exclusivity and high levels of personal service. But luxury hospitality can often be in conflict with responsible tourism, as high spending guests demand food that may not be available locally, alcohol and bottled water that needs to be imported, energy consuming air conditioning and so on.

While there are no official criteria for being an eco-resort, a number of websites make suggestions to help consumers make their choice of accommodation. One lists ten criteria which include:

- Ensuring the seafood served is ASI/MSC accredited
- Chocolate and coffee from Rainforest Alliance certified sources
- Paper and toilet paper from producers certified by the Forestry Stewardship Council
- Having food waste reduction initiatives in place
- A 'no single-use plastic' policy
- Carbon offsetting of the flights of guest flights
- Global Sustainable Tourism Council certified hotel operations
- Having renewable energy plans
- Leadership in Energy and Environmental Design certified buildings
- The resort is EarthCheck certified.

(www.greenisthenewblack.com, 2019)

This is not an official list, but it contains some important criteria that one would think should apply to any 'eco-resort'. However, when I checked out a number of properties in coastal locations around the world which used the term eco-resort, many did not meet these criteria. It is tempting to conclude that the term is being used in some cases as a deliberate marketing ploy or represents an exercise in 'green-washing'. Furthermore, it appears that guests are happy to buy into the term because it brings status for them and reduces any feelings of guilt they may have about flying halfway across the world to visit a resort by the ocean.

When considering whether something should be seen as an eco-resort in a coastal region or on an island the criteria should cover four elements, namely:

- The siting of the resort. Does it require the cutting down of trees and does it require motor vessels or even seaplanes to transport guests to it?

- The design of the buildings, spaces and landscapes. Will building materials need to be imported and have they been designed to encourage natural air conditioning rather than requiring electric air conditioning?
- The construction process. How disruptive will it be to marine and land-based wildlife and what effect will it have on sea water quality?
- The operation of the resort. How much will be done to minimise the carbon footprint of the resort, waste, and the use of energy and water resources?

Perhaps the final point is that if we are to make tourism more sustainable, then all coastal resorts will need to become 'eco-resorts', not just those which target high spending tourists who want luxury vacations in exotic destinations.

20: Lessons from the Indian Ocean Tsunami of 2004

On 26 December 2004, around 225,000 people were killed when a huge tsunami struck coastal communities in the countries around the Indian Ocean. The largest loss of life was seen in the Aceh province of Indonesia, but thousands also died in Sri Lanka, India and Thailand. The tourism industry in the Indian Ocean was seriously affected as a number of coastal resorts were devastated and left in ruins. Within days commentators were seeking to learn lessons from the disaster. In January 2005, a report from the UN Office for Disaster Risk Reduction (UNDRR) suggested these lessons included:

- *'We are all vulnerable to natural disasters.*
- *Coastal zones and small islands are often densely populated areas that increase people's risk and vulnerability.*
- *Public awareness and education are essential to protecting people and property.*
- *Early warning saves lives.*
- *Countries in the Indian Ocean need to develop a regional early warning system for tsunamis*
- *Reducing risk depends on communication ... between the scientific community and politicians*
- *Develop and enforce building codes where .... tsunamis are common.*
- *Humanitarian aid needs to invest more in prevention.*
- *International, regional and national organisations should ....be better coordinated.*
- *The media have a social responsibility to promote prevention.'*

(www.reliefweb.net, 2005)

The fact that some of these points needed making illustrates how ill-prepared the region was for such an event or for any other extreme weather event for that matter.

In 2014 on the tenth anniversary of the tsunami, the UNDRR looked at the lessons learned ten years on through a conference. Various speakers noted that:

- *'Ten years after the Indian Ocean tsunami the world has taken significant measures to make the world a safer place against disasters. We now have more efficient early warning systems and better evacuation procedures in place.*
- *One life-saving measure to emerge from the disaster was the Indian Ocean Tsunami Warning System...It is an efficient system which disseminated early warnings within eight minutes of the Banda Aceh earthquake in 2012.*
- *One great lesson ...is that we must build and plan our coastal urban areas in a more sustainable and responsible way.*
- *Hotel resilience can become a competitive advantage as the entire world will be affected by more severe and frequent weather-related disasters in the future.*
- *Indonesia has been ... investing a lot in decentralisation to empower local communities to deal with disaster risks.'*

(www.undrr.org, 2014)

Based on my travels and research, this seems a rather rosy picture of what has happened since the tsunami in 2004. Many homes have been rebuilt on the shore, particularly by those too poor to be able to build elsewhere, so that they remain at risk. And the tourism and hospitality sector has not always heeded the lessons of the tsunami, with hotels being built right on the beach despite it being clear that this maximises the risk for staff and guests.

However, some of the big companies have learned lessons. '*At Starwood, safety starts with the site of the hotel and a comprehensive risk assessment is conducted. Location is important, and assessing risk, is the key to better protect people and assets*' (www.undrr.org, 2014). Unfortunately the vast majority of hotels are small and medium sized enterprises, which lack the expertise and resources of these larger companies. On the other hand, UNESCO produced *A Guide to Tsunami for Hotels* which provides advice and a toolkit for hoteliers (www.greenhotelier.org, 2015).

In 2011 Erik Cohen claimed that '*the devastation left behind after the 2004 Indian Ocean tsunami ....led to numerous attempts to 'land grab', the displacement by powerful individuals and authorities of the weaker groups of local inhabitants from their land, in order to re-allocate it to tourism development*' (Cohen, 2011).

There is little evidence that lessons from the widespread corruption that accompanied the distribution of aid and the reconstruction work have been learned, such that it will not be seen after any future natural disasters. Perhaps inevitably, but nevertheless unfortunately, one lesson that was clear to many observers was that the international media paid a disproportionate amount of attention to tourist victims of disasters than it did to the victims who are local. The foreign media was understandably full of stories about their nationals who lost their lives, but I read no such stories about individuals from the local area, even those who rescued international tourists. The coverage also seemed to focus on the impact in tourist destinations, despite the fact that around three-quarters of those who died were from parts of Indonesia that receive hardly any tourists. Furthermore, tragically 5,000 international tourists were killed but this figure needs to be set against at least 220,000 local residents who died.

It remains to be seen whether the lessons learned from the 2004 tsunami will help reduce the risk for coastal communities from future natural disasters.

21: Hurricanes and the tourism industry in the Caribbean

We have already recognised that global warming is causing more frequent and increasingly severe extreme weather events across the planet. These events include typhoons, cyclones and hurricanes. In addition to the cost of such phenomena in terms of lives lost and homes destroyed there is also the damage done to local economies, and this is especially true in relation to the tourism industry. This can be seen clearly in the Caribbean, which is prone to major hurricanes and where the economies of many islands are highly dependent on tourism. Hurricanes can damage local tourism economies in several ways as follows:

- Cruise ships, on which Caribbean tourism relies heavily, have to modify itineraries to avoid hurricanes and that may mean them abandoning planned visits to ports of call.
- Tour operators may bring guests home early or cancel flights to destinations if a hurricane is forecast.
- Tourists who remain in a destination during a hurricane 'hunker down' in their hotels and so are not out and about in the destination spending money.
- Tourists may postpone or abandon plans for a trip during the hurricane season, particularly in light of media reports about hurricane risks.

During the 2017 Hurricane season, the worst hurricanes only badly affected around a third of the Caribbean, but they still resulted in an estimated loss of some 826,000 visitors who would have spent around 740 million US Dollars. Research suggested that it may take the affected islands up to four years to get tourism up to previous levels, during which time they would miss out on an estimated 3 billion US Dollars in revenue (www.wttc.org, 2018). In 2019, Hurricane Dorian devastated large parts of the Bahamas, a popular summer destination for tourists, where 60% of the economy is dependent on tourism. In addition to a large loss of life this storm devastated many hotels and other tourism businesses and seriously damaged the main international airport. It is expected to take years for the Bahamas to get back to where it was before the hurricane.

Recognising that hurricanes are becoming more frequent and severe, destinations in the Caribbean are focusing on building resilience in the tourism industry. An article published in 2019 reported a number of initiatives including:

- The work of the recently created Global Tourism Resilience and Crisis Management Center, based at the University of the West Indies in Jamaica. It is a regional centre for research covering the forecasting of natural disasters, mitigation measures and recovery strategies.
- A new sustainable draining system developed by the Grenada Hotel and Tourism Association.
- Efforts in St Lucia between the public and private sectors to improve the sharing of information on the effects of climate change to strengthen their ability to respond to crises.

(www.devex.com, 2019)

Resilience is now a 'hot topic' in tourism and seems likely to remain so as global warming leads to more extreme weather events that affect tourist destinations. Its about how to best prepare, so as to limit the impacts and how to ensure that recovery afterwards is as quick as possible.

22: The Cruise Lines International Association (CLIA) and responsible tourism

In Chapter 3 I suggested that, despite improvements in recent years, the cruise industry as a whole is not yet fully embracing the idea of responsible tourism. While it is taking action on a range of impacts which its activities have on the marine environment I believe that these actions are too often piecemeal and lack ambition and urgency. However, there have been some positive developments in recent years, not least amongst the professional associations in the cruise sector. The Cruise Lines International Association (CLIA) claims to be the world's largest cruise industry association, bringing together ocean and river cruise lines, travel agents, ports, destinations and industry suppliers (www.cruiseexperts.org, 2020). Its membership includes pretty well all the major cruise brands worldwide.

Its website recognises that '*no industry has a stronger interest in protecting the oceans we sail and the destinations we visit than. It is not simply our responsibility: operating sustainably is a business imperative* (www.cruiseexperts.org. 2020).

CLIA also has a separate section of its website that deals specifically with sustainability. including a set of fact sheets and infographics on issues such as technological innovations, the health of the oceans, and exhaust gas cleaning systems.

It communicates some key messages including:

- *We are reducing the rate of carbon emissions across the industry fleet by 40% by 2030 (compared to 2008 levels),*
- *We recycle 60% more than the average person on land,*
- *Our cruise lines never release untreated wastewater into the ocean,*

(www.cruiseexperts.org, 2020)

This sounds positive and is to be welcomed. However, the carbon emissions target relates to 2008 levels, not current levels so may not be as good as it appears. The recycling figure does not recognise that cruise ships may create more waste per cruise passenger than the average person generates at home. And, ironically in the context of the untreated wastewater message, on the day I am writing this (3 January 2020), the US media is reporting that a Carnival cruise ship has allegedly dumped 5,900 gallons of untreated grey water into the ocean at Port Canaveral in Florida against company policy and government regulations (www.triblive.com, 2020).

The CLIA website provides sustainability reports from the companies and brands that are members. These reports vary dramatically from brand to brand. Some are comprehensive and include measurable targets and information on performance against the targets. Others are more generic and talk of ambitions without measurable targets. A number focus largely on donations to environmental causes and projects, more than on how the company is reducing the impacts of its own activities on the marine environment. A few others say very little and in one case there was nothing at all about sustainability.

CLIA has tried to communicate the importance of sustainability to cruise operators and cruise consumers. Its website also includes useful industry-wide information on the progress being made on reducing the negative impacts of cruising on the marine environment. On the other hand, CLIA is very keen to balance what it says about sustainability with the fact that the cruise industry supports over a million jobs worldwide. This is perfectly valid but does not compensate for the damage cruise ships can do to the marine environment.

Hopefully, the CLIA will build on what it has done so far and try to raise the bar for its members so that progress can be made more rapidly across the whole sector. In doing this it could, perhaps, benefit from studying the experience of the International Tourism Partnership which is considered in Case Study 23.

23: Lessons from the International Hotels Environmental Initiative and the International Tourism Partnership

In Chapter 9 we looked at how the industry is responding to the negative impacts of tourism on the marine environment. With a few exceptions, the response has been fragmented and based on the actions of individual operators rather than coordinated across whole sectors or destinations. Where businesses are taking effective action to reduce negative impacts, it is being done in an attempt to gain competitive advantage. This means that actions can sometimes be seen as tokenistic gestures rather than evidence-based attempts to fundamentally reduce negative impacts. In the case of the cruise sector it appears the industry is yet to fully embrace responsible tourism and take effective action to address its impacts on the marine environment. This is all disappointing as the tourism industry has examples within it of self-regulation that could serve as a model if the industry is to effectively address its impacts on the marine environment.

The International Tourism Partnership (ITP), formerly known as the International Hotels Environmental Initiative, was set up as a direct response to the Rio Earth Summit in 1992. It is a sector-wide organisation designed to make the global hotel industry as a whole more responsible and sustainable based on cooperation and the sharing of good practice. From the beginning it has focused on practical help rather than strategies and mission statements. Its *Environmental Management for Hotels* manual and its *Green Hotelier Magazine* provide practical advice for hoteliers. In 2005, they published guidelines about the siting and design of new hotels and two years later introduced a guide to minimum environmental standards for sustainable hotels. In 2011, working with the World Travel and Tourism Council, the ITP launched the Hotel Carbon Measu rement Initiative, to develop an industry-wide way of calculating the carbon emissions of hotels. In 2020, its 18 corporate members managed 23,000 hotels with more than 3,400,000 rooms and 1,500,000 employees in over 100 countries (www.ihei.org, 2020).

Having begun with a focus on environmental issues the ITP has now broadened its focus to embrace the whole corporate social responsibility agenda and its goals now embrace six issues, namely:

- Carbon emissions
- Water resources
- Sustainable supply chains
- Youth unemployment
- Fair labour standards
- Youth unemployment

(www.tourismpartnership.org, 2020)

Despite its success, the ITP could be criticised potentially because of its focus on large hotel chains rather than privately owned hotels, which represent the vast majority of the hotel industry worldwide. However, many of their resources are available to non-members such as the *Green Hotelier Magazine.* ITP could also be criticised because it just focuses on hotels despite the fact that its name embraces the broader area of tourism. However, tourism does have its own umbrella organisation, the Global Sustainable Tourism Council (GSTC), which represents tour operators and destinations as well as accommodation. The GSTC currently offers training and accreditation and certification schemes for its members.

However, neither the ITP or the GSTC embraces the cruise sector and both tend to focus on land-based issues in tourism rather than the impacts of tourism on the marine environment. Perhaps, the Cruise Lines International Association could develop its own version of the ITP to improve responsible practice across the cruise industry. At the level of individual destinations, it would be good to see the creation of organisations bringing together transport, accommodation, visitor attractions and leisure activities of all kinds within a particular destination with the aim of making coastal tourism more sustainable.

24: Airlines and carbon offsetting

According to Emma Featherstone, '*carbon offsetting is the act of doing something to negate the impact of the CO2 your actions or travel have produced. It is seeking to balance out – or go some way to reduce – your carbon footprint to zero*' (www.telegraph.co.uk, 2019). This usually involves activities such as tree-planting. Carbon offsetting has been around some time, but it has come back into focus recently, not least because the rise of the 'flight shaming' movement (see Case Study 35) has highlighted the impact of air travel on the environment.

A number of tools exist to allow passengers themselves to find out the actual scale of their carbon footprint each time they fly. A random check on the myclimate website in January 2020 suggested that a single passenger making a return trip from Paris to Sydney in business class would generate 11.6 tons of CO2. The figure for a return from Copenhagen to Palma de Mallorca in economy class was 0.7 tons. The website then offered to offset the latter flights for the passenger at a cost of around US$ 34 for the business class flight and US$ 21 for the economy class trip (www.co2.myclimate.org, 2020).

Many airlines already have their own voluntary carbon offsetting schemes which passengers can buy into if they so choose. Table 11.3 presents the average cost of the schemes of ten airlines in US Dollars per 1,000 miles in 2019. It is based on economy class travel although some airlines charge more to passengers in premium economy and business class.

Table 11.3: Average rate to carbon offset per 1,000 Miles for Economy Class passengers in US Dollars

Airline	Average rate per 1,000 Miles in US$
Alaska	1.70
United	1.49
Air Canada	3.68 (International, domestic 2.86)
Austrian	4.63
Cathay Pacific	0.34
China Airlines	1.25
Eva Air	1.00
Japan Airlines	7.05
Lufthansa	2.21
Qantas	1.68

(www.thepointsguy.co.uk, 2019)

Although one might expect some variations based on the age of an airline fleet and the nature of the airline's operations, the fact that the figure for some airlines appear to be less than a tenth of others suggests that are some significant differences between the schemes. Presumably some are more ambitious than others and are offsetting a higher proportion of the carbon emissions than others? There is also concern over some of the projects funded by carbon offsetting. Some have been more effective and transparent than others. And while the passengers carbon footprint has an immediate effect, the projects funded by their offsetting donation may take months or even years to have an impact. Furthermore, some schemes simply do not have the results they promised.

Mindful of public concern over the impact of air travel on the environment, in 2019, British Airways and Air France announced they were to start offsetting all carbon emissions on domestic flights. Shortly after EasyJet announced that it would become '*the first carrier to start offsetting carbon emissions on flights across its whole European network*' (www.ft.com, 2019). However, many people do not believe that offsetting is a solution to the environmental impacts of air travel. Andrew Murphy of campaign group Transport and Environment said '*airlines paying others so they can go on polluting is not a solution*' (www.ft.com, 2019). Meanwhile, Justin Francis of Responsible Travel stated that '*carbon offsets are no substitute for carbon reduction. We are still advocating for a Green Flying Duty to reduce demand in the meantime, with the proceeds ring fenced for R and D into decarbonised aviation*' (www.telegraph.co.uk, 2019).

Whatever its merits, carbon offsetting looks set to remain part of the airline scene for years to come. Under the recently agreed UN Carbon Offsetting and Reduction Scheme for International Aviation (CORSIA), there will be an industry-wide carbon offsetting scheme in place on a voluntary basis from 2021 that will become compulsory for airlines from 2027 (www.ft.com, 2019).

I will leave the reader to decide if carbon offsetting represents a genuine solution to the carbon footprint of the airline industry, a way of letting passengers feel less guilty about their trip, or simple 'green-washing' by the industry.

25: The Great Barrier Reef Marine Park, Australia

The Great Barrier Reef in Australia is almost certainly the world's most famous coral reef. It is one of the major reasons for international tourists to visit Australia and is a major destination for domestic tourists. The Great Barrier Reef Marine Park was established in 1975 through the Great Barrier Reef Marine Park Act. It covers 344,000 square kilometres and in 2018 received 2.26 million visitor days, a figure derived from the number of visitors and their length of stay.

It has been estimated that the Great Barrier Reef contributes 5.6 billion Australian Dollars (AUD) to the national economy and supports around 70,000 jobs (www.gbrmpa.gov.au, 2019). It is the world's largest coral reef ecosystem and is 2,300 kilometres long and between 60 and 250 kilometres wide. It is home to 600 types of coral, 1625 types of fish, 100 types of jellyfish, 30 species of whales and dolphins, and 600 species of molluscs (www.gbrmpa.gov.au, 2019).

The primary purpose of the Marine Park is '*the long-term protection, ecologically sustainable use, understanding and enjoyment of the Great Barrier Reef for all Australians and the international community through the care and development of the Marine Park.*' It seeks to do this through four programme areas as follows:

1. Providing expert knowledge to influence and advise key decision makers on managing, reducing or avoiding significant threats to the reef.
2. Regulating and ensuring Marine Park use compliance.
3. Educating and fostering stewardship to enhance protection of the Reef.
4. Enhancing reef resilience through continuous improvement and new initiatives across all aspects of management.

The annual budget of the Marine Park in 2018-19 was 79.7 million AUD or around £42 million, of which just over two-thirds came directly from the government of the state of Queensland and the federal government of Australia (Great Barrier Reef Marine Protected Area Authority, 2019). However, revenue was also generated by the Environmental Management Charge which is levied on all tourists who visit the reef through a tourism industry operator. In 2020 this charge stood at 3.50 AUD for visitors staying less than three hours and 7.00 AUD for those staying for longer than three hours.

The park authority places a great emphasis on its governance and accountability, and it engages in a range of activities to safeguard the reef including:

- Research to increase its knowledge about the reef environment.
- Monitoring of wildlife populations including the corals and water quality.
- A huge amount of education and community engagement work both on site and digitally.
- Practical work 'in the field' to help conserve the reef.
- Working closely with stakeholders including indigenous people and the traditional owner groups.

However, perhaps its most important role is the granting of permissions for any activity that takes place within the Park. The system places permission applications into four types based on their scale and perceived level of risk. Only the highest category requires an Environmental Impact Statement to be produced. In 2018-2019, 369 applications for permits were received. None were refused because the Park Authority explained that it works closely with applicants to ensure that their proposals comply with Park policies. In the same year there were 136 alleged cases of non-compliance in relation to existing permissions. (Great Barrier Reef Marine Park Authority Annual Report for 2018-2019, 2019)

It is clear that the Park Authority is highly motivated and professional, but its task is a difficult one. It has to reconcile the need to protect the Great Barrier Reef with allowing visitors to enjoy it and being conscious of the value of reef tourism to the local economy. The tourism lobby is a powerful one in Queensland and one would imagine that its influence must be felt by the Park Authority in some way. The Authority has little or no influence over activities outside the Park which have an impact on it, whether that be global warming or pollution from agricultural and industrial run off from the land. As a result, despite the efforts of the Marine Park Authority the Great Barrier Reef continues to suffer deterioration including coral bleaching and damage from irresponsible diving.

26: Marine protected areas in Asia

Marine Protected Areas (MPAs), such as marine national parks and marine reserves, can play an important role in the conservation of the marine environment. However, in Case Study 25 we saw that even a well-resourced and managed MPA can struggle to protect vulnerable marine ecosystems from the threats they face. Many Asian countries have seen enormous growth in visitor numbers in recent years, due partly to the opportunities they offer for enjoying pristine beaches and marine leisure activities such as diving. It is interesting, therefore, to look at how countries in the region use MPAs to help them conserve the natural resource on which much of their tourism depends, whilst also protecting the marine environment from the negative impacts of tourism.

In 2020 it was reported that there were 16,977 MPAs in the world covering around 8% of the world's oceans. There has been a huge increase in MPAs in the past 20 years worldwide; in 2000 they covered 2 million square kilometres of ocean compared to over 27 million in 2020 (www.protectedplanet.net, 2020). The countries of Asia have been involved in the designation of MPAs from the early days of such designations and as far back as 2002 ASEAN was publishing a review of MPAs in nine ASEAN countries (Cheung et al., 2002). Given this fact I was surprised to see how little of the territorial waters of Asian countries is covered by MPAs, as can be seen in Table 11.4.

Table 11.4: MPAs as a percentage of the territorial waters of selected Asian countries, 2018

Country	MPAs percentage
Japan	8.2
China	5.4
Bangladesh	5.4
Indonesia	3.1
Thailand	1.9
Malaysia	1.5
Philippines	1.2
Pakistan	0.8
Vietnam	0.6
India	0.2

(www.data.worldbank.org, 2018)

To set this data in context, here are the equivalent figures for some other countries:

Table 11.5: MPAs as a percentage of the territorial waters of selected countries, 2018

Palau	83%
France	45%
USA	41%
Australia	41%
Jordan	36%
Brazil	27%
Mexico	22%
Dominican Republic	18%
Belize	10%
Bahamas	8%

(www.data.worldbank.org, 2018)

Even allowing for the fact that the data might not be wholly accurate or up to date, it is clear that Asian countries are using MPAs to protect their oceans less than most other countries across the world. However, what protects oceans is not the designation of MPAs alone but rather their regulation and management. A review of MPAs in Thailand conducted by a team from the University of Victoria in 2012 found significant weaknesses in the management of the MPAs. They made ten recommendations which included: updating the relevant legislation; addressing the threat of uncontrolled fishing; making funding more sustainable; developing partnerships with the tourism industry; and including local communities in decision-making (www.icriforum.org, 2012). These issues reflect those found in many other MPAs based on my personal investigations.

Academics have also not always been impressed with the performance of MPAs in Asia. Yunzhou and Fluharty in 2017 said that *'China has established more than 250 MPAs… but the overall management effectiveness is unimpressive'* (Yunzhou and Fluharty, 2017). In 2004 Christie was describing MPAs as *'biological successes and social failures in southeast Asia'* (Christie, 2004). In 2017 a review of literature review about MPAs by Kamil suggested that *MPA strategy has both positive and negative impacts on southeast Asia* (Kamil et al., 2017).

However, there are encouraging signs in Asia. In Indonesia, the WorldWide Fund for Nature is supporting the authorities with the designation and management of MPAs (www.wwf.org, 2019). And in 2019 the IUCN was able to report that the Tun Mustapha Marine Reserve in Malaysia was committed to joining the IUCN Green List, inclusion on which is based on *'good governance, sound design and planning, effective management and successful conservation outcomes'* (www.iucn.org, 2019).

27: Regulation of the cruise industry

The industry trade body, the Cruise Lines International Association (CLIA), says *'the cruise industry is one of the most heavily regulated industries in the world. The average ship undergoes dozens of … safety inspections per year involving … the implementation of thousands of specific requirements set by the International Maritime Organisation (IMO)'* (www.cruising.org, 2020).

The IMO sets international standards for all ships including cruise ships across a range of areas including labour protection, ship safety, security, and environmental practice.

There are other authorities involved in the regulation of cruise ships including:

- **Port states:** the countries with ports visited by ships, including cruise ships. They are responsible for ensuring that ships meet international, national and local regulations, usually by inspecting vessels when they are in port.
- **Classification societies**, which check compliance with regulations as well as setting their own safety and environment guidelines.
- **International Labour Organisation** regulates labour protection on cruise ships.

- **Countries of registration**, the country where a ship is registered has a responsibility to ensure that cruise ships comply with all international regulations.

Many cruise lines choose to register their vessels not in the country in which they are owned or operate from, but in countries whose registration regimes are often described as offering 'flags of convenience'. By registering in such countries cruise operators may enjoy preferential taxation rates, but critics claim that it means the cruise companies may benefit from less rigorous enforcement of the various regulations. The three leading 'flags of convenience' nations are Panama, Liberia, and the Marshall Islands, yet none of these countries is traditionally seen as a major maritime trading country. It appears that the majority of cruise ships are now registered in flag of convenience countries even when their operations are based in the USA.

The IMO has introduced a new regulation from 2020 which means that ships will have to reduce their sulphur emissions by some 85%. However, concern has been expressed that some port authorities currently lack the resources to enforce this new rule. It has also been suggested by some critics that the fines in some countries for breaching this new regulation seem modest and they add that the use of scrubbers to help achieve the reduction in emissions may have unintended and unwelcome consequences. (www.skiff.com, 2019)

A number of governments have been active in enforcing environmental regulations in respect of cruise ships. In April 2019, the Norwegian authorities imposed a fine on the Greek-owned *MS Magellan* cruise ship after its fuel was found to contain 0.17% sulphur compared to the 0.10% permitted at the time, before the new IMO regulation was introduced (www.worldmaritimenews.com, 2019). The US courts, as we saw in Chapter 3, have imposed fines on US based cruise lines which broke environmental protection rules. Indeed, in 2019, a judge in the USA threatened that the ships of one cruise line could be banned in future from US ports if they continued to infringe environmental regulations! While this is very unlikely to happen, it may reflect frustration on the part of the judiciary about alleged repeat offending by cruise companies.

We note in Case Study 28 that a number of ports of call have begun to act to limit the number of cruise ships that visit them for environmental reasons and because of the impact large cruise ships have on the local community. This trend seems likely to continue as politicians face increasing pressure from local communities and the climate change lobby.

However, I would argue that there is a need for further regulation of the environmental impacts of the cruise industry and less reliance on self-regulation by the industry. I believe that while progress is being made, it is not happening quickly enough given the crisis facing our planet from global warming. It is also my contention that some areas where the marine environment is particularly fragile should be made 'off limits' to cruise ships, if their operators do not voluntarily agree to do this themselves.

28: Cruise ship ports of call and over-tourism: Is the tide turning?

As we noted in Chapter 3, the cruise industry has been one of the most successful sectors of the tourism industry worldwide. The market for cruising has grown year on year and the cruise lines have invested more than ever before in bigger and better ships. It is now possible for tourists to cruise to virtually every place on earth including uninhabited wildernesses. For many port cities around the world, cruise ships inject hundreds of millions of US dollars into the local economy and support many thousands of jobs. However, there is increasing concern over the impacts which cruise ships can have on the coastal destinations they visit, and these concerns have played a major role in the growing debate over the concept of over-tourism. This focuses on the impacts which tourism can have on the lives of local residents and the point at which locals begin to perceive tourism as a negative rather than a positive influence in their communities.

This judgement can be based on the sheer volume of tourists, when they visit and the behaviour of the visitors. Demonstrations about over-tourism have been seen in recent years in a number of European cities including Barcelona, Venice, Amsterdam and Berlin. Academics have also begun to study over-tourism as something which appears to be gaining traction, not least due to the additional concern over the carbon footprint of the tourists visiting these cities, as well as their impact on the local community. The cruise sector has found itself at the centre of the over-tourism debate in the last few years because:

- Cruise ships are getting bigger and bigger with thousands of passengers.
- They disgorge thousands of people into the port for just a few hours, overwhelming the infrastructure and crowding the streets, shops and eating places.
- Most of the excursions involve little or no meaningful contact with the local community nor do they provide a real opportunity for visitors to have authentic experiences.

However, it was a single event in June 2019 that had perhaps the greatest impact, when a cruise ship and a river cruise vessel collided in the centre of the historic city of Venice. Photos of the cruise ship towering over the iconic city skyline made a stark point that perhaps things had gone too far, when huge cruise ships could travel right into the heart of this unique place. After this incident much media attention in summer 2019 was focused upon the impact of cruise ships on their ports of call and most of it was negative. In August 2019 the authorities in Venice announced that large cruise ships would no longer be able to dock in the city centre. They also announced that from 2020 a third of all cruise ships would be diverted away from the Venice Lagoon altogether. According to a UK newspaper report, the Chairman of the northern Adriatic Sea port authority wrote to eight other port authorities *'asking them to join forces to oblige cruise lines to launch ships compatible with our structures and the environment'* (www.telegraph.co.uk, 2019). The Venice incident also led to some calls for cruise ships to be banned altogether from destinations that were also World Heritage Sites.

In September 2019 the French Riviera destination of Cannes announced that it was to ban the most polluting cruise ships from 2020 in an effort to improve air quality in the town. Their mayor said, *'it's not about being against cruise ships, its about being against pollution'* (www.voanews.com, 2019). And these are not two isolated cases, as can be seen from the following news stories from 2019:

- In Dubrovnik from 2020 there will be a limit of 4,000 cruise passengers per day and each will pay a two Euro tourist tax (www.theguardian.com, 2019).
- The Belgian government has reduced the number of cruise ships that dock in Zeebrugge to protect Bruges from over-tourism. The Bruges mayor is reported to have said that *'we have to control the influx more if we don't want to become a complete Disneyland here'* (www.ship-technology.com, 2019).
- From 2019, cruise passengers visiting Amsterdam have been required to pay an eight Euro tax (www.ship-technology.com, 2019).
- From 2021, the number of cruise ships docking in Dublin will be limited to 80 compared to 160 scheduled to visit in 2019 (www.ship-technology.com, 2019).
- The Greek island of Santorini has imposed a daily limit of 8,000 cruise passengers (www.ship-technology.com, 2019).
- In Barcelona, a local councillor described cruise passengers as *'a plague of locusts who devour the public space and then leave'* (www.ship-technology.com, 2019).

The industry has made efforts to reach voluntary agreements with ports on ship sizes, passenger numbers and environmental impacts. Perhaps the pressure coming onto the cruise sector from some ports of call might be a vital catalyst in encouraging the cruise industry to improve its environmental performance more quickly than it had planned. Back in 2017 the head of Carnival Cruise Lines had acknowledged that the industry *'must listen to locals on over-tourism'*. He is reported to have said that *'the solution is a bit of regulation'* in an article that did seem to display an attitude that could have been seen as a little complacent (www.travelweekly.co.uk, 2017). However, in fairness, this comment was made two years before the June 2019 incident in Venice and the subsequent ban on large cruise ships.

The question of over-tourism also relates to the impact of cruise ships on wildlife and the environment when visiting uninhabited wilderness areas. One article about this issue in 2018 was headlined *'will cruise ships destroy the wonders their passengers claim to love'* (www.theguardian.com, 2018). The article focused on cruises to arctic regions to see polar bears and the natural environment and included a quote from Ricky Gervais, the British comedian, that *'lets get too close to a polar bear in its natural environment and then kill it if it gets too close. Morons.'* (www.theguardian.com, 2018).

There seems little doubt that the over-tourism debate will grow and spread further around the world and that the cruise industry will remain at the heart of this debate. I believe the answer will be more controls on the numbers of cruise passengers visiting ports of call each day and stricter environmental regulation of ships in ports, as well as 'no go' areas in uninhabited wilderness locations with fragile ecosystems. If the industry is smart it will do this voluntarily, if not it seems likely that public pressure will force politicians to introduce regulation.

29: The role of Environmental Impact Assessments in coastal tourism

We have seen that poorly designed coastal developments can have a negative impact on the marine environment. Of course, this should not happen if there is an effective development control and land use planning system in place. However, public sector decision makers need information in order to be able to evaluate the potential impacts of a proposed project. In many countries Environmental Impact Assessments (EIAs) are built into the planning process to provide information about these impacts, both positive and negative, for any large-scale proposed project or development, including environmental, economic and social impacts.

EIAs originated in the USA in the late 1960s and are well established in North America and Europe, through both a European Union Directive and national legislation. The use of EIAs has spread around the world although their quality can be variable. Each country has its own legislation for EIAs which can vary significantly. It is also important to recognise that often the decision as to whether an EIA will be required for a particular project can be left to the discretion of officials or politicians. It should be noted that the purpose of an EIA is often to improve the quality of a proposal rather than being used simply to refuse permission for the development. The EIA will propose changes to the proposal to mitigate potential negative impacts.

One also needs to remember that final decisions on proposals are taken usually by elected politicians rather than officials, so issues such as the economic benefits for the community or potential tax revenues a proposal may bring can influence decisions as much if not more than the EIA.

The EU believes that the following principles should underpin EIAs:

- Opportunities for participation by interested parties
- Transparency in decision-making
- The process and timescales agreed in advance
- Accountability of decision-makers
- Undertaken with professionalism and objectivity
- Cost effectiveness – environmental protection at the least cost to society
- Flexibility
- Information and outputs that are readily usable in decision-making

(www.ec.europa.eu, 2019)

In 2008 Jennifer Li highlighted areas of bad practice in some EIAs including:

- Inconsistent application to development proposals with some types of development not required to have an EIA
- Being a 'standalone' not clearly integrated into the decision-making process
- Weak follow up with a lack of monitoring and enforcement of conditions
- Public consultation that is perfunctory
- Results in reports which are descriptive and overly technical
- Providing irrelevant information for decision-makers
- Being costly in relation to the benefits delivered
- Understating and insufficiently mitigating the environmental impacts

(Li, 2008; Sadler, 1996; Glasson et al., 2005)

The Li paper is more than ten years old, but such criticism can still be found about EIAs in some countries today.

EIAs are technical documents and need to be produced by specialists, so developers may enlist experienced consultants. One such company, Atkins, offered three examples of EIA related work, on its website including:

- Undertaking a sub-tidal survey in connection with a proposal to strengthen sea defences at Newbiggin Bay in the UK
- Blue City in Oman where the company produced a holistic assessment of the proposed large new urban development to be built on the coast
- Mapping the coastal habitat in St Lucia to help inform future planning decisions

(www.atkinsglobal.com, 2020)

In relation to tourism and the marine environment, the main use of EIAs relates to infrastructure development, including cruise ship terminals, marinas, artificial islands and large resort hotels. Recent high-profile examples involving EIAs for coastal projects include:

- The World Islands project in the UAE which critics claim will damage marine habitats and disrupt ocean currents.
- Malaysia, where a proposed new artificial island as part of the Penang South Reclamation Project was rejected following a negative EIA; the government has since apparently pressed on with the plan having made improvements to the design.
- Proposed new cruise terminals in Liverpool in the UK and Falmouth in Jamaica.
- A plan to improve cruise ship dock facilities in the Cayman Islands which was rejected following a critical EIA which noted that the development would damage the local coral reef. The project has been redesigned and the authorities still plan to go ahead with it, saying that corals will be relocated before the dredging commences.
- A number of proposed new marinas in Scotland.

With a few notable exceptions, EIAs are rarely used to allow authorities to make decisions relating to tourism industry operations, including wildlife-watching boat trips, scuba diving or granting cruise ships permission to visit a port. That is not to say that the potential impacts of these operations are not considered, but it tends to be done informally without a formal EIA being conducted. Furthermore, EIAs only focus on development proposal for specific sites and do not take into account the impacts on the wider environment of the journeys tourists will make to visit the new development.

30: Zoning tourism uses in the marine environment

Zoning, whereby activities or uses are permitted or prohibited in certain areas of the sea, is a well-established technique for managing marine environments. It can be based on the need to conserve rare resources so, for example, it is widely used to control commercial fishing to protect fish stocks. Zoning is also used to avoid problems where particular activities or uses conflict with each other. In the UK, for example, zoning is used in territorial waters to manage fisheries, shipping, military activities, underwater archaeological remains, nature conservation, and oil and gas exploration (Gubbey, 2005).

It is a technique used in the growing field of Marine Spatial Planning (MSP) which seeks to apply some techniques used in spatial planning on land. MSP is a '*public process of analysing and allocating the spatial and temporal distribution of human activities in marine areas to achieve ecological. Economic and social objectives that are usually specified through a political process*' (www.mspglobal2030.org, 2019). Therefore, zoning is usually based on government regulation or legislation, although in tourism there may be local schemes based upon a voluntary agreement amongst different user groups and local authorities. One particular form of zoning is the designation of Marine Protected Areas (MPAs). These are designed to protect vulnerable or exceptionally important marine environments from human activities including tourism. Ironically though, these designations can actually make the areas even more attractive to tourists which can increase pressure on them.

In the context of tourism and the oceans there are two main dimensions to zoning, namely:

- Zoning which allocates particular areas of the sea to tourism rather than other economic activities or seeks to exclude tourism or certain tourist activities from areas, usually to protect vulnerable marine environments or ecosystems.
- Zoning based on the fact that some marine leisure activities are incompatible and need to be separated, such as jet skis and swimming, or fishing and powerboating.

The former is more common, but the latter is found in locations where there is a fear that conflicting uses may be a safety risk to tourists themselves.

In terms of the zoning of tourism specifically, it is primarily seen in inshore waters where marine leisure activities and wildlife-watching are concentrated rather than in the 'the open sea'. This is because these areas are under the control of national governments and are where conflicts between uses or activities are most common and affect the largest number of people.

In the state of South Australia in 2012, the following four categories of marine zone were being used:

- General Managed Use Zone
- Habitat Protection Zone
- Sanctuary Zone
- Restricted Access Zone

In the General Managed Use Zone all marine leisure activities were permitted while in the Restricted Access Zone all leisure activities were prohibited, including diving, boating, surfing and swimming. In the Restricted Access Zone, in addition to leisure uses, all fishing was prohibited – apart from traditional fishing practiced by aboriginal people – together with the anchoring of vessels, marina construction, renewable energy infrastructure, and aquaculture (www.portvictoria.org.au, 2012). Also in Australia, the Great Barrier Reef Marine Park uses a system which features eight different types of zones with each having both a name and a colour which is used on maps (www.gbrmpa.gov.au, 2020).

The approach on Canada's Northern Pacific coast in 2016 was based on two matrices that focused on the compatibility of uses with the natural environment and on the vulnerability of the marine environment or rather of particular elements within the marine ecosystem including kelp, seagrass, and deep sea corals and sponges. This resulted in the creation of three types of zone. (www.mappocean.org, 2016)

I was intrigued by a draft report produced by a working group in the Seychelles in 2014 which talked in terms of *'sustainable tourism zones'*, suggesting that when zoning, different types of tourist activity might be treated differently depending on whether they were considered to be sustainable tourism or not (www.seychellesmarinespatialplanning.com, 2014). Interestingly, this working group report also drew upon the experiences of other destinations including St Kitts and Nevis which was discussed in a paper published in 2014 (Agostini et al., 2014)

In general, it can be said that most zoning of marine environments covers all activities and uses rather than just tourism. However, in many coastal destinations there are zoning policies limiting where certain leisure activities can take place, notably jet skiing and powerboating. Beaches may also be zoned with specific areas set aside for everything from exercising dogs to naturist bathing!. The common denominator for such zoning appears to be the perceived incompatibility of certain uses or activities. The final point about zoning relates to enforcement, which can be logistically difficult, expensive and unpopular with tourists who resent limitations on their activities.

31: The concept of carrying capacity in tourism and the oceans

The concept of carrying capacity in tourism dates back to the 1980s and has been widely studied by academics. While several definitions exist, it is basically the idea that every destination has a measurable limit to the number of tourists it can receive before negative impacts are experienced. There are several types of carrying capacity including:

- **Physical** – the area cannot physically accommodate any more tourists.
- **Environmental** – the point after which further tourists will lead to damage to the physical damage.
- **Ecological** – any increase in tourist numbers will cause harm to the ecosystem.
- **Socio-cultural** – an increase in tourists will harm the society or culture of the host community.
- **Economic** – an increase in visitors will damage the functioning of the local economy.
- **Visitor satisfaction** – the point at which the visitor experience begins to be adversely affected by the number of tourists.
- **Perceptual** – the threshold beyond which the perception is that the sense of place and what makes it special is harmed.

From the idea of carrying capacity developed the concept of 'limits to acceptable change' which suggests that a point may be reached when intervention is needed to prevent impacts occurring which are seen as unacceptable. These two related concepts have come to the fore recently with the debate over 'over-tourism'.

In Figure 11.1 I identify the contexts in which the idea of carrying capacity might apply to tourism and the marine environment.

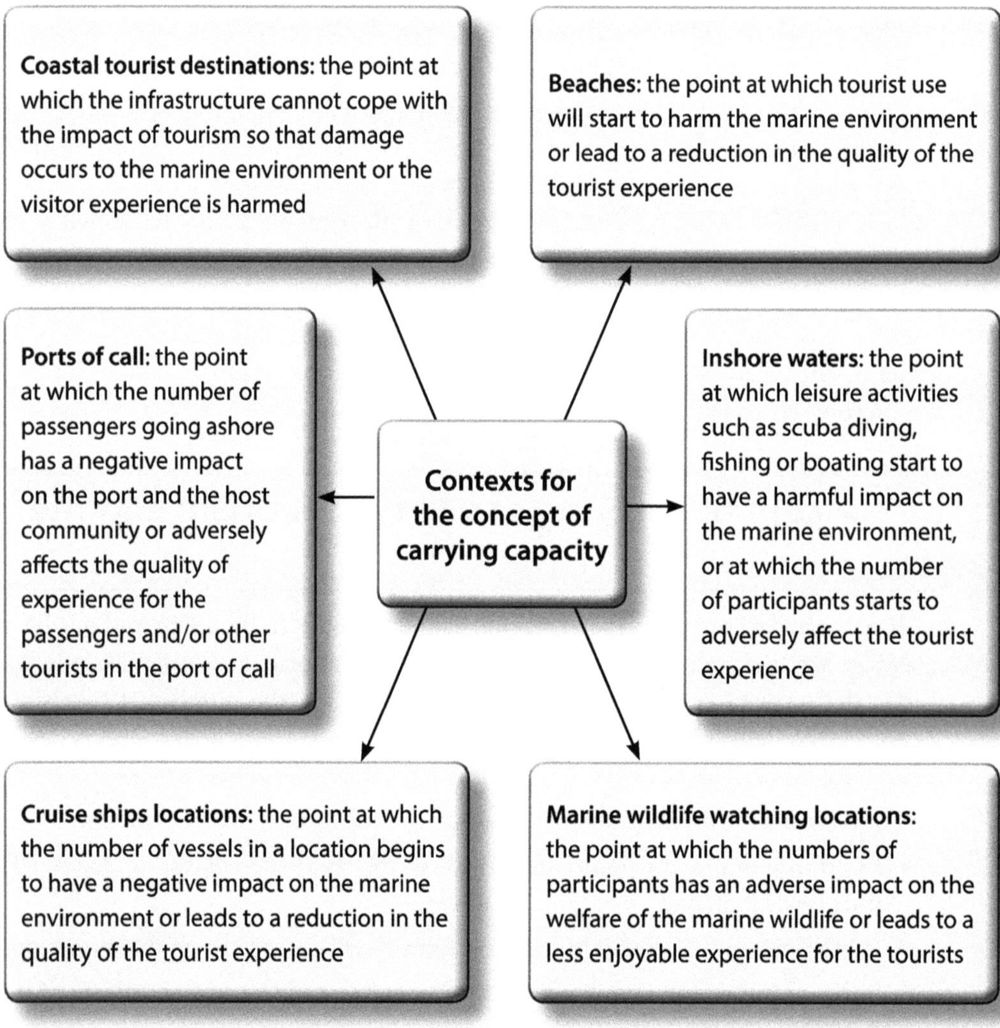

Figure 11.1: The different contexts for the concept of carrying capacity in relation to the marine environment

Although most of the work on carrying capacity has tended to be based in protected areas on land, some academics have looked at carrying capacity in relation to the marine environment. Most studies have focused on tourism in marine protected areas, activities such as scuba diving and the use of beaches. However, with the growing debate on the phenomenon of over-tourism we will probably see work being done on measuring the carrying capacity of the ports of call of cruise ships.

There are several observations to be made about carrying capacity in relation to tourism and the marine environment.

- First, it assumes that up to a certain number of tourists no harm will occur, yet in fragile marine environments even the presence of one tourist may well have a detrimental impact which may be irreversible.
- Second, the impacts of tourism are not just a result of the number of visitors, they also depend on their behaviours and the activities they are pursuing.
- Third, deciding how many tourists a location can carry is incredibly complex, based on the resilience or fragility of the environment, which may change from season to season, based on everything from weather systems to the migration and breeding behaviour of marine creatures.
- Fourth, at a time when the oceans are in crisis, perhaps our priorities should be focused on the impact of tourism on the marine environment rather than on the quality of the visitor experience.
- Fifth, given its focus on a particular coastal destination or area of ocean, the idea of carrying capacity ignores the impact of the tourist journey to the destination on the wider environment.

Perhaps the key point is that even if the carrying capacity of a location could be measured it is extremely difficult to implement in most locations in the open sea or in coastal destinations. It is perhaps, therefore, more useful as a theoretical idea than as a management technique for tourism and the marine environment.

32: The impact of tourism in Antarctica

Until the 1960s, the only visitors to Antarctica, which has no native population, were research scientists. Since then Antarctica has found itself increasingly on the tourist map, thanks almost exclusively to cruise ships. With no culture, a harsh climate, a limited variety of wildlife and no infrastructure, the only reasons to visit are the austere beauty of the scenery and perhaps the status of visiting somewhere that is unlikely to have been visited by the tourist's acquaintances. The main reason for visiting appears to be simply 'because it is there'. While some tourists call it an adventure, the fact that they are normally sleeping in luxurious cabins on board ship, seems to rather undermine this argument, although the voyage itself may be an adventure given the rough seas!

In 2016-2017, 44,202 tourists visited Antarctica, and 36,907 of these actually went ashore. Of those who landed 30% came from the USA, 14% from China, and

10% from Australia. Of those who land most simply sightsee or take small boat trips to explore with a guide. For those who undertake leisure activities, the most popular are kayaking, walks, and swimming (www.coolantarctica.com, 2019).

Interestingly, and perhaps worryingly, tourism in Antarctica is currently self-regulated by the industry in the form of the International Association of Antarctic Tour Operators which was set up in 1991. Tourism has a number of impacts on the marine environment in Antarctica including:

- Disturbance of breeding birds that may result in nest sites being abandoned.
- Inadvertent introduction of invasive species on the clothes of tourists and in food they bring ashore.
- Erosion of fragile environments due to trampling by large numbers of feet.
- Accidental oil spills by ships and boats.
- Debris left behind by small private group expeditions, which is not always removed as required under the Antarctic Treaty.

(www.coolantarctica.com, 2019)

The impacts noted above are made worse because they are concentrated in small geographical areas and take place over short periods of time when ships disgorge their passengers ashore for a limited period of time.

Furthermore, there are no local native communities to benefit from tourist expenditure to set against the environmental costs of tourism.

There is also always the risk that accidents will cause devastating damage to this fragile marine environment. In 2007 the cruise ship M/V Explorer sank and in 1979 an Air New Zealand sightseeing flight crashed on Mount Erebus. Neither caused an environmental catastrophe but the risk is real. And, as British researcher John Durban says, '*The rate of climate change around the Antarctic Peninsula is rapid. Increasing tourism* obviously *means burning more fossil fuels here*' (www.nationalgeographic.co.uk, 2018). A worrying development in the context of global warming is the reported launch of flights to Antarctica by an Argentinian airline and plans for a service from China after a successful experiment in December 2017 (www.nationalgeographic.co.uk, 2018) .

I believe that given the effects of global warming on Antarctica and its fragile environment, together with the huge number of other vacation opportunities open to affluent travellers, perhaps tourists should just not go to Antarctica at all. As we have noted throughout this book, tourism is not a necessity of life and no one will suffer through not visiting Antarctica. It is hard to see how any kind of tourism on that continent could be seen as responsible at the moment.

33: The World-Wide Fund for Nature and the impact of tourism on the marine environment

Based in the UK, and founded in 1961, the World-Wide Fund for Nature is probably the largest independent conservation organisation in the world. They rely on donations, membership revenue and income from activities such as 'adopt an animal' schemes. WWF are active all over the world and have become an influential lobbying group on the world stage. Their mission is to help create a world in which *'people and wildlife can thrive together'* (www.wwf.org.uk, 2019). They have been involved for many years in endeavouring to protect the marine environment and have become increasingly involved in *'promoting the adequate use of marine and coastal resources and increased marine protected areas, while creating economic benefits for local communities'* (www.panda.org, 2019). On their website, they stress their desire to work with industry as partners and recognise that marine and coastal tourism worldwide generates 200 million jobs, but they also say that cruise ships in the Caribbean produce more than 70,000 tons of waste water per annum. Their priorities for tourism and the marine environment include:

- *Encouraging smart coastal planning and zoning of critically important areas*
- *Providing support for the establishment and management of marine protected areas and key species*
- *Providing the skills for people to rely on tourism activities that are not destructive*

(www.panda.org, 2019)

WWF operates all over the world, with offices in many countries, which undertake their own projects. For example, in 2010, five national WWF offices combined to create the Mediterranean Marine Initiative which covered tourism, commercial fishing, and plastic and other forms of pollution.

Over the past decade WWF has undertaken a wide variety of initiatives aimed at reducing the negative impacts of tourism on the marine environment including:

- Working in partnership with Royal Caribbean Cruises Ltd to set targets for environmental sustainability for the cruise line including carbon emissions and sustainable seafood procurement, minimise the effects of marine leisure activities organised by operators and try to ensure that accommodation is not built too close to the shoreline.
- An 'adopt a sea turtle' scheme to raise funds for turtle conservation projects.
- A Global Marine Programme that encouraged the tour operators to play a pro-active role in reducing the negative impacts of tourism on the marine environment. They wanted them to work with accommodation that had some form of environmental practice certification, and provided financial assistance for protected areas.

- Launching in 2019 a 'Coral Reef Rescue Programme' in partnership with Vulcan Inc and the governments of Fiji and Indonesia.

WWF has developed an approach which tries to build mutually beneficial bridges between the conservation movement, industry and governments. It works quietly behind the scenes to influence industry practice and government policy, as well as publicly raising awareness of the issues. This seems to be producing results, but critics might argue that cooperation with WWF gives businesses and governments something of an umbrella when criticism is rained down upon them.

34: Crisis planning for tourist destinations in Australia

At the time of writing, parts of Australia are being ravaged by wildfires on a scale not seen before. While these have mostly been impacting inland areas, they have had a devastating impact on tourism across much of the country. The wildlife, which is the main reason for tourists to visit the unique ecosystem of Kangaroo Island in South Australia, has been devastated and air quality in the coastal city of Sydney has been amongst the worst in the world due to smoke from the fires.

Given the subject of this book it is interesting to note that there are already suggestions that the fires will lead to the oceans becoming more acidic due to the CO_2 produced by the fires but absorbed by the sea. This can harm shellfish and coral while the ash contains heavy metals and other materials which harm phytoplankton which is vital to the marine food chain.

Natural disasters, from catastrophic floods to wildfires to droughts, are common in Australia and despite the views of a few people, there is a general consensus that such crises will increase in frequency and severity due to climate change. This could have a devastating impact on the tourism sector, so it is little wonder that state governments have been developing strategies and advice for businesses. *Victoria's Tourism Strategy 2020*, published in 2013 by the state government, said '*It is becoming increasingly common for the tourism sector to be adversely affected by natural disasters such as bushfires, floods, drought, blue green algae outbreaks and locust plagues…There will be an increasing need for the industry to consider crises as a normal part of business.*' (www.ecotourism.org.au, 2020). The state government cites four types of crises which are focused on the ocean, including tsunamis, water safety, shark sightings and storms. It also offers a range of resources to help tourism businesses with their crisis planning (www.business.vic.gov.au, 2020).

In 2018 the state government of Queensland published *Building a Resilient Tourism Industry: Queensland Tourism Climate Change Response Plan*. This states that '*for Queensland's tourism business community climate change is both a risk and an opportunity too big to ignore. A strategic plan, underpinned by ambitious actions, will support Queensland's tourism industry to respond to climate risks and opportunities …*

A recent report by the Climate Council concluded that Australia's tourism industry is the most vulnerable and the least prepared to manage climate risks' (State of Queensland, 2018) The report is detailed and based on impressive research, although it remains to be seen how fully and quickly its many suggestions will be implemented.

The risks for the tourism industry were said to include heatwaves and sea-level rise, biodiversity loss and ocean acidification. At the same time as recognising the need to do more to reduce global warming, the report went on to say climate change might also provide opportunities to develop new products and diversify visitor experiences. Barriers to tackling climate risk were identified as a lack of specialist knowledge, short termism in planning and business decision-making, and inadequate resources and investment.

By 2019 the state government was providing detailed advice for tourism businesses in Queensland to help them prepare for emergency events including those caused in some way by climate change. This advice covers the following:

- *Assess the likelihood and impact of risks*
- *List key contacts and communication strategies*
- *Create an evacuation plan or emergency shut-down procedure*
- *Build an emergency kit*
- *Prepare a crisis response plan*
- *Minimise risks*
- *Form a crisis management team*

(www.business.qld.gov.au, 2020)

The buzz words in crisis planning in tourism appear to be *resilience* and *recovery* and while crisis planning in Australia is well established and professional it does seem to be based on the idea that crises will be generally short-lived and that recovery will be possible within a relatively short period. In terms of climate change, that may not be the case. Rising sea level may make flooding of existing coastal communities the norm, while some extreme weather events may so devastate an area that recovery could take years or be impossible or not cost-effective.

35: The 'flight-shaming' movement in Europe

'Flight shaming' is about encouraging people to fly less by making them feel guilty about the environmental impact of their flights. It has its origins in Sweden in 2017 and has become associated with the Swedish climate activist, Greta Thunberg, who in 2019 travelled by boat rather than air to New York to address the United Nations, to highlight the impact of air travel on global warming. In Sweden in 2019 some 23% of the population of Swedes claimed not to have set foot on a plane in the past year to reduce their impact on the environment, with many switching to train for journeys within Scandinavia (www.whatsorb.com, 2019). And in January 2020 it was reported that '*Sweden has seen a 4% drop in the number of people flying via its airport, a rare decrease in recent years for a European country'* (www.bbc.co.uk, 2020).

In October 2019, Reuters reported the results of a survey of 6,500 travellers in France, Germany, the UK and the USA, saying that 27% of respondents indicated that they were thinking about reducing their personal consumption of air travel. (www.reuters.com, 2019). The flight shaming movement has begun to have an impact on governments and the airline industry in Europe. In June 2019, the French government introduced an 'eco-tax' on all flights taking off from France, with the revenue being used to support other forms of transport. The Dutch airline KLM has launched a sustainability campaign that urges travellers to fly less over short distances where the train is a viable alternative (www.washingtonpost.com, 2019).

While the movement has made progress in Europe, it is yet to make a real impact in much of the world. Hannah Simpson in the Washington Post in June 2019 said '*Europe's* 'flight shame' *movement doesn't stand a chance in the U.S.'* (www.washingtonpost.com, 2019). Her view was largely based on the size of the country and the fact that high speed rail travel, which is common in Europe, is in the USA '*a dream that may never come true*' (www.washingtonpost.com, 2019). There is no doubt that the airline industry as a whole sees the flight shaming movement as a threat and there are rumours that the International Air Transport Association is planning a campaign to counter the effects of the movement. If true this may be understandable, as a report by CiTi in October 2019 suggested that flight shaming was '*gaining traction and could cost airlines billions*' (www.cnbc.com, 2019).

At the beginning of 2020 a debate had begun that did involve the USA as it revolved around celebrities flying to work and award ceremonies. The actions of celebrities seem to have a big impact on consumer behaviour these days, so a few film stars saying they will cut down their use of air travel could influence a large number of consumers. But if we are to reduce the impact of air travel on global warming, we need fewer flights rather than just fewer people flying on the same number of flights.

Fight shaming is an example of responsible tourism where consumers change their own behaviour voluntarily and make sacrifices for the benefit of the planet. It remains to be seen whether it is the start of a truly global phenomenon that will have a long-term impact on the world of air travel.

36: The International Panel on Climate Change 2019 Report on the Oceans

In September 2019, the International Panel on Climate Change produced a major report entitled, *The Ocean and Cryosphere in a Changing Climate* which received enormous media attention across the world. Produced by a team of international experts its findings made uncomfortable reading. The report stated that:

- *It is virtually certain that the global ocean has warmed unabated since 1970 and has taken up more than 90% of the excess heat in the climate system.*
- *Since 1993, the rate of ocean warming has more than doubled.*
- *By absorbing more CO_2, the ocean has undergone increasing surface acidification.*
- *Global mean sea level is rising, with acceleration in recent decades due to increasing rates of ice loss from the Greenland and Antarctic Ice Sheets as well as continued glacier mass loss and ocean thermal expansion.*
- *Increases in tropical cyclone winds and rainfall, increases in extreme waves, combined with relative sea level rise, exacerbate extreme sea level events and coastal hazards.*
- *Coastal ecosystems are affected by ocean warming including marine heatwaves, acidification, salinity intrusion, and sea level rise, in combination with adverse effects from human activities on ocean and land.*
- *Changes in the oceans have impacted marine ecosystems and ecosystem services with regionally diverse outcomes, challenging their governance. The impacts on ecosystem services have negative consequences for health and well-being and for indigenous peoples and local communities dependent on fisheries.*
- *Coastal communities are exposed to multiple climate-related hazards, including tropical cyclones, extreme sea levels and flooding.*
- *The Greenland and Antarctica Ice Sheets are projected to lose mass at an increasing rate throughout the 21st century.*
- *Sea level continues to rise at an increasing rate. Extreme sea level rates that are historically rare are projected to occur frequently at many locations by 2050. Extreme sea levels and coastal hazards will be exacerbated by projected increases in tropical cyclone intensity and frequency.*
- *A decrease in global biomass of marine animal communities, their production, and fisheries catch potential.... The rate and magnitude of decline are projected to be highest in the tropics.*
- *Warm-water corals are at high risk already and are projected to transition to very high risk even if global warming is limited to 1.5°C.*
- *Increased mean and extreme sea level, alongside ocean warming and acidification, are projected to exacerbate risks for human communities in low-lying areas.*

(www.ipcc.ch, 2020)

While the predictions are accompanied by caveats relating to levels of confidence, the overall message is clear – the situation is serious and is getting worse. The report also notes that the worst impacts are likely to be felt most strongly in two types of location, namely:

- 'The tropics', areas of the world where some of the poorest and most vulnerable people on Earth live in communities which are often increasingly dependent on tourism
- The Arctic where indigenous peoples are likely to pay a high price for what humanity as a whole has done

37: United Nations Sustainable Development Goal 14

UN Sustainable Development Goal 14 is to '*conserve and sustainably use the oceans, seas and marine resources for sustainable development*' (www.sustainabledevelopment.un.org, 2019). It is one of 17 goals agreed by the UN in 2015 as part of its 2030 Agenda for Sustainable Development. This is the latest stage in a process that began with the Earth Summit in 1992 in Rio de Janeiro. There are a number of target indicators to help monitor progress, although most of them do not include measurable performance indicators, but they do have target dates. For example, three of the target indicators are as follows:

- *By 2020 effectively regulate harvesting and end over-fishing.*
- *By 2025 prevent and significantly reduce marine pollution of all kinds.*
- *By 2020 sustainably manage and protect marine and coastal ecosystems to avoid significant adverse effects.*

  (www.sustainabledevelopment.un.org, 2019)

Each year, the UN reports progress on the goal as a whole and on the individual target indicators. These can make dismal reading as in reality there seems little progress to report! The 2019 report focuses on the problems that still remain and the only three sets of figures used to show where progress has been achieved are as follows:

- In December 2018, 17% of waters under national jurisdiction (those waters 0-200 nautical miles from a national border), were covered by a protected area compared to 12% in 2015.
- The global mean percentage of each marine key biodiversity area covered by protected areas increased from 31% in 2000, to 45% in 2015, and 46% in 2018.
- The number of parties to the 2016 Agreement on Port State Measures to Prevent, Deter and Eliminate Unreported and Unregulated Fishing stood at 58 in February 2019.

  (www.sustainabledevelopment.un.org, 2019)

Clearly, any improvements in the protection of our oceans are to be applauded but progress appears very limited to date. Still, 83% of waters under national jurisdiction are not part of protected areas. And, in terms of the percentage of key biodiversity areas covered by protected areas the rate of growth has actually fallen since SDG 14 was agreed! Finally, just because nations sign agreements does not mean they will be implemented. It is easy to be critical of the lack of progress on SDG 14, but in fairness the United Nations is in a difficult position. It has to adopt a 'lowest common denominator' approach to ensure that all members will agree things and in general it has no real power to ensure that governments pursue any of the SDGs.

That is why I believe that only a bottom up, rather than a top down, approach can bring about the rapid changes needed if we are to protect our oceans effectively.

38: Towards more responsible tourist behaviour

Throughout this book we have focused on the tourism industry and the public sector and their responsibilities in relation to the impacts of tourism on the marine environment. However, it could be argued that we should focus on the tourists themselves, for they are the key to making tourism more responsible because:

- It is their decisions and behaviour which actually creates the negative impacts of tourism in the marine environment.
- It is the consumers who buy the products and experiences offered by the tourism industry; if no one bought them the industry would not offer them.
- The tourists are also voters who elect local, regional national politicians so they have the power, potentially, to influence public policy.
- Tourists are also citizens who can become activists and campaigners based on their experiences as tourists.

Unfortunately, most tourists seem to be part of the problem rather than part of the solution currently. Perhaps this is because, psychologically, people see vacations as a time when you can escape from routine and get away from the responsibilities of work and daily life for a week or two. Whatever the reason, few tourists can argue that they were not aware of the plight of the oceans, global warming and the impacts of tourism, for the global media covers these stories regularly, in the main tourist generating countries at least. Irresponsible behaviour is everywhere though despite this media attention including tourists:

- Damaging coral reefs while diving
- Disturbing marine wildlife by jet skiing
- Disposing of plastic waste from boats and on the beach

- Making too much noise or getting too close when on wildlife watching trips.
- Buying souvenirs that are not sustainable and/or are made from marine creatures.
- Taking cruises to uninhabited places with fragile marine environments where every tourist has a negative impact on the environment simply by being there.
- Making long-haul flights to vacation destinations so their trip has a substantial carbon footprint.

Some observers would say that tourists need to be 'educated' to improve their behaviour, but I am not convinced that this will work, and it seems patronising in the extreme. Perhaps it is better to suggest that behaving more responsibly makes the tourist feel good and can thus enhance their enjoyment of their trip. This means encouraging them to ask questions before they buy. However, I believe that there will always be people, unfortunately, who do not respond to such an approach and in those cases strong regulation, backed by substantial penalties, will be required for those who are not willing to behave more responsibility. The crisis facing our oceans and the contribution of tourism to this crisis is so great and the need for action so urgent, that everyone needs to be on board as soon as possible. On the other hand, I believe there is reason for hope that things may improve because of developments such as:

- The rise of the 'flight shaming' movement in Europe.
- Tourists who become activists such as 'Surfers against Sewage in the UK.
- The rise of 'catch and release' in sea angling.
- The growing willingness of passengers to contribute to carbon offsetting schemes when taking flights, although many would argue these are not a solution but merely assuage the guilty feelings of the tourist.
- Individual tourists using social media to highlight examples of irresponsible tourism.

However, I would like to see things go further, with more tourists:

- Exercising self-control and self-censorship and getting away from the idea that tourists need to try to go everywhere and see everything, regardless of the impact this may have on the marine environment.
- Just enjoying wildlife rather than being obsessed by photographing it and sharing the images on social media. Much of the disturbance of wildlife is caused by getting too close because of the desire to take a good photograph.
- Joining together more with like-minded people to campaign to influence industry practice and boycott those operators who continue to behave irresponsibly.

Casual observation might lead one to suggest that levels of interest in responsible tourism vary between countries, with the highest level of interest being seen in

Scandinavia, Germany, and the UK. I believe that this is partly a function of history and the fact that some nationalities have been travelling longer than others, and their national tourist markets are at different stages of maturity as a result.

The sector where I see least evidence of increasing consumer interest in responsible tourism is the cruise sector. I have spent hours watching television channels in the UK which focus on cruising. In all that time I have never seen anything that encourages cruise passengers to try to reduce food waste or water use, move away from single-use plastic, or make sure they buy things which contribute directly to the local economy. The main motivation for buying cruises appears to be simply the seductive notion that it is all about hedonism and luxury with no thought to the consequences or the impact on the marine environment. Given the continued rapid growth in the cruise market this is a major concern.

Ultimately, all any individual can really control is their own behaviour and choices, and if everyone made more responsible choices the negative impacts of tourism on the marine environment could be greatly reduced. This will probably only happen if:

- Industry advertising is controlled more rigorously to eliminate false or unproven claims made by some businesses that risk making tourists cynical about the idea of responsible or sustainable tourism.
- Credible accreditation schemes, including independent verification, are developed, so tourists who want to 'do the right thing' can buy products which are genuinely more responsible.

39: 'What's in the news this month'

I thought it would be an interesting exercise to look at the media over a month and see what was being written that related to tourism and the marine environment in any way, to see how many stories there were and what kind of issues they covered. I intended to do this for a whole month but actually found that I had more than enough stories after just the first nine days! Below the reader will find a selection of stories I found, which makes it clear that there is a lot happening in our oceans that is related in some way to tourism. By way of context readers should note that during this month the global media was primarily focusing on the emerging Coronavirus crisis in China and the upcoming US Presidential election, both huge stories for the media.

- February 1st. CBS News in the USA reported that *'scientists alarmed to discover warm water at "vital point" under Antarctica's "doomsday glacier".'* (www.cbsnews.com, 2020) This story claimed that for the first time, warm water had been discovered underneath one of the fastest melting glaciers. It suggested that this might mean that the retreat of this glacier is now unstoppable which would have huge implications for global sea level rise. It went on to say that Dr Daniel Holland from New York University had suggested that if this glacier system were to melt completely, global sea levels could rise by nearly one metre.

- February 2nd. An online industry news website reported that the cruise line MSC had launched a new programme of shore excursions under the name of 'Protectours'. This was said to be *'part of the cruise line's mission to offer passengers an environmentally conscious vacation while making a positive contribution to the planet'* (www.travelpulse.com, 2020). It is for the reader to decide if this is a genuine attempt to make shore excursions more responsible or an exercise in green washing. It is interesting that this story appeared at a time when cruise passenger numbers were predicted to rise to 32 million in 2020 compared to 23 million just five years earlier (www.statista.com, 2020). The story also noted that a survey had shown that for some 54% of Generation Z consumers, the environmental impact of travelling affected their vacation choices.
- February 4th. The BBC reported that *'the UK's aviation industry is promising to reduce its net carbon emissions to zero by 2050'*. It suggested the industry believed this could be achieved through technological developments such as cleaner engines, new fuels and tree planting to offset emissions. However, this claimed was criticised by environmental campaigners who stated that the *'only way to cut airline emission is by reducing air travel'* (www.bbc.co.uk, 2020).
- February 4th. US broadcaster CNN had a story under the heading of *'nearly $1 million worth of shark fins seized by wildlife inspectors in Florida'* (www.cnn.com, 2020). They claimed that around 600 kilograms of shark's fins had been intercepted on their way to Asia from South America. It is against the law in Florida to trade shark's fins. The report claimed that Hong Kong is the world's largest shark's fin trading hub and that the populations of some species of sharks had fallen by 90% largely as a result of the trade.
- February 5th. The Seychelles News Agency stated that *'the World Bank will help the Seychelles with the provision of experts and financial support to help the island nation deal with coastal erosion as a result of climate change'* (www.seychellesnewsagency.com, 2020). The article also stressed the importance of the terrestrial and marine ecosystems to the economic well-being of the country in terms of attracting tourists.
- February 6th. CNN reported on progress being made on the conservation of mangroves in Sri Lanka through the work of volunteers and local communities. The value of mangroves was recognised after the 2004 tsunami, when it was noticed that casualties were fewer in areas protected by mangroves. In 2015 the government announced it would conserve all of its mangroves (www.edition.cnn.com, 2020).

- February 6th. A New Zealand news source had a headline which stated, '*divers to move marine life before Waikawa marina expansion*' (www.i.stuff.co.nz, 2020). This was a community-led project to relocate starfish, sea cucumbers, tube worms and other marine creatures to a nearby bay in Queen Charlotte Sound, because local people were worried that they would not survive the construction process as the Waikawa marina was expanded. This followed an Environmental Impact Assessment for the project which suggested that the project would have significant adverse effects on marine wildlife in the area. The local volunteers apparently believed that they could not save all the marine life that would be impacted by the development, but they wanted to save as many as they could.
- February 6th. The US Space Agency NASA reported that '*Arctic ice melt is changing ocean currents ... a major ocean current in the Arctic is faster and more turbulent as a result of rapid sea ice melt*'. The report suggested that, ultimately, this process could lead to a modification of ocean currents worldwide which would in turn have a significant impact on climate and weather patterns worldwide (www.climate.nasa.gov, 2020).
- February 9th. Reporting a study by scientists at Herriot Watt University, The Times newspaper, in the UK, reported that *every blade of seagrass in ocean habitats that are havens for fish and other marine life has been contaminated by tiny pieces of plastic.* The report claimed that plastic particles cling to the seagrass blades and are consumed by fish, sea snails and shrimps. (www.thetimes.co.uk, 2020)

From this selection I believe we can make some useful observations.

- First, the global media appears to have bought into the idea that our oceans are in crisis and that climate change is at the heart of this crisis.
- Second, the stories come from across the world showing that the problems facing our oceans and the communities that live around their fringes are a truly global issue.
- Third, the reports featured show examples of citizens taking action themselves rather than relying on governments to do what needs to be done.
- Fourth, the stories show that industry recognises that it is under pressure to respond to environmental concerns, but that it remains to be seen whether this will result in meaningful action or in initiatives that could be seen as 'green-washing'.

I have mixed feelings about the plethora of media stories about the oceans and tourism, for it indicates growing media interest in these issues which is raising public awareness, but this growing level of interest reflects the increasing severity of the challenges we face.

References

Agostini, VN, Margles, SW, Knowles, JK, Schill, SR, Bovino, RJ, and Blyther, RJ (2015) Marine zoning in St Kitts and Nevis: a design for sustainable management in the Caribbean. *Ocean and Coastal Management* **104**, 1-10.

www.amnh.org (2019) www.amnh-org/explore/videos/biodiversity/mangroves-the-roots-of-the-sea/mangrovethreats-and-solutions [18 Dec 2019]

www.asean.cruising.com (2018) asean.cruising.com/royal-caribbean-scouts/private-islands-asia 2018 [22 Oct 2019]

www.bali.com (2019) www.bali.com/diving-code-of-conduct.html [20 Jan 2020]

www.bbc.co.uk (2020) www.bbc.co.uk/news/world-europe-51067440 [15 Jan 2020]

www.bsee.gov (2019) www.bsee.gov/faqs/what-is-the-national-artificial-reef-plan [8 Dec 2019]

www.business.vic.gov.au (2020) www.business.vic.gov.au/tourism-industry-resources/Business-Tools-and-Support/crisismanagement-guide/plan/stay-informed [2 Feb 2020]

www.charroa.org.au (2019) www.charroa.org.au/codes_of_practice.html [22 Jan 2020]

Cheung, C, Aliño, PF, Uychiaoco, AJ and Arceo, HO (2002) *Marine Protected Areas in Southeast Asia*. ASEAN Regional Centre for Biodiversity Conservation, Department of Environment and Natural Resources, Los Baños, Philippines

Christie, P (2004) Marine protected areas as biological successes and social failures in southeast Asia. *American Fisheries Society Symposium* **44**, 155-164.

Cohen, E (2011) Tourism and land grab in the aftermath of the Indian Ocean Tsunami. *Scandinavian Journal of Hospitality and Tourism* **11**(3) 224-236.

www.coolantarctica.com (2019) www.coolantarctica.com/Antarctica%20fact%20file/science/threats_tourism.php [28 Jan 2020]

www.co2.myclimate.org (2020) www.co2.myclimate.org/en/flight-calculators/new [25 Jan 2020]

www.crew-center.com (2019) www.crew-center-com/juneau-alaska-criose-ship-schedule-july-october-2019 [3 Jan 2020]

www.cruisecheap.com (2019) www.cruisecheap.com/cruises/private-islands.html [19 Nov 2019]

www.cruiseexperts.org. (2020) www.cruiseexperts.org/clia-agents/what is clia/ [Feb 2020]

www.cruising.org (2020) www.cruising.org/about-the-the-industry/policy-priorities/cruise-industry-regulation [8 Feb 2020]

www.data.worldbank.org (2018) www.data.worldbank.org/indicators/ER.MRN.PTMR.25. [13 Jan 2020]

www.dec.alaska.gov (2019) www.dec.alaska.gov/air/air-monitoring/juneau-cruise-ship-monitoring-project/ [15 Jan 2020]

www.devex.com (2019) www.devex.com/news/opinion-how-to-build-resilient-tourism-in-the-caribbean-95570 [22 Jan 2020]

www.ec.europa.eu (2019) www.ec.europa.eu/environment/cites/info_souvenirs_en.htm [25 Oct 2019]

www.fishfishme.com (2019) www.fishfishme.com/blog/top-ten-fish-south-africa [2 Dec 2019]

www.ft.com (2019) www.ft.com/content/8bceef94-86cd-11e9-a028-86cea8523dc2 [3 Feb 2020]

www.ftnnews.com (2019) www.ftnews.com//38299-how-tourists-and-operations-can-protect-Egypt-s-coral-reefs [22 Dec 2019]

www.gbrmpa.gov.au (2019) www.gbrmpa.gov.au/access-and-use/zoning/about-zoning [18 Feb 2020]

www.gbrmpa.gov.au (2020) www.gbrmpa.gov.uk/the-reef-facts [22 Jan 2020]

Glasson, J, Therivel, R, and Chadwick, A (2005) *Introduction to Environmental Impact Assessment*. Routledge. London.

www.greenfins.net (2019) www.greenfins.net/about-green-fins [12 Jan 2020]

www.greenglobetravel.com (2020) www.greenglobetravel.com/eco-lodges-eco-hotels [19 Feb 2020]

www.greenhotelier.org (2015) www.greenhotelier.org/best-practice-sub/talking-point/tourism-ten-years-on-the-role-for-hotels [10 Feb 2020]

www.greenisthenewblack.com (2019) www.greenisthenewblack.com/10-things-to-consider-when-looking-for-real-eco-resorts [14 Feb 2020]

Great Barrier Reef Marine Protected Area Authority (2019) *Great Barrier Marine Park Authority Annual Report 2018-2019*. GBRMPA. Townsville

Gubbay, S (2005) *Marine Protected Areas and Zoning in a System of Marine Spatial Planning*. A Discussion Paper for the WWF. WWF

www.humpbackswims.com (2020) www.humpbackswims.com/about [8 Feb 2020]

www.icriforum.org (2019) www.icriforum.org/about-coral-reefs [20 Jan 2020]

www.ihei.org (2020) www.ihei.org/about-us/history[29 Jan 2020]

www.independent.co.uk (2018) www.independent.co.uk/travel 19 Dec. [4 Jan 2020]

www.ipcc.ch (2020) www.ipcc.ch/srocc [17 Jan 2020]

www.iucn.org (2019) www.iucn.org/news/protected-areas/201912/iucn-green-list-gets-down-earth-and-sea-Malaysia. [5 Jan 2020]

www.iwc.int (2020) www.iwc.int/wwguidelines. [20 Feb 2020]

Kamil, KA, Hailu, A, Rogers, A and Pandit, R, (2017) An assessment of marine protected areas as a marine management strategy in southeast-Asia: a literature review. *Ocean and Coastal Management* **145**, 72-81.

www.knoema.com (2019) www.knoema.com/atlas/Maldives/topics/Tourism/Travel-and Tourism-Total-Contribution-to-GDP [12 Jan 2020]

www.kuoni.co.uk (2019) www.kuoni.co.uk/inspiratio/the-best-places-in-the-world-to-see-turtles [8 Jan 2020]

Li, J (2008) *Environmental Impact Assessment in Developing Countries: An Opportunity for Greater Environmental Security?* Working Paper No. 4. USAID and the Foundation for Environmental Security and Sustainability

www.maldives.net.mv (2019) www.maldives.net.mv/tag/tourist-arrivals [12 Jan 2020]

www.mappocean.org (2016) www.mappoean.org. Presentation 6 Oct [11 Dec 2019]

www.mspglobal2030.org (2019) www.mspglobal2030.org/wp-content/uploads./2019/03/Marine-Spatial-Planning-a step-bystep-approach.pdf [12 Feb 2020]

www.nationalgeographic.co.uk (2018) www.nationalgeographic.co.uk/travel/2018/03/tourism-antarctica-what-does-it-mean-worlds-1st-wilderness [22 Oct 2019]

www.newscientist.com (2017) www..newscientist.com/article/2125198-on-front-line-of-climate-change-as-maldives-fight-rising-seas [14 Oct 2019]

www.oceanservice.noaa.org (2019) www.oceanservice.noaa.gov/facts/artificial-reef.html/ [22 Dec 2019]

www.oceanwealth.org (2019) www.oceanwealth.org/ecosystem-services/recreation-tourism/ [22 Jan 2020]

www.onlinelibrary.wiley.com (2019) www.onlinelibrary.wiley.com/doi/abs/10.1002/jtr.442 [30 Dec 2019]

www.panda.org (2019) www.panda.org/our_work/oceans/problems/pollution [29 Nov 2019]

Parsons, ECM (2012) The negative impacts of whale-watching. *Journal of Marine Biology*, Article ID 807294.

www.portvictoria.org.au (2012) www.portvictoria.org.au/files/marineparksactivitiesuses.pdf [2 Oct 2019]

www.reliefweb.net (2005)www.reliefweb.net/report/Bangladesh/10-lessons-learned-south-asia-tsunami-26-Dec-2004 [1 Feb 2020]

Sadler, B (1996) *Environmental Assessment in a Changing World: Evaluating Practice to Improve Performance. International Study on the Effectiveness of Environmental Assessment*. Canada Environmental Assessment Agency. Ottawa

www.sas.org.uk www. sas.org.uk/plastic-free-communities [2 Jan 2020]

www.seacc.org (2019) www.seacc.org/cruise-ship-impacts [3 Jan 2020]

www.seychellesmarinespatialplanning.com (2014) www.seychellesmarinespatialplanning.com/wp-content/uploads/2014/06/SeychellesZoningDraft_v.1.1.pdf [6 Feb 2020]

www.ship-technology.com (2019) www.ship-technology.com/features/cities-who-banned-cruise-ships [5 Jan 2020]

www.skiff.com (2019) www.skiff.com/2019/10/31/the-cruise-industry-will-soon-face-its-strictest-pollution-standard-yet-can-anyone-enforce-it [2 Feb 2020

www.solentskiers.org.uk (2014) www.solentskiers.org.uk/2014/09/04/environmental-impacts-jet-skiing [2 Feb 2020]

www.sportquestholidays.com (2019) www.sportquestholidays.com/fishing-region/Africa [22 Dec 2019]

www.sustainabledevelopment.un.org (2019) www.sustainabledevelopment.un.org/sdg14 [21 Dec 2019]

www.telegraph.co.uk (2019) www.telegraph.co.uk/travel/739287/flying-three-times-cleaner-than-cruising.html [19 Oct 2019]

www.theculture.trip.com (2018) www.theculture-trip-com/central-america-articles/watch-sea-turtles-hatch-central-america/ [18 Jan 2020]

www.theguardian.com (2018) www.theguardian.com/environment/2018/nov/28/diving-force-experts-join-forces-to-savethe-worlds-coral-reefs [19 Nov 2019]

www.theguardian.com (2019) www.theguardian.com/business/2019/sep/16/a-rising-tide-overtourism-and-the-curse-of-the cruise-ships [8 Jan 2020]

www.thepointsguy.co.uk (2019) www.thepointsguy.co.uk/guide/everything-you-need-to-know-carbon-offsetting-flights [18 Jan 2020]

www.thesevenseas.net (2017) www.thesevenseas.net/wp/wp-content/uploads/2017/06/DiversCode.pdf [8 Jan 2020]

www.triblive.com (2020) www.triblive.com/news/carnival-cruise-ship-dumped-5900-gallons-of-grey-water-into-the-ocean-at-Port-Canaveral [4 Feb 2020]

www.undrr.org (2014) www.undrr.org/news/decade-after-indian-ocean-tsunami-lessons-learned [2 Feb 2020]

www.voanews.com (2019) www.voanews.com/europe/frances-cannes-ban-polluting-cruise-ships [22 Dec 2019]

Walker, R (2019) *Evaluating Marine Mammal Watching Legislation, Regulations and Codes of Conduct*. Winston Churchill Memorial Trust Report

www.washingtonpost.com (2019) www.washingtonpost.com/travel/2019/07/09/europes-flight-shaming-movement-doesnt-stand-chance-in-the-usa [31 Jan 2020]

www.worldcetaceanalliance.org (2018) www.worldcetaceanalliance.org/wp-content-uploads/WCA-Global-Best-Practice-Guideline-Whale-Watch-low-res.pdf [20 Jan 2020]

www.worldmaritimenews.com (2019) www.worldmaritimenews.com/archives/276946/cruise-ship-fined-for-violating-norwegianfjords-sox-limit [15 Feb 2020]

www.wwf.org (2019) www.wwf.org.id/en/about_species_how_we_work/marine_protected_areas [21 Jan 2020]

www.wttc.org (2018) www.wttc.org/-/media/files/-reports/2018/caribbean-recovery-report- - - executive-summary. pdf [2 Dec 2019]

Yunzhou, L and Fluharty, DJ (2017) Marine protected area networks in China; challenges and prospects. *Marine Policy* **85**, 8-16

12 Conclusions and the Future

In this book, I have attempted to cover a very broad subject, namely the impact of tourism on the marine environment. The text and the case studies together present the reader with a picture of the situation at the time of writing, which is late 2019 and early 2020. However, the situation is changing rapidly, not least due to the pace at which global warming is affecting the oceans and the rapid growth of the tourism industry in many parts of the world. In this final chapter, I will try to bring everything together and make a number of points by way of a conclusion although I hope that for many readers this chapter will stimulate them to find out more and undertake research that takes our knowledge of this fascinating subject further. Given how important the oceans are to the tourism industry and tourists it is highly surprising, and perhaps almost shocking, how little of the tourism literature focuses on the relationship between tourism and the marine environment.

Before we look specifically at this relationship, though, I believe that it is important to start by setting the scene by looking at the 'big picture' that provides the context for the relationship between tourism and the marine environment.

The 'big picture'

This book has been written at a time when the news media is full of stories about global warming and, for the first time, it appears that these stories are increasingly focused on the impact of climate change on the oceans rather than on land. This increased prominence in the media makes sense because the warming of the oceans is a major factor in the growth of extreme weather events that have been experienced on land in recent years, particularly in coastal communities. Furthermore, the melting of glaciers and polar ice caps as a result of global warming is leading to rising sea levels, which means the threat of more frequent and severe flooding in coastal communities and eventually the destruction of low-lying coastal settlements altogether. Rises in water temperatures and sea levels pose a major threat to both the marine environment across the world and the lives of people in low-lying coastal communities.

These issues also pose a particular threat to the tourism industry, given the important role which coastal destinations play in the vacation experiences around the world. Given the dependence of much of the tourism industry on clean beaches and unpolluted seas, it is very worrying that in the past few years a number of major threats to the both have been identified from pollution of the ocean and beaches by plastic, including microplastics, as well as pollution from industrial and agricultural run-off coming from inland sources. So, the crisis facing our oceans is having an increasing impact on tourism, and this can only get worse unless global warming and plastic pollution are tackled effectively, which frankly seems unlikely, at the time of writing. Rather than being seen as a potential victim of global warming, tourism needs to be seen as a significant contributor to the problem, not least but not solely, through the effects of air travel. Leisure car use is also a significant contributor to global warming and every part of the tourism industry has a carbon footprint, no matter how small they may be; no part of tourism is currently carbon neutral!

Tourism is also responsible directly for a number of negative impacts on the marine environment and our oceans which we will now consider in terms of the cruise industry and tourism in the coastal zone.

The main impacts of the cruise industry on the marine environment

We have seen that until recently relatively little attention had been paid to the impacts of the cruise industry but that has now all changed. As the cruise market has grown and cruise ships have become larger and larger, the spotlight has begun to be focused on the sector. This has also been stimulated by media coverage of legal cases against cruise companies relating to them being accused of breaking rules which exist to limit the environmental impact of cruise ships. Photos of huge cruise ships towering over small ports of call and the impacts of thousands of cruisers being disgorged in destinations for just a few hours has started a debate over whether cruising is sustainable in the longer term. This has been reinforced by the fact that cruise ships now seen to be everywhere, even to fragile marine environments such as Antarctica. It is clear that the cruise industry contributes to global warming through its emissions and also causes pollution from fuel spills and waste disposal as well as injuring and killing marine wildlife accidentally. While cruise companies are beginning to take sustainability more seriously, not least due to the effects of the court cases referred to earlier, they still need to do more and be more ambitious in their targets. They also need to do more to make their consumers aware of the impacts cruising can have and encourage their passengers to behave more responsibly.

The main impacts of tourism in coastal destinations on the marine environment

In Chapters 4 to 7 we explored the impacts of tourism on the marine environment in the coastal zone, which covers coastal communities, beaches, and inshore waters – the area of sea which is close to the land. This is where most of the impacts of tourism on the marine environment are concentrated, including:

- Wildlife disturbance and harm caused by irresponsible wildlife-watching.
- Wildlife disturbance, pollution of the seas, and noise pollution from marine leisure activities including diving, boating, jet skiing, and sea angling, for example.
- The depletion of marine wildlife through over-fishing and the taking of souvenirs such as dried sea horses and pieces of coral, for instance.
- Short- and long-term harm to the marine environment caused by coastal infrastructure projects including the construction of artificial islands, beach nourishment and desalination.
- Plastic pollution of the marine environment from coastal destinations and potential marine pollution from inadequately treated sewage.

These impacts tended to be highly concentrated in small geographical areas around destinations although their effects will be felt further afield given that the ocean is an open system. We did see evidence that some voluntary initiatives by groups of activists, enlightened entrepreneurs and representative bodies of some marine leisure activities are beginning to make a difference, albeit a modest one.

The ocean bites back!

The ocean can 'bite back', particularly when Man mistreats the planet and the marine environment specifically. Most extreme weather events have their origins in and above the oceans, but their destructive power is largely felt when they make landfall and often for hundreds of kilometres inland. The frequency and severity of such events appears to be increasing due to the effects of global warming.

There seem to be a growing number of shark attacks on humans, despite the fact that shark populations are under pressure, some of which is being blamed on the warming of the sea and the fact that more sharks re coming close to shore, sometimes because marine protected areas are creating rich concentrations of prey fish. At the same time the global spread of tourism and growth in various water-sports is bringing sharks into contact with more and more people. Over-fishing may be making it harder for sharks to find food in their traditional hunting grounds forcing them to come closer to shore, while some suggest that urbanisation and run off into inshore waters is creating silty conditions that sharks favour for hunting.

The growth of attacks at a time when the global population of sharks is declining suggests that man-made factors are likely to be playing a role in the increase. Whatever the reasons, there had been three fatal shark attacks on the coast of the small Indian Ocean island of La Reunion between January and October 2019. By comparison, between 1958 and 2018, there were a total of 19 fatal attacks in La Reunion, which averages just under one every three years.

A more subtle way in which the marine ecosystem can hit back against humans, albeit unintentionally of course, is through the food chain. Scientists are becoming increasingly concerned that the micro-plastics that are ingested by fish and shellfish will end up in the stomachs of humans. This may seem like poetic justice, but it is a significant and little-researched concern for public health officials.

Of course, sometimes the ocean can harm humans in ways which have little or nothing to do with anything that human beings have done. The most extreme example of this is the tsunami, one of which killed over 200.000 people around the Indian Ocean in December 2004.

For me, two things are worthy of note in relation to this idea that the oceans can 'bite back' against humanity.

- First, the apparent increase in examples of natural disasters which have their origins in or over the sea but affect the land suggests a need for every coastal zone to have an effective system of emergency planning, crisis management and the facility to provide early warnings to local residents.
- Second, the fact that the growing frequency and severity of some so-called 'natural disasters', notably extreme weather events are believed to be partly due to global warming caused by human beings, suggests the term 'natural disaster' may no longer be appropriate in these cases!

Tourism is not a necessity of life

If we are to try to halt global warming and tackle the crisis facing our oceans it is going to require radical changes to lifestyles and how industries operate, particularly in so-called developed countries and those which are developing at a rapid pace today. However, whereas industries such as agriculture and energy can argue that they are essential to human survival, tourism can, surely make no such argument. It is an industry that is about quality of life not survival, and it is not necessity of life. After all, you do not die if you do not get an annual vacation and if the planet is at risk it is surely easier to stop travelling than it is to stop eating or wearing clothes! Furthermore, taking a vacation it is not something that everyone on the planet is able to enjoy today. Perhaps, the majority of the world's population and certainly more than a billion people are too poor to take a vacation. Global warming due to air travel is largely due to a minority of people making multiple numbers of flights for business or leisure.

Therefore, it could easily be argued that tourism should take more than its fair share of the changes required to reduce global warming and protect our oceans. This could mean tourists being limited to one international trip per annum, restrictions on the length of flights that can be taken, or incentives or extra taxes being used to reduce the use of private cars for leisure purposes. While many would argue that leisure travel is less justified than business travel, technology has certainly already provided us with tools that could reduce the need to travel for business meetings and conferences.

There is a potential downside to reducing tourism globally that cannot be ignored. Millions of people worldwide rely on tourism for their livelihood and many governments rely on it for tax revenue. It is a cruel twist of fate that many of the countries which are most dependent on tourism for income and jobs are amongst the poorest on the planet. They often rely on long-haul flights from affluent source markets, and if long-haul flying were to be reduced, they would stand to lose most. While northern Europeans, for example, might face nothing worse than having to take a train to another European country rather than a flight to Asia for their vacation, people in the Asian country could lose their source of income with a devastating impact on their whole family. Solutions need to be found for this, for it would surely be unfair for those who are already the most disadvantaged to pay a disproportionate price for saving the planet.

■ Tourism is not an island!

I hope readers will forgive the pun given the subject of this book, but this phrase is a perfect way of making the point that tourism does not exist in isolation. Tourism is only one source of impacts on the marine environment alongside industries such as agriculture, commercial fishing, mineral, oil and gas extraction and shipping. The marine environment is also negatively impacted by pollution which is caused by those who simply live by the coast as well as well as from tourists. Some of the pollution that affects our oceans has its origins hundreds of kilometres from the ocean in inland towns and cities. At the same time, tourism itself is affected by the actions of some other industries, and resident populations which may harm the marine environment and beaches.

In the opposite direction, though, coastal tourist destinations are often the victims of extreme weather events that have their origins in and above the oceans. However, these events also affect coastal settlements that have nothing to do with tourism as well as inland towns and villages. What is required, therefore, is a holistic approach towards planning and managing that recognises the links between what happens in the marine environment and what happens on land as well as recognising the relationships between tourism and other industries

'One size fits all' solutions will not work!

One thing that has been clear throughout this book, or at least after Chapter 2, is that the idea of a single marine environment is simply not true. Every part of the ocean is different and in reality, there are a myriad of marine environments. They vary based on a range of criteria including: the salinity of the water; its depth; its temperature; the material on the seabed; the amount of sunlight which filters through the water; and the nature and size of the tidal flows. They vary based on whether they are open sea, or inshore and coastal zone environments. In the case of the latter they also vary in terms of their relationship with the land, which can mean everything from wide-open bays to narrow rocky inlets, coral reefs and lagoons or sandbanks, mudflats or estuaries, and so on. And, of course, because of the inter-play of all these factors, they also vary enormously in terms of the marine wildlife that lives within a particular environment. Finally, they vary substantially in terms of the direct impacts which human activities and industry have upon them.

We therefore need to ensure that planning policies and management plans take these differences into account and make sure that the solutions proposed are not generic ones but, rather, are tailormade to the specifics of an individual location. This is easier said than done because marine environments are not closed systems and their interactions with other marine environments need to inform planning and management. There is also a need to recognise that not all marine environments are equally vulnerable in the face of pressure from tourism and other human activities and industry. Some are more resilient and some recover from damage quicker than others. This needs to be borne in mind when planning and managing tourism in marine environments.

Finally, it is important to bear in mind that the impacts of tourism specifically, on the marine environment, tend to be concentrated on relatively small geographical areas of the marine environment. These areas need to be our priorities for pro-active planning and management.

De-regulation or regulation?

A constant theme throughout this book has been the issue of the balance between voluntary industry self-regulation and the regulation of industry by the public sector. To date, the tourism industry, with the exception perhaps of the cruise sector, has been quite proactive and has voluntarily undertaken initiatives to reduce the negative impacts of tourism on host communities and destinations. Some of this has been undertaken by industry bodies while individual corporations have also taken action on their own initiative. In doing so they have gone well beyond anything required by government regulators and have led, rather than responded to, consumer demand in respect of more responsible tourism.

One can speculate on the motives for the industry being so proactive and taking so much voluntary action. I believe there have been three main motivations:

- A wish to improve the image of the company or sector in the market amongst consumers, the media and the regulators, as part of a broader corporate social responsibility strategy.
- A desire to head off the need for government regulation by demonstrating that the industry is responsible and capable of 'putting its own house in order' without the need for public sector regulation that might prove burdensome and expensive.
- An attempt to 'get ahead of the game' and ensure they do not suffer competitive disadvantage if and when consumers start to take ethical issues and sustainability into account when making vacation decisions, at some point in the future.

Some of those industry bodies and individual corporations which have been at the leading edge of driving change are beginning to tacitly accept the need for further regulation, largely for commercial reasons. Those that are 'doing the right thing' are seeing themselves lose consumers to operators who cut corners and give consumers what they want with no concern for the impacts of their actions. They are usually able to undercut the prices of more responsible operators, which is an issue in a market as price sensitive as tourism. They are now beginning to support the idea of further regulation, if only to create a more 'level playing field' in relation to competition. However, as we have also seen, regulation itself is not enough, the regulations have to be strictly and fairly enforced and that is still rare in most tourist destinations.

We have seen that most sectors of the industry and some destinations themselves have voluntary codes of conduct aimed at encouraging operators and tourists to behave more responsibly. I think in the vast majority of cases it is crystal clear that such voluntary codes are ineffective and are often ignored by less scrupulous operators and tourists who are intent on just enjoying their vacations with little thought about the impact of their behaviour. I therefore believe that we need to regulate behaviour more rather than just relying on goodwill and the morality of operators and tourists. We need everyone to behave more responsibly not just some operators and tourists. In terms of tourist behaviour in particular I also think that peer group pressure can help improve behaviour. Surfers, divers and sea anglers, for example, are more likely to take notice of other surfers, divers and sea anglers than they are of well- meaning authorities.

Perhaps the main reason why we cannot rely on industry self-regulation and voluntary codes of conduct is simple. They bring positive change too slowly given the crisis we face in terms of global warming and the plight of our oceans.

Nowhere is this more evident than in the cruise sector where things are improving but not quickly enough. Industry targets for reducing emissions and pollution from vessels are not ambitious enough and cruise companies are not doing enough to raise consumer awareness of the impacts of their behaviour and how they can become more responsible. If this does not change then there will be no alternative to strict regulation, which may even need to restrict where cruise ships can go if the sector does not practice some self-regulation so it steers clear of offering cruises to particularly vulnerable marine environments; Antarctica comes to mind in this respect. At the same time, we surely cannot continue to permit huge cruise ships to dock right in the middle of fragile ports such as Venice.

Protecting 'the goose that lays the golden egg'!

Hopefully, albeit reluctantly, the tourism industry will accept the need to be further regulated, because ultimately it is in its own interests to ensure that marine environments, both the open seas and the coastal zones are protected. Without healthy oceans and clean beaches, tourism may lose much of its appeal and the tourism industry will find itself facing huge reductions in income. It is virtually impossible to imagine tourism that does not involve the ocean in some way or other. Industry action to reduce the negative impacts of tourism on the marine environment is therefore simply a case of enlightened self-interest.

But this idea of self-interest also extends to the public sector in coastal tourist destinations. If local government planning and management of the coastal zones fails to protect the marine environment, on which its local tourism industry depends, then we could see tourism collapsing in some destinations. Given how dependent many coastal communities have become on tourism in recent years this would be disastrous for the economies of these places and the local residents who rely on tourism for their livelihoods.

Perhaps one of the biggest and most complex challenges for the tourism industry remains how effectively it tackles the contribution it makes to global warming, whether that is due to carbon emissions and other pollutants from air travel, private cars being used for leisure trips and cruise ships or the carbon footprint of the global hospitality industry. Most of these effects are far beyond the control of the authorities in individual coastal tourist destinations. However, the effects of global warming, including rising sea levels and increasing water temperatures have a disproportionately large impact on exactly these places. It is in these coastal destinations that the extreme weather events, made more frequent and severe by the effects of global warming, are primarily felt and it is these communities which experience the gradual inundation, and regular flooding caused by rising sea levels.

Can the public sector deliver sustainable tourism?

I have argued that the negative impacts of tourism on the marine environment can only really be reduced if the public sector intervenes in terms of land-use planning policies, effective control of new development, the management of tourism operations and the regulation of industry and tourist behaviour. How likely is this to happen? On the basis of current evidence there seems little basis for optimism, whether we are talking about local, regional and national governments or supra-governmental bodies such as the EU and even the United Nations.

First, large areas of open sea, the Areas Beyond National Jurisdiction (ABNJ), have no public sector agency that has responsibility for them. These areas are crucial given that oceans are open systems and what happens in the coastal zone eventually affects the open seas and vice-versa.

Second, the Economic Exclusion Zones (EEZ) appear to be seen as areas within which national governments are able to oversee the economic exploitation of the ocean in their national interest. Conservation seems very low down the priorities for most governments in the EEZs.

Third, public sector bodies in tourist destinations appear to be wary of upsetting the local tourism industry through taxes and increased regulation because of the contribution it makes to the local economy. They also seem unwilling to do anything that may make tourists less likely to visit in future, whether it be the introduction of tourist taxes or regulation of tourist behaviour.

Fourth, governments in the generating countries, from which most tourists travel internationally, do not usually want to do anything that might upset their citizens who take international vacations, as these people are also voters. While introducing environmental taxes on air travel might make sense in terms of reducing the impacts of aviation on the environment, governments fear that voters will show their anger at such taxes at the ballot box.

Fifth, while the designation of marine protected areas is a step forward, the reality is that many of them exist on paper only, lacking the powerful regulation and financial resources they require to make them effective.

Sixth, many countries do not have the highly developed planning and management systems needed to effectively manage the impacts of tourism on the marine environment. This is particularly true of those less affluent countries which are often the countries which are most dependent on tourism, economically.

Penultimately, corruption is a reality in most countries, although in some places it is more subtle than others. This can distort public sector decision-making even where planning and management systems do exist.

Finally, the sad fact is that the research data available to government decision-makers is often lacking or unreliable and may be challenged by the tourism industry. This makes government nervous about making decisions which will have far-reaching consequences. The fact is that governments in general will only act if sufficient pressure is placed upon them by voters, the media, industry or influential non-governmental organisations. There is currently in most countries simply not enough such pressure, especially from voters. This will only change if public awareness of the issues is raised by the media. This has happened to some extent in recent years in terms of issues such as plastic pollution of the oceans and the impact of global warming on the polar ice caps and glaciers and the effects that is having on sea levels globally. In general, even when it does act the public sector often moves too slowly and too timidly to adequately respond to the scale of the challenges faced by our oceans.

The problems are not just related to local and national governments. The United Nations, despite its commitment to sustainability, does not help with its rather laissez-faire attitude towards the ABNJs and the EEZs. Then there is its 'law of the sea', which leaves cruise companies free to send their vessels anywhere they like, even to places which are blatantly unsuitable for large cruise ships.

Someone reading this critique of the current performance of the public sector could reach the chilling conclusion that the best chance for reducing the negative impacts of tourism on the marine environment rests with countries with totalitarian governments rather than those with democratically elected governments. Such totalitarian governments often own or at least control much of the tourism industry themselves and are nowhere near as accountable to their population as is the case in a democratic country. They therefore have the freedom of movement and the resources to act if they so wish. I am not condoning dictatorships for one moment nor suggesting that non-democratic governments represent the best chance of achieving sustainable tourism. Rather I am seeking to use this point to highlight the current desperate failure of democratic governments across the world to effectively tackle the impacts of tourism on the marine environment, or on climate change.

Are we talking responsible tourism or sustainable tourism?

At this point I would like to set the issue of the impact of tourism on the marine environment in a broader context that I believe is relevant. That context is the debate over whether we should talk about 'sustainable tourism' or 'responsible tourism'. There is no doubt that at the present time the former term is much more widely used, but commentators such as Harold Goodwin have argued strongly that the latter term is more appropriate in relation to tourism. I have become increasingly convinced in recent years that if we want to achieve change, rather than just talking about it, then the idea of responsible tourism is most likely to achieve that. At the same time, I believe that responsible tourism seems the approach most suited to the world today with its growing mistrust of political elites and the 'establishment'.

Sustainable tourism has been around, as a concept, for several decades and has been embraced by supra-governmental organisations such as the United Nations and the European Union, as well as by the public sector at national, regional and local levels around the world. It has also been embraced by academia and has become the subject of endless books, refereed journal papers and conferences, usually in exotic locations which can only be reached by taking a long flight on an aeroplane! I will now try to explain why I believe responsible tourism rather than sustainable tourism is a more pragmatic approach to trying to solve something which is a real rather than a theoretical challenge, and a situation where things need to happen quickly across a broad front. I will begin by offering a critique of the way in which the idea of sustainable tourism has developed in practice.

From the beginning, sustainable tourism has followed a top down approach, with supra-governmental bodies such as the UN World Tourism Organisation (UNWTO) and the European Union taking the lead. In many ways, its development has paralleled the evolution of the broader issue of sustainable development. The UNWTO was present at the crucial global conference on sustainability in Rio de Janeiro in 1992 and at the World Summit on Sustainable Development in Johannesburg in 2002. The UNWTO is meant to support the implementation of the 17 'sustainable development goals', developed by the UN, in terms of the contribution of tourism towards the realisation of these goals. However, the UNWTO does not have the resources to make this happen on its own so it relies on national governments to implement the goals in their respective countries.

Sustainable tourism, like sustainable development as a whole, focuses on the destination rather than the journey and it is based on the idea that sustainability is something that can be achieved if we do certain things. This does not recognise that the concept of sustainable tourism is not an absolute and is likely to evolve over time. And it fails to acknowledge the uncomfortable possibility that it could be something that human beings cannot accomplish – an impossible dream.

The supra-governmental bodies and national governments have promoted the idea of sustainable tourism through innumerable summits, declarations, conferences, policies, voluntary codes of conduct, and one-off funded projects. I would argue that the impact of all of this has been negligible, because:

- The strategic level policies and voluntary codes of conduct have rarely been backed up by legislation or regulation or substantial amounts of funding or the use of taxation to modify industry activity and tourist behaviour.
- The focus on macro-level policy and actions rather than small scale local actions has made many citizens feel that sustainability is something that happens above their heads and that they personally cannot do anything about.

This helps account for the failure of the public sector to engage tourists in the development of sustainable tourism

- Tourism is a private sector industry, and although the public sector plays a vital role in it through the provision of infrastructure and tourism policy and planning, the simple fact is that it is largely the private sector which has direct lines of communication with tourists and provides the services for them. The public sector cannot make sustainable tourism happen without the active support of industry and it has failed to engage effectively with it, because the approach has been abstract and long-term, rather than concentrating on practical actions which can be seen to yield results in the short-term.
- Ironically, the one-off funded sustainable tourism projects supported by bodies such as the European Union, for example, have generally not been sustainable themselves! Most have been funded for a fixed term only and they have tended to collapse once the funding runs out.
- After years of talk and report writing there is still no internationally recognised standard for sustainable tourism products that tourists can use to help them make purchase decisions, so those who want to support sustainability are left frustrated by not knowing what they should do.
- The focus is on technocratic solutions rather than seeing sustainable tourism as a fundamentally political issue and framing solutions appropriately.

Furthermore, this top-down approach runs counter to a strongly emerging theme across the world in recent years, namely a loss of faith in elites and political establishments. Why would they respond positively to their messages about the need to develop sustainable tourism when they do not trust anything these people say?

Because bodies like the UNWTO have to work on consensus, progress tends to be slow and piecemeal, but the crisis we face with global warming and our oceans is developing quickly and requires radical action. This is something that organisations like UNWTO are simply unable to provide and while most governments could do this if they had the will, most have chosen not to for fear of alienating powerful industry lobbies or the tourists who are also their electorate.

I believe that things could have been different, and that the idea of sustainable tourism is still valid. My argument is that the way it has been led by public sector bodies has meant that it has become devalued, but more importantly it will fail to achieve more sustainable tourism and will not ensure that tourism plays its full part in tacking the crises of global warming and the plight of our oceans.

Let us now turn our attention to the concept of responsible tourism, which is in many ways, the direct opposite of what has been said about sustainable tourism.

The first point to make is that it is about taking a 'bottom up' rather than a 'top down' approach. It is based on the idea that every tourist and every tourism business can do something to make tourism a little more sustainable through their own actions. It is about creating change and reducing the negative impacts of tourism through the combined effects of a myriad of such voluntary initiatives. It is also about recognising that every tourism business, tourist destination and individual tourist is responsible for the negative impacts of tourism and as such has a responsibility to take action to reduce these impacts. As such, as far as businesses at least are concerned, it fits neatly with the growth of Corporate Social Responsibility, although one always has to ensure that this is not just used to enhance the reputation of an organisation, of course.

The focus on action, rather than talking and policymaking, appeals to industry and to tourists who can see concrete results from their activities. It also means that everyone feels there is something they can do, however small, which takes away the sense of helplessness that many feel. As we noted in earlier chapters, most of the progress made in making tourism more sustainable has been led by industry, on a voluntary basis, limited though that progress may have been.

Responsible tourism is based on the idea that sustainable tourism is not an absolute and that it may be unattainable whatever we do. It does not see sustainable tourism as a road with a specific destination at the end of it. Instead it sees the journey itself as what matters most, and believes that the actual destination will change over time as circumstances change and the destination might not actually be reached. But it believes that we can, and should, make tourism more sustainable with every step we take on the journey. Responsible tourism is an inclusive philosophy that believes that any citizen can, and should, have a say in what happens and can contribute towards making tourism more sustainable in many different ways. It is not just about political elites and government officials making decisions, but rather about citizens taking as much responsibility as they are willing and able to take. This idea seems to resonate with what we are seeing at the time of writing, where activists such as 'Extinction Rebellion' and individuals such as Greta Thunberg are leading the debate over the need for urgent action to tackle global warming and other environmental challenges. These campaigners have also shown the value importance of the media, and social media in particular, in spreading examples of good practice and putting pressure on public bodies to take action against bad practices, including their own.

We have seen in various chapters examples of responsible tourism in practice and the case studies that form part of the book offer further such examples. However, responsible tourism does not mean the public sector has no role in reducing the negative impacts of tourism on the marine environment and making tourism more sustainable, but it does mean it is not seen as the leader in this process but just as an important co-creator. It will only gain recognition in that role if it shows it has the political will to regulate where required, provide adequate funding for marine protected areas and use taxation to influence the activities of the tourist industry.

It seems to me that the idea of responsible tourism, incorporating an effective public sector alongside all the other stakeholders, is the most effective way to tackle the big challenges tourism faces today. These challenges include: the effects of air travel on global warming; the growing backlash against perceived over-tourism in some destinations; and of course, or the impacts of tourism on the marine environment.

While apologising for the length of this section I believe it has dealt with a question which is crucial to whether or not we can make tourism, wherever it occurs, more sustainable.

Everyone needs to play their part

This section is based on the idea that the negative impacts of tourism on the marine environment can only be reduced if all stakeholders work constructively together. If not, then the future outlook is bleak for humanity whether we are talking about climate change as a whole or the impacts of tourism on the marine environment specifically. And, picking up on the theme of the last section we need every stakeholder group to recognise that they have a responsibility in relation to tourism and its effects on the marine environment.

Partnership is the key word and these partnerships will be many and varied and will include:

- The tourism industry and national and local government authorities.
- The tourism industry and communities in coastal tourist destinations.
- Communities in coastal tourist destinations and the cruise sector.
- The public sector and non-governmental organisations, pressure groups and activists.
- The tourism industry and non-governmental organisations, pressure groups and activists.
- Government regulators and the representative bodies for different marine leisure activities.

These partnerships will not be easy ones and there will be disagreements and conflicts between partners. But we will only manage the impacts of tourism on marine environments if we can build and maintain these partnerships, based on mutual respect.

Perhaps, most important of all, success will not be achieved by any of these partnerships unless we can engage tourists themselves as active partners, so let us now focus on the role of tourists in achieving more sustainable tourism in the marine environment.

■ Responsible tourist behaviour is the key

This is a crucial issue because, ultimately, all of the impacts of tourism on the marine environment are the result of demand from, and decisions made by, tourists. New infrastructure is developed because more tourists visit a destination and unsustainable marine leisure activities only continue if tourists decide to participate in them. Therefore, it is vital that tourists become active partners, in making tourism more sustainable. The first step is getting tourists to think consciously about the decisions they make and the impacts of these decisions on the marine environment. This includes decisions about where to go on vacation, when they go, how they travel to and around the destination, the activities they do on vacation, what they eat in the destinations and the souvenirs they take home.

I believe that both destinations and tourism businesses have a responsibility to provide information to consumers so that they can choices which reduce the negative impacts of their trip on the marine environment. Governments also have a clear duty to work together, and with the tourism industry, to develop industry standards and internationally recognised labels so that tourists can make purchase decisions in the knowledge that what they are buying will genuinely contribute towards more sustainable tourism in coastal zones and on the oceans. At the same time, sooner rather than later, governments, researchers and industry need to evaluate whether carbon offsetting is something that is beneficial and should be encouraged for all flights, car journeys for leisure and cruises. If this is the case, then guidelines need to be provided around the best types of carbon offset schemes so tourists can make informed choices. Or, the decision might be made that such schemes should become compulsory for all tourist travel. Alternatively, it may be decided that carbon offsetting is not a solution to the problem of carbon emissions from tourist travel and is being used just to reduce tourist guilt or make tourism companies look more responsible than they really are. Either way tourists need clear information about the value or otherwise of carbon offsetting.

The representative bodies for marine leisure activities such as diving, sea angling, and sailing need to encourage clubs and societies and experienced participants to apply gentle peer group pressure on all participants in their activity to help reduce irresponsible behaviour.

Destinations and industry need to encourage more responsible behaviour by tourists by ensuring that they get the opportunity to 'feel good' as a result of making choices which promote more sustainable tourism.

I believe that most tourists will respond positively and modify their behaviour if they feel they are being engaged with and actively involved as partners in trying to find more sustainable ways of having a great vacation. This will be more effective than being 'educated', that terribly patronising term that implies that those doing the 'educating' have the moral right to do so and know what is best. The literature on tourist compliance with voluntary codes of behaviour is mixed but contains enough positive experiences to give us some hope for the future, but there have also been cases where such codes have been ignored or not complied with by tourists. (Waayers et al., 2006)

Therefore, encouraging voluntary behaviour changes by tourists needs to exist alongside regulation to protect the marine environment from damage caused by irresponsible tourist behaviour. Hopefully the need for regulation will reduce over time, but there will always be a minority who refuse to behave responsibly, and regulation is needed to deal with them.

Perhaps the most difficult challenges facing us are those where we want tourists to make significant sacrifices to directly or indirectly protect the marine environment. This includes:

- Making certain fragile or endangered marine environments 'no go' areas which tourists are not permitted to visit either temporarily or permanently.
- Getting tourists to take fewer flights, particularly long-haul, to help tackle global warming. The phenomenon of 'flight shaming' may turn out to be a passing fad or it may represent a change of mindset on the part of some tourists at least.

We may also need to ensure that tourists pay a fair price for their vacation experience and help fund marine conservation to reduce the negative impacts of tourism on the marine environment. There are a number of studies of this subject in which tourists say they would be prepared to pay some form of marine conservation levy to help fund marine conservation, such as the work of Stithou and Scarpa (2012) in Greece and Batel et al. (2014) in Croatia. However, there may be a big difference between a hypothetical question about willingness to pay and what happens in reality when a charge is introduced.

I strongly believe that if we want tourists to acknowledge their responsibilities, it should also be acknowledged that they should be seen to have rights, including the right not to be ripped off and the right to be safe when on vacation. If tourists are prepared to behave more responsibly then they need to be respected.

There also needs to be a recognition that the damage is not always caused by 'outsiders', by foreign tourists. Domestic tourists in some places cause more harm, simply because there are more of them or because they are used to behaving in a particular way in places they visit regularly.

So far, we have focused on tourists as individuals, but it will be a game changer if tourists can be persuaded to act together in various ways, including:

- Using social media to spread messages about responsible tourism and the marine environment to friends and acquaintances, in a systematic way.

- Establishing groups of tourist-activists who campaign on issues around the impact of tourism on the marine environment.
- Working with local communities to protect the marine environment.
- Applying peer group pressure to fellow beach users, cruise ship passengers or divers, for example, to encourage them to behave more responsibly.
- Putting pressure on the tourism industry about the impact of tourism on the marine environment, and organising boycotts of irresponsible operators and lobbying local authorities who are not managing tourism effectively.

I do not underestimate the scale of the task of getting tourists to behave more responsibly. People fortunate enough to be able to take a vacation work hard for the whole year to afford a week or two of escapist fun in a vacation destination, and it will never be an easy thing to get people to take responsibility for the impacts of their vacation on the marine environment. But I believe that if it is done in the right way, it can be done, and can even enhance the experience by giving the tourist the feeling that they have done something positive through their vacation.

Further consideration of the issue of responsible tourist behaviour is to be found in Case Study 38.

The need for holistic thinking

When looking at the relationship between tourism and the marine environment we are looking at a complex system of inter-relationships. To manage it effectively therefore requires a much more holistic approach than we have seen to date. We need to recognise and understand the key relationships between:

- The different elements within the marine ecosystem.
- The open sea, inshore waters and the beach.
- Different types of tourist use of the marine environment and ecosystem.
- Planning and management decisions made by public bodies and their impacts on the marine environment.
- The impacts on the marine environment of tourism and other industries and economic activities.
- The marine environment, what happens in coastal communities and what happens inland.
- Cruise ships and the marine environment and the ports of call.
- Air travel, global warming, changes in the ocean and extreme weather events

I also contend that the holistic thinking required for the management of the relationship between tourism and the marine environment can only be achieved through partnerships that represent all the relevant stakeholders. However, we need to ensure that such partnerships enable the voice or voices of the marine environment to be clearly heard, however that is facilitated. There also needs to be a clear recognition that human beings and communities are part of the global ecosystem rather than being separate from it or above it in some way.

Towards an effective research agenda

Holistic thinking is only possible if it can be underpinned by sound research. While much excellent research has been, and is being, conducted, that is relevant to the relationship between tourism and the marine environment, much more needs to be done. Before we look at some specific research priorities, we need to recognise that the research needs to be:

- Truly inter-disciplinary or multi-disciplinary to reflect the complexities of what needs to be researched.
- Focused on identifying actions and potential solutions in a timely manner.
- Research where the findings are targeted at decision-makers, as well as the scientific communities, and are presented in a form which is 'user-friendly.'

There are clearly a number of priority areas for research based on what we have seen in the book so far.

First, we need empirical studies which identify the impacts of particular types of tourism on particular types of marine environment so that we can start to develop different approaches to managing tourism tailored to particular locations.

Second, we need more longitudinal studies to show how impacts develop over time and how long the effects of tourism last.

Third, we need research projects that objectively monitor the effectiveness or otherwise of public sector planning and management of tourism in the marine environment in general and of individual projects specifically.

Fourth, much more research is needed on the impacts of the cruise sector including emissions and marine pollution, the effects of docking in fragile marine environments, collisions with wildlife, and the physical and temporal concentration of impacts of cruise passengers in ports of call. Much more work is also needed on understanding the attitudes of cruise passengers towards responsible tourism and their awareness of the impacts of cruising on marine environments.

Fifth, we need to better understand the factors that lie behind successful partnerships so lessons can be learned that will help make the success of future partnerships more likely.

Finally, more research is required to help us identify the most effective ways of engaging tourists in responsible tourism and making tourist behaviour more responsible in relation to the marine environment.

Achieving these things will require not only substantial extra funding but a change in research culture, not least in tourism academia. Researchers will need to reach out to decision-makers to help ensure that their research can make a difference to what happens, as well as contributing to the development of knowledge within the scientific community. Perhaps presenting research findings in a popular newspaper or on the television news or at an industry conference needs to be recognised as being as important as having a paper published in a top peer refereed journal or speaking at an academic conference.

And finally ...

The context in which this book has been written is the threat which our planet faces from climate change. We have seen that global warming is having a massive and detrimental impact on the marine environment across the planet. Furthermore, the increases in water temperatures and rises in sea level caused by climate change are in turn causing extreme weather events that are causing death and destruction, particularly in coastal communities, although their effects are being felt hundreds of kilometres inland too.

Sadly, we have seen that tourism is contributing significantly to global warming through emissions from air travel and leisure use of private cars, as well as from the carbon footprint of the hospitality and cruise sectors. At the same time, we have seen that coastal zone tourism is having a range of negative impacts on the marine environment due to irresponsible wildlife watching, marine leisure activities and consumption of marine resources, in addition to major infrastructure developments, beach nourishment schemes and desalination. Unfortunately, unlike most communities on land that suffer the negative impacts of tourism, the ocean has no voice with which to complain and demand action. And the marine environment sees no benefits from tourism unlike the communities on land for whom tourism brings income and employment to balance against negative environmental impacts.

The marine environment can only be protected if all the interests work together, including the public sector, industry, tourists, local communities, and voluntary organisations. We, as human beings, are causing the problems that the marine environment is experiencing so it is our responsibility to try to solve them. And it is in our own interests to do so because what happens in the oceans affects us directly in terms of everything from extreme weather events, to whether or not coastal tourism and the cruise industry are sustainable in the long term.

Furthermore, tourism is not an island, so we also need to work holistically with other industries which are also harming the marine environment through their activities. Perhaps, the most worrying aspect of the current response to the challenges facing our oceans, or the whole planet for that matter, is the apparent lack of urgency. Things are being done but not quickly enough and the window of opportunity for action is closing! We cannot plead ignorance as an excuse for our inadequate response to the crises we are facing, for the media is increasingly filled with stories about the plight of the oceans and the relationship between tourism and the marine environment as can be seen in Case Study 39.

And a final thought. Next time you hear people talking about the future of the planet, just remember that the oceans make up 70% of its surface!

References

Batel, A, Basta, J and Mackelworth, P (2014) Valuing visitor willingness to pay for marine conservation - the case of the proposed Cres-Lošinj marine protected area, Croatia. *Ocean and Coastal Management* **95**, 72-80.

Stithou, M and Scarpa, R (2012) Collective versus voluntary payment in contingent valuation for the conservation of marine diversity: an exploratory study from Zakynthos, Greece. *Ocean and Coastal Management* **56**, 1-9.

Waayers, D, Newsome, D and Lee D (2008) Research note. Observations of non-compliance behaviour by tourists to a voluntary code of conduct: a pilot study of turtle tourism in the Exmouth region, Western Australia. *Journal of Ecotourism* **5**(3) 211-222.

12

Questions and exercises

1. Using the scientific literature, produce a report on the carbon footprint of the tourism industry as a whole. In your report note how easy or difficult you found this task.
2. Compare and contrast the concepts of responsible tourism and sustainable tourism
3. Discuss which of the following can, in your opinion, could or should make the greatest contribution towards responsible tourism:

   ☐ governments

   ☐ the tourism industry

   ☐ tourists

   Please provide evidence and examples to support your view
4. Critically evaluate the statement that 'Tourism needs the oceans, but the oceans do not need tourism'.
5. Working in a group, design a survey to use with a representative sample of the population with the objective of seeing how likely it is that tourists might be willing to modify their behaviour to reduce the negative impacts of tourism on the marine environment. Present your results in a report that also offers suggestions for ways of getting tourists engaged in responsible tourism, based on your results.
6. On a scale of 1 (not at all) to 10 (totally) how responsible is your behaviour as a tourist? Please use examples to support your assessment of your own behaviour.

P Postscript: The COVID-19 Pandemic

I completed the main text of this book a few days before Coronavirus, as it was called at the beginning, started to become a major story in the news in Europe. Now, just over three months later, as the book is about go for printing it seems as if the COVID-19 pandemic, as it is now called, is about the only story in the world's media. In the circumstances, it seems important that I say something about the virus and its potential impact on the subject of this book.

As I write these words, in early May 2020, the pandemic has killed at least 264,000 people worldwide and some 3.8 million people are confirmed to have been infected, although the actual number is likely to be significantly higher as many people who have had the virus may not have had it confirmed through testing. To put this in context, the Spanish Flu pandemic in 1918 killed an estimated 50 million people, while the highly publicised outbreak of SARS in 2003 killed fewer than 1,000 people. The 2014-2016 Ebola outbreak in Africa resulted in the deaths of an estimated 11,300 people. So COVID-19 is far and away the largest pandemic, in terms of deaths, to hit the world in just over a century. Of course, we do not yet know the final death toll from it, for as I write it is still continuing. Furthermore, unlike SARS and Ebola this virus is a true pandemic, affecting virtually every part of the planet where human beings live.

It is important to recognise that this is the first global pandemic the world has seen since the development of mass tourism, and it has quickly become apparent that there has been a powerful two-way relationship between travel and tourism and COVID-19. It was international travel and tourism which took the virus from its point of origin in China to begin its journey to infect the whole world. Business travellers took COVID-19 home with them after they attended conferences in Asia and some of those taking skiing holidays in the Italian Alps picked up the virus there and returned home to spread it to friends and relatives around Europe. And cruise ships became floating incubators for the virus and lax quarantine restrictions in some ports meant that some infected passengers went ashore and spread the disease as they made their way home.

But then, in a development which seemed almost like poetic justice after tourism facilitated its spread, the virus brought international tourism – and also domestic tourism – to a standstill. Air travel has virtually ceased in many countries and there are hardly any contrails from jet aircraft to be seen in our skies. Those living next to airport can hear the birds singing for the first time ever. The carbon footprint of tourism has been dramatically reduced almost overnight, not by some amazing new technological innovation, but by a virus which has forced governments to close their borders and impose lockdowns on their citizens. It is ironic that a virus has achieved what decades of inter-governmental summits, endless policies and strategies and pressure group activity have failed to achieve, namely reversing the growth of air travel and tourism and reducing its carbon footprint! It remains to be seen whether this is a purely temporary 'blip' before things return to 'business as usual' or whether the virus, and the fear of others that may follow, may change tourism forever. This is being vigorously debated by academics and media commentators, literally, as I write these words.

Those who hope or believe tourism will bounce back quickly after the worst of the pandemic is over will be relying on the fact that people who have been enjoying international travel for years will not suddenly have lost the taste for it. They may even think that as people are being denied the chance to travel because of lockdowns and border closures, there is pent up demand that will explode once travel restrictions are relaxed. And they might also point to the fact that many airlines and tour operators have not refunded travellers for their cancelled trips, instead allowing them to postpone their trip to a future date. So, unless that changes or governments force the industry to give refunds, which would probably bankrupt many tourism organisations, people will need to travel or lose the money they have spent already.

However, those who think tourism will be changed for ever by this virus can also point to factors that support their argument. Even today, while the virus is still raging, it has been estimated that the virus will cause a recession that will probably be the most severe in living memory. Unemployment is rising and many businesses are saying they are unlikely to survive because of the COVID-19 pandemic. Economic downturns always have a negative impact on tourism demand so this one could affect the recovery of the tourism industry significantly. Second, it could be argued people have become nervous about travelling with others due to the virus, and the idea of social distancing has become ingrained in people's minds. It is very difficult to reconcile social distancing with tourism, which involves sitting centimetres away from others on a plane or sharing space on crowded beaches or at theme parks, or queuing at check-ins at hotels. They can also point to the fact that many airlines, hotels and tour operators may not survive the COVID-19 crisis, which may well lead to a reduction in capacity for a period following the end of the virus, whenever that may be. Only time will tell which prediction for the long-term future of tourism is the right one.

As I am writing this, some countries are beginning to ease their lockdowns and travel restrictions on their citizens. It seems that this will happen in a phased manner and that tourism will not be something that is prioritised as governments relax restrictions step by step. However, they will be under pressure from the tourism industry and those destinations within their own countries which depend heavily on tourism. Because of the fact that air travel has been seen to spread the virus internationally, it may well be that domestic tourism trips are permitted before international travel. Ironically, this could lead to something of a boom in domestic tourism in some counties, at least for a short period, as those who might normally take an international trip settle for a vacation in their own country instead.

In the meantime, we have been seeing some interesting developments in relation to the link between tourism and climate change. Climate change activists and some politicians are arguing that state aid for struggling airlines should be tied to action by these airlines to reduce their impact on climate change. The French government recently announced state aid for Air France but said this would not cover their domestic flights because alterative means of transport could be used which are less environmentally damaging, namely the railways.

It remains to be seen whether the 'overtourism' movement will lose support in European cities whose economies have been badly affected by the impact of COVID-19 on tourism.

It also remains to be seen what the long-term consequences of the virus will be for tourism; whether it will bring change to the way the industry operates or have any long-term effect on the decisions made by tourists about whether to travel, when to travel, where to travel to, and how to get to their chosen destination. The answers to these questions may not emerge for months or even years. They will also depend on whether COVID-19 turns out to be an isolated 'incident' or the first of a number of pandemics over the years to come. If it is the latter, then the travel and tourism industry will come under increasing scrutiny, given its potential role as the main means by which viruses can move between countries and spread.

I would like to turn my attention to what all this might mean for the marine environment in the short, medium and long terms. Before that, though, it is important to look at the bigger picture as our oceans are negatively impacted by economic activities apart from tourism and by the behaviour of consumers in their everyday lives not just when they are on vacation. It is, therefore, important to say something about how COVID-19 might on the impact which other forms of human activity have on our oceans.

One major effect of the virus we have seen to date has been the increase in people working at home, where that is possible, to avoid them travelling to work and potentially spreading the virus. This has led to fewer cars on the roads and thus less carbon emissions and a reduction in air pollution. Indeed, social media and the news has been full of satellite images and 'before and after' images showing the tangible impact of the virus on visible pollution in big cities and in the sky. At the same time, the closing of all but non-essential shops during the crisis has led some consumers to question materialism and ask if they need to buy as much 'stuff' as they have up to now. Many people have started sourcing groceries from local shops and buying locally produced food, due to the problems with travelling during the lockdowns, which again should have reduced car use.

If at least some of these changes continue to some extent after the virus crisis is over, then the impact of humanity on climate change should come down, so that we may be able to slow down the process of global warming. However, it is a big 'if', and if that were to happen it would have a significant short-term economic impact, most notably in terms of unemployment.

There can be little doubt that in the short term the COVID-19 virus will have set in chain a sequence of events that will ultimately benefit the oceans, albeit probably only in a marginal way. Less carbon emissions should slow global warming and thus both rising sea levels and increasing sea water temperatures, which are both having a serious impact on the marine environment. Furthermore, the fact that most cruises have currently been suspended means that the oceans have at least a temporary respite from not only the effects of the carbon emissions from cruise ships but also from the pollution and waste problems they cause to our marine environment. At the same time, the huge reduction in tourism will mean a substantial reduction in the pollution of the seas caused by the waste, plastic and otherwise, that comes from tourism. The marine environment, and certainly the creatures that live in it, will be befitting from reduced tourist activity in coastal resorts too. Fewer whales will be being chased by irresponsible trip boats, fewer divers will be damaging coral reefs and fewer jet skiers will be disturbing the wildlife. There will be fewer people littering the beaches with plastics that the receding tide will take out into the sea, and fewer people on beaches taking shells and seaweed away with them.

Some of these benefits for the marine environment arising from the reduction in tourism worldwide are being felt now, others will take much longer to bring tangible improvements to the wellbeing of the oceans.

We also need to recognise that if the current reduction in tourism helps slow down global warming that will, in time, benefit the land too in terms of less frequent and less severe extreme weather events. However, for that to happen the current temporary reduction in tourism would need to be maintained and/or ways found to reduce the carbon footprint of travel and tourism.

Sadly, what limited evidence there is at present suggests that this may not be the case. Already the cruise lines are reporting that advance bookings for 2021 are looking very good and there is nothing on the horizon at the moment in terms of technological innovation that seems likely to reduce the carbon footprint of air travel dramatically any time soon. And even if such an innovation were to come along, the airlines that survive this crisis are unlikely to have the financial capacity, on their own, to make the levels of investment that would be required to introduce new, less environmentally damaging, technologies.

Therefore I believe that the only hope of something good for the marine environment coming out of the COVID-19 pandemic is if governments, and hopefully the UN, are prepared to take decisive action rather than just talking, making the right noises, but not backing them up with effective action. The industry and tourist behaviour itself may need to be regulated, or at least encouraged to behave more responsibly, through a mixture of incentives and penalties. Furthermore, governments also need to fully understand that what happens in the oceans has a direct impact on the quality of life of their citizens, and work together to protect marine environments. They need to recognise the immense implications of the fact that 70% of the surface of our planet consists of oceans and marine environments, and that they are essential to our survival as a species.

The simple fact is that this appalling virus crisis may be our last opportunity to change direction, a final wakeup call before it is too late. If we do not take this opportunity, the crisis of our oceans will just get progressively worse until it becomes too late to reverse the damage.

There are two other final points to be made. We seem to be facing more and more crises, not just the COVID-19 virus. Some are a direct result of the climate change, such as extreme weather events, to which tourism is contributing while others are viruses that are spread by international travel and tourism. If we do not take action to manage tourism more effectively and reduce its environmental impacts one day we may face a 'perfect storm' when a major global crisis emerges at a time when our failure to tackle climate change means we are so weakened that we cannot deal with any new crisis which may emerge.

If the COVID-19 virus, and the devastation it has caused, make us reflect and change our behaviour, as consumers, business people and politicians, so that we put the environment, and particularly the marine environment, at the top of our agenda, then hundreds of thousands of innocent people may not have lost their lives in vain.

B Bibliography and Further Reading

You will find below a selection of books, journal papers and online sources which should encourage and help you to further explore some of the issues covered in this book. Given that the subject of this book is one which is evolving at a rapid pace I have deliberately made great use of reliable and trustworthy online sources, as well as on books and journal papers, for much of the data used as I wanted you to have the most up-to-date information possible.

In reality, I struggled to find enough material in books and journal papers to provide you with topical coverage of the wide variety of issues discussed in this book. For every excellent paper or book I found there were probably ten I would like to have found but they did not exist. This reflects the pace at which things are moving in relation to the crisis facing our oceans and the impact of tourism on the marine environment but it is also probably a function of the time it takes for a conventional book or traditional journal paper to be published.

Finally, it remains disappointing that there is still a paucity of genuinely trans-disciplinary and multi-disciplinary academic literature about the relationship between tourism and the marine environment that brings together the expertise of tourism academics and specialists from marine sciences, planning, psychology, and geography. We need more joint work between scientists and tourism academics to strengthen the quality of the academic literature in respect of the impacts of tourism on the marine environment.

Books, Book Chapters and Reports

Apostpolpoulos, Y, Apostolopoulos, Y and Gayle, DG (Eds) (2002) *Island Tourism and Sustainable Development: Caribbean, Pacific and Mediterranean Experiences*. Praeger. Westport, USA

Becken, S and Hay JE (2007) *Tourism and Climate Change: Risks and Opportunities*. Channel View Publications. Clevedon

Bigano, A, Goria, A, Hamilton, JM and Tol, RSJ (2009) *The Effect of Climate Change and extreme Weather Events on Tourism*. FEEM Working Paper No.30.05. CMCC Research Paper No. 1

Carlsen, J and Butler, R (2011) *Island Tourism: Towards a Sustainable Perspective*. CAB International. Wallingford

Cater, C and Cater, E (2007) *Marine Tourism: Between the Devil and the Deep Blue Sea*. CAB International. Wallingford

Clark, JR (Ed) (1996) *Coastal Zone Management Handbook*. CRC Press. Boca Raton

Cruise Lines International Association (2019) *2019 Cruise Trends and Industry Outlook*. CLIA. Washington

Dowling, RK (ed) (2006). *Cruise Tourism*. CAB International. Wallingford

Dowling, RK (2009) Coastal Tourism Development: Planning and Management Issues. CAB International. Elmsford, USA

Garrod, B and Gössling, S (2008) *New Frontiers in Marine Tourism*. Elsevier. Oxford and Amsterdam

Gössling, S and Hall, CM (eds) (2017) *Tourism and Global Environmental Change: Ecological, Social, Economic and Political Inter-relationships*. Routledge. London

Harriott, VJ (2002) *Marine Tourism Impacts and their Management on the Great Barrier Reef*. CRC Reef Research Centre Technical Report No. 46. Townsville

Haslett, SK (2009) *Coastal Systems*. Routledge. London

Higham, J and Lück, M (2008) *Marine Wildlife and Tourism Management*. CAB International. Wallingford

Honey, M and Krantz, MA (2007) *Global Trends on Coastal Tourism*. Marine Program, World Wildlife Fund, Washington.

Kay, R and Alder, J (2005) *Coastal Planning and Management*. Second edition. Taylor and Francis. London

Laffoley, D (1995) Techniques for managing marine protected areas: zoning. In: Gubbay, S (Ed) *Marine Protected Areas. Conservation Biology*, Volume 5. Springer, Dordrecht

Luck, M (2007) *Nautical Tourism: Concepts and Issues*. Cognizant Communications Corporation. Elmsford

Markwell, K (Ed) (2015) *Animals and Tourism: Understanding Diverse Relationships*. Channel View Publications. Bristol

Orams, M (1999) *Marine Tourism: Development, Impacts and Management*. Routledge. London

Palumbi, S (2003) *Marine reserves: a tool for ecosystem management and conservation*. Pew Oceans Commission, Arlington, USA.

Parry, ML, Canziani, OF, Palutikof, JP, van der Linden, PJ and Hanson, CE (eds) (2007) *Climate Change 2007, Impacts, Adaption and Vulnerability. Contribution of Working Group II to the Fourth Assessment Report of the Intergovernmental Panel on Climate Change*, Cambridge University Press, Cambridge, UK

Rosendahl Appelquist, L, Balstrøm, T and Halsnaes, K (2016) *Managing Climate Change Hazards in Coastal Areas – The Coastal Hazard Wheel Decision-Support System*. UNEP

Sawkar, K, Noronha, L, Mascarenhas, A, Chauhan, OS and Saeed, S (1998) *Tourism and the Environment: Case Studies on Goa, India, and the Maldives*. The International Bank for Reconstruction and Development/The World Bank. Washington

Stonehouse, B and Snyder, J (2010) *Polar Tourism: An Environmental Perspective*. Channel View Publications. Bristol

Thevenon, F, Carroll, C and Sousa, J (eds) (2014) *Plastic Debris in the Ocean: The Characterization of Marine Plastics and their Environmental Impacts, Situation Analysis Report.* IUCN. Gland

United States Department of Commerce and National Oceanic and Atmospheric Administration (2007) *National Artificial Reef Plan: Guidelines for Siting, Construction, Development, and Assessment of Artificial Reefs.* USDCNOAA

UP-MSI, ABC, ARCBC,DENR, ASEAN (2002) *Marine Protected Areas in Southeast Asia.* ASEAN Regional Centre for Biodiversity, Conservation, Department of Environment and Natural Resources, Los Baños, Philippines.

Wong, PP (1991) *Coastal Tourism in Southeast Asia.* Association of Southeast Asia Nations/ United States Coastal Resources Management Project. Education Series 8

Journal Papers

Abed, MH, Monaviri, M, Karbasi, A, Farshchi, P and Abedi, Z (2011) Site selection using analytical hierarchy process by Geographical Information System for sustainable coastal tourism. *2011 International Conference on Environmental and Agricultural Engineering,* **15.**

Baker, DM (2016) The cruise industry: past, present and future. *Journal of Tourism Research* **127**, 141-153.

Belle, N, and Bramwell, B (2005) Climate change and small island tourism; policy maker and industry perspectives in Barbados. *Journal of Travel Research* **44**, 32-41.

Bennett, NJ and Dearden, P (2014) Why local people do not support conservation: community perceptions of marine protected area livelihood impacts, governance, and management in Thailand. *Marine Policy* **44**, 107-116.

Bigano, A, Bosello, F, Roson, R and Tol, RSJ (2008) Economy-wide impacts of climate change: a joint analysis for sea level rise and tourism. *Mitigation and Adoption Strategies for Global Change* **13**, 765-791.

Borch, T (2004) Sustainable management of marine fishing tourism: some lessons from Norway. *Tourism in Marine Environments* **1**(1) 49-57.

Buckley, R (2002) Surf tourism and sustainable development in Indo-Pacific islands. I. The industry and the islands. *Journal of Sustainable Tourism.* **10**(5) 405-424.

Caffyn, A and Jobbins, G (2003) Governance capacity and stakeholder interactions in the development and management of coastal tourism: examples from Morocco and Tunisia. *Journal of Sustainable Tourism* **11**(2-3) 224-245.

Christie, P (2005) Observed and perceived environmental impacts of marine protected areas in two southeast Asia sites. *Ocean and Coastal Management* **48**(3-6) 252-270.

Davenport, J and Davenport, JL (2006) The impact of tourism and personal leisure transport on coastal environments: a review. *Estuarine, Coastal and Shelf Science* **67**(1-2) 280-292.

De Miguel-Molina, B, De Miguel-Molina, M and Rumiche-Sosa, M (2011) Does luxury indicate sustainability: an analysis of the Maldives. *Electronic Journal of Business Ethics and Organizational Studies* **16**(1) 21-31.

Diedrich, A (2007) The impacts of tourism on coral reef conservation awareness and support in coastal communities in Belize. *Coral Reefs* **26**(4) 985-996.

Elliott, M, Cutts, ND and Trono, A (2014) A typology of marine and estuarine hazards and risks as vectors of change: a review for vulnerable coasts and their management. *Ocean and Coastal Management* **93**, 88-99.

Fang, Y, Yin, J and Wu, B (2016) Flooding risk assessment of coastal tourist attractions affected by sea level rise and storm surge, a case study in Zhejiang Province, China. *Natural Hazards* **84**, 611-624.

Ghaderi, Z, Som, APM, and Henderson, JC (2015) When disaster strikes: the Thai floods of 2011 and tourism industry response and resilience. *Asia Pacific Journal of Tourism Research* **20**(4) 399-415.

Gladstone, W, Curley, B, and Shokri, MS (20130 Environmental impacts of tourism in the Gulf and the Red Sea. *Marine Pollution Bulletin* **72**(2) 375-388.

Gössling, S, Borgström Hansom, C, Hörstmeier, O and Saggel, S (2002) Ecological footprint analysis as a tool to assess tourism sustainability. *Ecological Economics* **43**(2-3) 199-2011.

Gouveia-Souza, CL de, and Suguio, K (2003) The coastal erosion risk zoning and the São Paulo State Plan for *Coastal Management Journal of Coastal Research*, **35**, 530-647.

Graci, S (2013) Collaboration and partnership development for sustainable tourism. *Tourism Geographies*. **15**(1) 25-42.

Hall, CM (2001) Trends in ocean and coastal tourism: the end of the last frontier? *Ocean and Coastal Management* **44**(9-10) 601-618.

Hamilton, JM, Maddison, DJ and Tol, RSJ (2005) Climate change and international tourism: a simulation study. *Global Environmental Change* **15**(3) 253-266.

Hawkins, JP (1994) The growth of coastal tourism in the Red Sea: present and future effects on coral reefs. *Ambio* **23**(8) 503-508.

Howes, L, Scarpaci, C, and Parsons, ECM (2012) Ineffectiveness of a marine sanctuary zone to protect burranan dolphins (Tursiops Australia sp nov.) from commercial tourism in Port Phillip Bay, Australia. *Journal of Ecotourism* **11**(3) 188-201.

Hoyt, E (2005) Sustainable tourism on Atlantic islands, with special reference to whale watching, marine protected areas and sanctuaries for cetaceans. *Biology Environment* **105B**(3), 141-154 .

Imran, S, Alam, K, and Beaumont, N (2014) Environmental orientations and environmental behaviour: perceptions of protected area tourism stakeholders. *Tourism Management* **40**, 290-299.

Irandu, EM (2006) sustainable tourism development on Kenya's coast: a hospitality sector view. *Anatolia* **17**(2) 189-209.

Jennings, S (2004) Coastal tourism and shoreline management. *Annals of Tourism Research* **31**(4) 899-922.

Johnson, D (2002) Environmentally sustainable cruise tourism: a reality check. *Marine Policy* **26**(4) 261-270.

Kennington, R (1993) Tourism in coastal and marine environments – a recreational perspective. *Ocean and Coastal Management* **19**(1) 1-16.

Kurniawan, F, Adrianto, L, Bengen, DG and Prasetyo, LB (2016) Vulnerability assessment of small islands to tourism: the case of the marine tourism park of the Gili Matra Islands,

Indonesia. *Global Ecology and Conservation* **6**, 308-326.

Lamb, JB and Willis, BL (2011) Using coral disease prevalence to assess the effects of concentrating tourism activities on offshore reefs in a tropical marine park. *Conservation Biology* **25**(5) 1044-1052.

Lewis, A and Newsome, D (2003) Planning for stingray tourism in Hamelin Bay, Western Australia: the importance of stakeholder perspectives. *International Journal of Tourism Research* **5**(5) 331-346.

Lück, M and Jiang, Y (2007) Keiko, Shamu and friends: educating visitors to marine parks and aquaria? *Journal of Ecotourism* **6**(2) 127-138.

Luna, B, Pérez, CV and Sánchez-Lizaso, JL (2009) Benthic impacts of recreational divers in a Mediterranean marine protected area. ICES *Journal of Marine Science* **66**(3) 517-523.

McNamara, KE and Gibson, C (2008) Environmental sustainability in practice? A macro-scale profile of tourist accommodation facilities in Australia's coastal zone. *Journal of Sustainable Tourism* **16**(1) 85-100.

Marafa, LM and Chau, KC (2014) Framework for sustainable tourism development on coastal and marine zone environment. *Tourism, Leisure and Global Change.* 1. Papers from the 6th Tourism Outlook Conference, April 2013. Kota Kinabalu, Sabah, Malaysia.

Marzuki, A (2008) Impacts of tourism development. *Anatolia* **20**(2) 450-455.

Mayes, G and Richins, H (2008) Dolphin watch tourism; two differing examples of sustainable practices and proenvironmental outcomes. *Tourism in Marine Environments* **5**(2-3) 201-214.

Miller, ML (1993) The rise of coastal and marine tourism. *Ocean and Coastal Management* **20**(3) 181-199.

Miller, ML, Orams, M, Lück, M, Auyong, J and Gräupl, A (2009) Editorial. A field taking shape: papers from the 5th International Coastal and Marine Tourism Congress. *Tourism and Marine Environments* **5**(2-3) 75-87.

Miller, ML and Auyung, J (1991) Coastal zone tourism: a potent force affecting environment and society. *Marine Policy* **15**(2) 75-99.

Mkono, M (2020) Eco-anxiety and the flight-shaming movement: implications for tourism. *Journal of Tourism Futures*. Ahead of print.

Moreno, A and Amelung, B (2009) Climate change and coastal and marine tourism: review and analysis. *Journal of Coastal Research*. Special issue No. 56. Proceedings of the 10th International Coastal Symposium ICS 2009. Vol II, 1140-1144.

Moscado, G, Pearce, PL, Gren, D and O'Leary, JT (2001) Understanding coastal and marine tourism demand from three European markets: implications for the future of ecotourism. *Journal of Sustainable Tourism* **9**(3) 212-227.

Musa, G and Dimmock, G (2012) Scuba diving tourism: introduction to special issue. *Tourism in Marine Environments* **8**(2) 1-5.

Nadzir, NM, Ibrahim, M and Mansor, M (2014) Impacts of coastal reclamation to the quality of life: Tanjung Tokong community, Penang. *Procedia – Social and Behavioural Sciences* **153**, 159-168.

Nara, P, Mao, G-G and Yen, T-B (2014) Applying environmental management policy for sustainable development of coastal tourism in Thailand. *International Journal of Environmental Protection and Policy* **2**(1) 19-23.

Nathan, A (2017) The law of the seas: a barrier to implementation of Sustainable Development Goal 14. *Sustainable Development Law and Policy* **16**(2), Article 8.

Neumann, DR and Orams, MB (2006) Impacts of ecotourism on short-beaked common dolphins (Delphinus Delphis) in Mercury Bay, New Zealand. *Aquatic Mammals* **32**(1) 1-9.

Ong, LTJ and Smith, RA (2014) Perception and reality of managing sustainable coastal tourism in emerging destinations: the case of Sihanoukville, Cambodia. *Journal of Sustainable Tourism* **22**(2) 256-278.

Parsons, ECM (2012) The negative impacts of whale-watching. *Journal of Marine Biology*, Article ID 807294.

Petrosillo, I, Zurlini, G, Corlianò, ME, Zaccarelli, N, and Dadamo, M (2007) Tourist perception of recreational environment and management in a marine protected area. *Landscape and Urban Planning* **79**(1) 29-37.

Pini-Fitzsimmons, J, Knott, NA, and Brown, C (2018) Effects of food provisioning on site use in the short-tail stingray Bathytoshia Brevicaudata. *Marine Ecology Progress Series 600*, 99-110.

Powell, RB, Cuschnir, A and Peiris, P (2009) Overcoming governance and institutional barriers to integrated coastal zone, marine protected area, and *Tourism Management* in Sri Lanka. *Coastal Management* **37**(6) 633-655.

Prabhakaran, S, Nair, V, and Ramachandran, S (2013) Marine waste management indicators in a tourism environment: exploring possibilities for Semporna District, Sabah, *Worldwide Hospitality and Tourism Themes* **5**(4) 365 - 376.

Priskin, J (2003) Tourist perceptions of degradation caused by coastal nature-based recreation. *Environmental Management* **32**, 189-204.

Rees, SE, Foster, NL, Langmead, O, Pittman, S and Johnson, DE (2018) Defining the qualitative elements of Aichi Biodiversity Target 11 with regard to the marine and coastal environment in order to strengthen global efforts for marine biodiversity conservation outlined in the United Nations Sustainable Development Goal 14. *Marine Policy* **93**, 241-250.

Rodger, K, Smith, A, Newsome, D and Moore, SA (2011) Developing and testing an assessment framework to guide the sustainability of the marine wildlife tourism industry. *Journal of Ecotourism* **10**(2) 149-164.

Rouphael, A, and Inglis, GJ (1997) Impacts of recreational scuba diving at sites with different reef topographies. *Biological Conservation* **82**(3) 329-336.

Showalter, GR 91995) Cruise ships and private islands in the Caribbean. *Journal of Travel and Tourism Marketing*. **3**(4), 107-118.

Simpson, M (2013) Impacts of climate change on tourism and marine recreation). *MCCIP Science Review 2013*, 271-283.

Smith, RA (1992) Coastal urbanization: tourism development in the Asia Pacific. *Built Environment* **18**(1) 27-40.

Southon, MP and van der Merwe, CD (2018) Flooded with risks or opportunities, exploring flooding impacts on tourist accommodation. *African Journal of Hospitality, Tourism and Leisure* **7**(1) 1-27.

Stewart, E and Draper, D (2006) Sustainable cruise tourism in Arctic Canada: an integrated coastal management approach. *Tourism in Marine Environments* 3(2) 77-88.

Stewart, MC (1993) Sustainable tourism development and marine conservation regimes. *Ocean and Coastal Management* **20**(3) 201-217.

Steyn, JN and Spencer, JP (2012) Climate change and tourism; implications for South Africa and adventure. *African Journal for Physical Health Education, Recreation and Dance* **18**(1) 1-19.

Stonich, S (1998) Political ecology of tourism. *Annals of Tourism Research* **25**(1) 25-54.

The, L, and Cabanban, AS (2007) Planning for sustainable tourism in southern Pulau Banggi: as assessment of biophysical conditions and their implications for future tourism development. Journal of *Environmental Management* **85**(4) 999-1008.

Tisdell, C and Wilson, C (2005) Perceived impacts of ecotourism on environmental learning and conservation: turtle watching as a case study. *Environment, Sustainability and Development* **7**, 291-302.

Trave, C, Brunnschweiler, J, Sheaves, M, Diedrich, A and Barnett, A (2017) Are we killing them with kindness? Evaluation of sustainable marine wildlife tourism. *Biological Conservation* **209**, 211-222.

Wall, G (1998) Climate change, tourism and the IPCC. *Tourism Recreation Research.* **23**(2) 65-68.

Warnken, J and Buckley, R (1998) Scientific quality of tourism environmental impact assessment. *Journal of Applied Ecology*. **3**(1) 1-8.

Weatherdon, LV, Magnan, AK, Rogers, AD, Sumaila, R and Cheung, WWL (2016) Observed and projected impacts of climate change on marine fisheries, aquaculture, coastal tourism and human health: an update. *Frontiers in Marine Science*. **3**.

Weaver, D (2010) Can sustainable tourism survive climate change? *Journal of Sustainable Tourism* **19**(1) 5-15.

Wiener, CS, Needham, MD, and Wilkinson, PF (2009) Hawaii's real-life marine park: interpretation and impacts of commercial marine tourism in the Hawaiian Islands. *Current Issues in Tourism* **12**(5-6) 489-504.

Wilhelmsson, D, Öhman, MC, Stähl, H, and Shlesinger, Y (1998) Artificial reefs and dive tourism in Eilat, Israel. *Ambio* **27**(8) 764-766.

Wilson, SP and Verlis, KM (2017) The ugly face of tourism: marine debris pollution linked to visitation in the southern Great Barrier Reef, Australia. *Marine Pollution Bulletin* **117**(1-2) 239-246.

Wong, P (1998) Coastal tourism development in southeast Asia: relevance and lessons for coastal zone management. *Ocean and Coastal Management* **38**(2) 89-109.

I Index

www.ingramcontent.com/pod-product-compliance
Ingram Content Group UK Ltd.
Pitfield, Milton Keynes, MK11 3LW, UK
UKHW051847100926
8556IPUK00005B/503

9 781911 635574